Y0-AAF-215

mylar bnab

CHAPMAN & HALL/CRC

Research Notes in Mathematics Series

Submission of proposals for consideration
Suggestions for publication, in the form of outlines and representative samples, are invited by the Editorial Board for assessment. Intending authors should approach one of the main editors or another member of the Editorial Board, citing the relevant AMS subject classifications. Alternatively, outlines may be sent directly to the publisher's offices. Refereeing is by members of the board and other mathematical authorities in the topic concerned, throughout the world.

Preparation of accepted manuscripts
On acceptance of a proposal, the publisher will supply full instructions for the preparation of manuscripts in a form suitable for direct photo-lithographic reproduction. Specially printed grid sheets can be provided. Word processor output, subject to the publisher's approval, is also acceptable.

Illustrations should be prepared by the authors, ready for direct reproduction without further improvement. The use of hand-drawn symbols should be avoided wherever possible, in order to obtain maximum clarity of the text.

The publisher will be pleased to give guidance necessary during the preparation of a typescript and will be happy to answer any queries.

Important note
In order to avoid later retyping, intending authors are strongly urged not to begin final preparation of a typescript before receiving the publisher's guidelines. In this way we hope to preserve the uniform appearance of the series.

CRC Press UK
Chapman & Hall/CRC Statistics and Mathematics
Pocock House
235 Southwark Bridge Road
London SE1 6LY
Tel: 0171 407 7335

Situ Rong

Reflecting stochastic differential equations with jumps and applications

CHAPMAN & HALL/CRC

Boca Raton London New York Washington, D.C.

Library of Congress Cataloging-in-Publication Data

Situ, Rong
    Reflecting stochastic differential equations with jumps and applications / Situ Rong
        p. cm.--(Chapman & Hall research notes in mathematics series)
    Includes bibliographical references and index.
    ISBN 1-58488-125-9 (alk. paper)
    1. Stochastic differential equations. 2. Jump processes.
  I. Title II. Series.
  QA274.23.S525 1999
  519.2—dc21   99-33103
                                                   CIP

International Standard Book Number 1-58488-125-9
Library of Congress Card Number 99-33103
Printed in the United States of America   1 2 3 4 5 6 7 8 9 0
Printed on acid-free paper

Contents

Preface

This book is written for scientists who are interested in dealing with stochastic dynamic systems.

The components of many important physical variables can only take non-negative values, and the variables usually can be considered to satisfy some dynamic evolution systems. Therefore, one can examine them by studying such dynamic systems. In the deterministic case, one can put some conditions on the coefficients to make all components of the solutions to the dynamic systems non-negative. However, as the system is disturbed by a random process, e.g. a Wiener process, even in the simplest case the components of solutions to stochastic differential equations (SDE) can still be changing between arbitrary large positive and negative values. This prevents research going any further. To overcome this difficulty, we are motivated to examine the reflecting stochastic differential equation (RSDE) with the coordinate planes as its boundary, or with a more general boundary. In this book, the general theory and applications of RSDEs with jumps, which can be used as a stochastic population control model and a neurophysiology model etc., are systematically studied. In particular, for the case that the coefficients of RSDEs are discontinuous, with greater than linear growth, that may include jump reflection, the existence, uniqueness, comparison, convergence, and stability of strong solutions to such RSDEs are examined. The non-linear filtering and Zakai equations, the maximum principle for stochastic optimal control and the necessary and sufficient conditions for the existence of optimal control are derived. The applications to the stochastic population control problem and the neurophysiological control problem, etc., are also developed. Most parts of this book are new. It includes many new works of the author, both in stochastic differential equations with reflection and without reflection, and much of it appears here for the first time. More precisely, the book is organized as follows: Chapter 1 presents some recent results on SDE with jumps and without reflection obtained by the author, and also some preparation, e.g , Tanaka formula etc., for the RSDEs discussed later. In Chapter 2, we discuss the existence and uniqueness results for some general Skorohod problems in two cases: one is for some general reflection domain but the reflection is without jumps, which basically is adopted from Liptser and Shiryayev, 1989; another is for a half space as a reflection domain but the reflection may have jumps, and this is a generalization of Chaleyat-Maurel et al., 1980. In Chapter 3, we examine RSDE with jumps. The Yamada-Watanabe theorem, the Tanaka type formula, the Krylov estimate, the Girsanov theorem and the existence and uniqueness theorems for solutions under non-Lipschitzian conditions are all obtained for RSDEs with jumps. It seems that all such results are new. In Chapter 4, we discuss the convergence, stability, and comparison theorems for solutions to RSDE with jumps. In Chapter 5, we derive the non-linear filtering equation and a Zakai equation for the partially observed RSDE system with jumps by using the martingale representation theorem. In Chapter 6, we discuss and derive the maximum principle and necessary and sufficient conditions for the stochastic optimal control problem subject to some RSDEs with jumps by using the Girsanov theorem, variational technique, and the martingale method. Finally, we used the results obtained from the previous chapters to discuss the stochastic population control and neurophysiological control problems.

The whole book, except 2.1 and 2.2 of Chapter 2, which are adopted from Liptser and Shiryayev, 1989, is the work of the author, as is the case with reflection and jumps. However, there are still some lemmas and theorems drawn from other references, which are pointed out there.

Acknowledgement

This work is supported in part by The Foundation of Chinese National Natural Science No. 79790130. The author also would like to express his sincere thanks to Professor Alan Jeffrey for his kindly recommending the book and offering many valuable and helpful suggestions.

Chapter 1

Some Recent Results on SDE with Jumps in 1-Dimensional Space

This Chapter will give some recent results obtained by the author on stochastic differential equations (SDE) with Poisson jumps, mainly in 1-dimensional space. Actually, some of them are new.

1.1 Local Time and Occupation Density Formula

Consider a real semi-martingale x_t as follows:
(1.1) $x_t = x_0 + A_t + M_t + \int_0^t \int_Z c(s,z,\omega)q(ds,dz), t \geq 0$
where A_t is a real continuous, finite variational process with $A_0 = 0$, M_t is a real continuous, local square integrable martingale with $M_0 = 0$, and $q(dt,dz)$ is a Poisson random martingale measure such that

$q(dt,dz) = p(dt,dz) - \pi(dz)dt,$

where $p(dt,dz)$ is a Poisson point process with the compensator $\pi(dz)dt$, $\pi(\cdot)$ is a non-random $\sigma-$finite measure in some measurable space $(Z, \Re(Z))$. Obviously, we may assume that x_t is a right continuous with a left limit process (denote it by a cadlag process). To make sense of (1.1) let us assume that

(1.2) $c \in \mathbf{F}_p^{2,loc} = \left\{ \begin{array}{c} f(t,z,\omega) : f \text{ is } \Im_t - \text{predictable such that} \\ \int_0^t \int_Z |f(t,z,\omega)|^2 \pi(dz) < \infty, P-a.s. \text{ for all } t \geq 0. \end{array} \right\}$

1.1 Definition. The two-variable function $L_t^a(x)$ is called the local time of semi-martingale x_t, if it satisfies: for $\forall t \geq 0$ and $a \in R$
(1.3) $(x_t - a)^+ = (x_0 - a)^+ + \int_0^t I_{(x_{s-}>a)} dx_s + \frac{1}{2} L_t^a(x)$
$+ \int_0^t \int_Z ((x_{s-} + c(s,z,\omega) - a)^+ - (x_{s-} - a)^+ - I_{(x_{s-}>a)} c(s,z,\omega)) p(ds,dz).$

1.2 Remark. It is easily seen that the last term in the right side of (1.3) equals

$$(1.4)\ \int_0^t \int_Z (I_{(x_{s-}>a)}(x_{s-}+c(s,z,\omega)-a)^- + I_{(x_{s-}\le a)}(x_{s-}+c(s,z,\omega)-a)^+)p(ds,dz)$$
$$= \sum_{0<s\le t} I_{(x_{s-}>a)}(x_s-a)^- + \sum_{0<s\le t} I_{(x_{s-}\le a)}(x_s-a)^+.$$

Hence (1.3) can be rewritten as

$$(1.3)'\ (x_t-a)^+ = (x_0-a)^+ + \int_0^t I_{(x_{s-}>a)}dx_s + \tfrac{1}{2}L_t^a(x)$$
$$+\sum_{0<s\le t} I_{(x_{s-}>a)}(x_s-a)^- + \sum_{0<s\le t} I_{(x_{s-}\le a)}(x_s-a)^+.$$

This is the definition of local time $L_t^a(x)$ for a general real semi martingale x_t given by Meyer, 1976.

In the following we will show that the last two terms on the right side of (1.3)' are absolutely convergent for all $t \ge 0, P-a.s.$ Hence, definitions (1.3) and (1.3)' make sense. For simplicity, from now on we will always use the phrase "by localization" to simplify our proofs or statements. It means that we let

$$(1.5)\ \tau_n = \inf\left\{t\ge 0 : |x_t|^2 + |A|_t^2 + |M_t|^2 + |N_t|^2 + \langle M\rangle_t + \langle N\rangle_t > n\right\},$$

where

$|A|_t =$ the total variation of A on $[0,t]$,

$N_t = \int_0^t \int_Z c(s,z,\omega)q(ds,dz)$,

$\langle N\rangle_t = \int_0^t \int_Z |c(t,z,\omega)|^2 \pi(dz)$,

and discuss our problem for processes $x_{t\wedge\tau_n}, A_{t\wedge\tau_n}, M_{t\wedge\tau_n}$, and $N_{t\wedge\tau_n}$. After obtaining the result we then let $n\to\infty$. The advantage of such a localization technique is that $|A|^2_{t\wedge\tau_n}, |M|^2_{t\wedge\tau_n}, \langle M\rangle_{t\wedge\tau_n}$, and $\langle N\rangle_{t\wedge\tau_n}$ are bounded by n for each n,and

$$E|N_{t\wedge\tau_n}|^2 \le n,\ E|x_{t\wedge\tau_n}|^2 \le 4E|x_0|^2 + 12n.$$

1.3 Lemma. (Meyer, 1976). $L_t^a(x)$ defined by (1.3) exists, and it is a continuous increasing process.

Proof. Take $0 < a_n \downarrow 0$ and continuous functions $0 \le g_n(x)$ such that

$\int_0^{a_{n-1}} g_n(x)dx = 1,\ g_n(x) = 0$, as $x \notin (0, a_{n-1})$; $n = 1,2,\cdots$

Let

$$h_n(x) = \int_0^{(x-a)^+} dy \int_0^y g_n(u)du, n = 1,2,\cdots$$

Then

$$h_n'(x) = I_{(x>a)}\int_0^{(x-a)^+} g_n(u)du,\ h_n''(x) = g_n((x-a)^+).$$

Hence

$$h_n(x)\in C^2(R),\ |h_n'(x)|\le 1;\ h_n(x)\to (x-a)^+, h_n'(x)\to I_{(x>a)},\text{ as } n\to\infty.$$

Applying Ito's formula, we get

$$(1.6)\ h_n(x_t) - h_n(x_0) = \int_0^t h_n'(x_{s-})dx_s + A_t^n,$$

where

$$A_t^n = \sum_{0<s\le t}(h_n(x_s) - h_n(x_{s-}) - h_n'(x_{s-})\triangle x_s) + \tfrac{1}{2}\int_0^t h_n''(x_s)d\langle M\rangle_s.$$

Since $h_n''(x) \ge 0, h_n(x)$ is convex, and A_t^n is an increasing process for each n. Now let

$$\widetilde{A}_t = (x_t-a)^+ - (x_0-a)^+ - \int_0^t I_{(x_{s-}>a)}dx_s.$$

By the localization technique we may assume that M_t and N_t are martingales, moreover, $M_t, \langle M\rangle_t$, and $E\langle N\rangle_t$ are bounded. Then it is easily seen that as $n\to\infty$ by (1.6)

$A_t^n \to \widetilde{A}_t$, in probability for each $t \geq 0$.

In fact, as $n \to \infty$

$h_n(x_t) \to (x_t - a)^+$, $P - a.s.$ for each $t \geq 0$,

and by Lebesgue's domination convergence theorem as $n \to \infty$

$$E\left|\int_0^t h_n'(x_{s-})dx_s - \int_0^t I_{(x_{s-}>a)}dx_s\right|^2 \to 0,$$

From this it yields easily that $\widetilde{A}_t$ is also an increasing process. Note that

$$\triangle\widetilde{A}_t = (x_t - a)^+ - (x_{t-} - a)^+ - I_{(x_{t-}>a)} \triangle x_t, \forall t \geq 0.$$

Hence $\forall t \geq 0$

$$\triangle\widetilde{A}_t = I_{(x_{t-}>a)}(x_t - a)^- - I_{(x_{t-}\leq a)}(x_t - a)^+.$$

Therefore

$$\widetilde{A}_t \geq \sum_{0<s\leq t} \triangle\widetilde{A}_s = (1.4),$$

$$\widetilde{A}_t - \widetilde{A}_s \geq \sum_{s<u\leq t} \triangle\widetilde{A}_u, \text{ as } s < t.$$

This means that (1.4) is absolutely convergent , $P - a.s.$, and

$$\tfrac{1}{2}L_t^a(x) = \widetilde{A}_t - (1.4) = \widetilde{A}_t - \sum_{0<s\leq t} \triangle\widetilde{A}_s$$

is a continuous increasing process. ■

1.4 Corollary. Under assumption (1.2) we have that (1.4) is absolutely convergent, $P - a.s.$, i.e. $\forall t \geq 0$

$$\int_0^t \int_Z ((x_{s-} + c(s,z,\omega) - a)^+ - (x_{s-} - a)^+ - I_{(x_{s-}>a)}c(s,z,\omega))p(ds,dz)$$

$$(= \int_0^t \int_Z (I_{(x_{s-}>a)}(x_{s-} + c(s,z,\omega) - a)^-$$

$$+ I_{(x_{s-}\leq a)}(x_{s-} + c(s,z,\omega) - a)^+)p(ds,dz)) < \infty, \; P - a.s.$$

The following occupation density formula on local time is very useful.

1.5 Lemma. (Meyer, 1976). For any bounded Borel measurable function f

(1.7) $\int_0^t f(x_s)d\langle M\rangle_s = \int_R f(a)L_t^a(x)da.$

Proof. By approximation we can assume that $f \in C_0^\infty(R)$, the totality of functions with compact support. Let

$$F(x) = \int_R f(a)(x-a)^+ da, \forall x \in R.$$

Then it is easily seen that

$$F'(x) = \int_{-\infty}^x f(a)da, \; F''(x) = f(x).$$

By Ito's formula

(1.8) $F(x_t) - F(x_0) - \int_0^t F'(x_s)dA_s - \int_0^t F'(x_s)dM_s$

$$- \int_0^t \int_Z F'(x_{s-})c(s,z,\omega)q(ds,dz)$$

$$- \int_0^t \int_Z (F(x_{s-} + c(s,z,\omega)) - F(x_{s-}) - F'(x_{s-})c(s,z,\omega))p(ds,dz))$$

$$= \tfrac{1}{2}\int_0^t f(x_s)d\langle M\rangle_s.$$

By localization we can suppose that $|A|_t$, M_t, $\langle M\rangle_t$ and $\langle N\rangle_t$ are bounded. Denote the left side of (1.8) by $\sum_{i=1}^6 I_i$. Obviously,

$$I_1 = \int_R f(a)(x_t - a)^+ da,$$

$$-I_3 = \int_R f(a)(\int_0^t I_{(x_s>a)}dA_S)da,$$

$$-I_6 = \int_R f(a) \int_0^t \int_Z ((x_{s-} + c(s,z,\omega) - a)^+ - (x_{s-} - a)^+$$

$$- I_{(x_{s-}>a)}c(s,z,\omega))p(ds,dz)da,$$

where we have applied the Fubini's theorem for Lebesgue-Stieltjes integrals. Now we apply it for the stochastic integral (Ikeda & Watanabe, 1989). Then we get

$\int_0^t F'(x_s)dM_s = \int_R f(a)\int_0^t I_{(x_s>a)}dM_S da.$

Note that

$E|\int_0^t \int_Z(\int_R f(a)I_{(x_{s-}>a)}da)c(s,z,\omega))q(ds,dz)$
$-\int_R f(a)\int_0^t \int_Z I_{(x_{s-}>a)}c(s,z,\omega))q(ds,dz)da|^2$
$= E[\int_0^t \int_Z(\int_{R^2} f(a_1)f(a_2)I_{(x_{s-}>a_1)}I_{(x_{s-}>a_2)}da_1da_2)c(s,z,\omega)^2\pi(dz)ds$
$+\int_{R^2} f(a_1)f(a_2)\int_0^t \int_Z I_{(x_{s-}>a_1)}I_{(x_{s-}>a_2)}c(s,z,\omega)^2\pi(dz)dsda_1da_2$
$-2\int_{R^2} f(a_1)f(a_2)\int_0^t \int_Z I_{(x_{s-}>a_1)}I_{(x_{s-}>a_2)}c(s,z,\omega)^2\pi(dz)dsda_1da_2] = 0,$

by the Fubini's theorem for Lebesgue-Stieltjes integrals again. Hence, for each $t \geq 0$ we have that $P-a.s.$

$I_5 = \int_0^t \int_Z F'(x_{s-})c(s,z,\omega)q(ds,dz)$
$= \int_R f(a)\int_0^t \int_Z I_{(x_{s-}>a)}c(s,z,\omega))q(ds,dz)da.$

Since I_5 and $\int_0^t \int_Z I_{(x_{s-}>a)}c(s,z,\omega))q(ds,dz)$ are right continuous processes with left limit it is not difficult to derive that $P-a.s.$ the above equality for I_5 holds for all $t \geq 0$. Hence (1.7) is only a rewritten form of (1.8). ■

1.6 Remark. By (1.1) we get

(1.1)' $x_t - a = x_0 - a + \int_0^t dx_s.$

(1.3) minus (1.1)' yields that

(1.9) $(x_t - a)^- = (x_0 - a)^- - \int_0^t I_{(x_{s-}\leq a)}dx_s + \frac{1}{2}L_t^a(x)$
$+\int_0^t \int_Z((x_{s-} + c(s,z,\omega) - a)^- - (x_{s-} - a)^- - I_{(x_{s-}\leq a)}c(s,z,\omega))p(ds,dz)$
$= (x_0 - a)^- - \int_0^t I_{(x_{s-}\leq a)}dx_s + \frac{1}{2}L_t^a(x)$
$+\sum_{0<s\leq t} I_{(x_{s-}>a)}(x_s - a)^- + \sum_{0<s\leq t} I_{(x_{s-}\leq a)}(x_s - a)^+.$

(1.3) plus (1.9) becomes

(1.10) $|x_t - a| = |x_0 - a| + \int_0^t sgn(x_{s-} - a)dx_s + L_t^a(x)$
$+\int_0^t \int_Z(|x_{s-} + c(s,z,\omega) - a| - |x_{s-} - a| - sgn(x_{s-} - a)c(s,z,\omega))p(ds,dz),$

where we define $sgn0 = -1$. (1.9) and (1.10) define the same local time $L_t^a(x)$ as (1.3).

1.7 Remark. By (1.7) one has that for a.e. $a \in R$

$L_t^a(x) = \lim_{\varepsilon\downarrow 0}\varepsilon^{-1}\int_0^t I_{[a,a+\varepsilon)}(x_s)d\langle x^c\rangle_s,$

where $x^c = M$. Hence $L_t^a(x)$ can be viewed as the "measure" of occupation at the point a by the process $x(\cdot)$ during the time interval $[0,t]$.

1.2 A Generalization of Ito's Formula

Note that formula (1.3) can be rewritten as

(2.1) $f(x_t) = f(x_0) + \int_0^t f'_-(x_{s-})dx_s + \frac{1}{2}\int_R L_t^u(x)\mu_f(du)$
$+\int_0^t \int_Z(f(x_{s-} + c(s,z,\omega)) - f(x_{s-}) - f'_-(x_{s-})c(s,z,\omega))p(ds,dz)),$

where we denote $f'_-(x)$ as the left derivative of f at the point x, and μ_f as the second derivative measure of f defined by

$\mu_f\ ([a,b)) = f'_-(b) - f'_-(a)$,

which satisfies $\forall \varphi \in C_0^\infty(R)$ (the totality of smooth functions with compact supports)

$\int_R \varphi(u)\mu_f(du) = \int_R \varphi''(u)f(u)du.$

Obviously, (2.1) holds for the special convex function $(x-a)^+$. (Recall that $f(x)$ is convex, if $\forall 0 \le \lambda \le 1$, $f(\lambda x + (1-\lambda)y) \le \lambda f(x) + (1-\lambda)f(y)$). It is a natural question to ask: is (2.1) still true for any convex function? Indeed, we have the following

2.1 Lemma. (Meyer, 1976). If f is a real convex function, then (2.1) holds, provided that (1.2) is satisfied.

To prove Lemma 2.1 we need the following

2.2 Proposition. If f is a real convex function, then $f'_-(x)$ exists, it is increasing, left continuous, and satisfies

$f(b) - f(a) = \int_a^b f'_-(x)dx,\ \forall a < b.$

Proof. By the convexity of f as $x_1 < x_2 < x < x_0$

$$\frac{f(x_1) - f(x)}{x_1 - x} \le \frac{f(x_2) - f(x)}{x_2 - x} \le \frac{f(x) - f(x_0)}{x - x_0}.$$

Hence $f'_-(x)$ exists, and it is increasing. Now set a subdivision in $[a,b]$:

$a = x_0 < x_1 < \cdots < x_n = b.$

Note that

$f(b) - f(a) = \sum_{k=1}^n (f(x_k) - f(x_{k-1}))(\triangle x_k)^{-1} \triangle x_k,$

where $\triangle x_k = x_k - x_{k-1}$. Hence

$\sum_{k=1}^n f'_-(x_{k-1}) \triangle x_k \le f(b) - f(a) \le \sum_{k=1}^n f'_-(x_k) \triangle x_k.$

Now let $\text{Max}_{1\le k\le n} \triangle x_k \to 0$. Then,

$f(b) - f(a) = \int_a^b f'_-(x)dx.$

Therefore $f(x)$ is continuous. Finally, for any given x_0 and $\varepsilon > 0$ by the definition of $f'_-(x_0)$ one can take a $x_1 < x_0$ such that

(2.2) $0 \le f'_-(x_0) - (f(x_1) - f(x_0))/(x_1 - x_0) < \varepsilon/2.$

By virtue of the continuity of $f(x)$ at x_0 there exists a $\delta > 0$ such that as $x_1 < x_0 - \delta < x < x_0$

(2.3) $(f(x_1) - f(x_0))/(x_1 - x_0) - f'_-(x)$

$\le (f(x_1) - f(x_0))/(x_1 - x_0) - (f(x_1) - f(x))/(x_1 - x) < \varepsilon/2.$

By (2.2) and (2.3) it follows that $f'_-(x)$ is left continuous at x_0. ■

Now let us apply (1.7) (Lemma 1.5) to show Lemma 2.1. Actually, the proof is also similar to that of Lemma 1.5. (cf. Meyer, 1976; and He, et al., 1992).

Proof. 1) If f is convex, and μ_f has a compact support in $[-N_0, N_0]$, let

$g(x) = \int_R (x-a)^+ \mu_f(da) = \int_{-\infty}^x (x-a)\mu_f(da),$

then

$g'_-(x) = \int_R I_{(x>a)}\mu_f(da) = \int_{-\infty}^x \mu_f(da) = f'_-(x) - f'_-(-N_0).$

Hence

$$f(x) = a + bx + g(x), \forall x \in R,$$

where a and b are constants. Since (2.1) is obviously true for $a + bx$ and

$$\mu_f(da) = \mu_g(da).$$

Hence, one only needs to show that (2.1) holds for g. However, integrating (1.3) with respect to measure $\mu_g(da)$ on R and using Fubini's theorem as the proof of Lemma 1.5, one finds immediately that (2.1) holds for g.

2) Assume that f is a general convex function. Let

$$f_N(x) = \begin{cases} f(N) + f'_-(N)(x-N), \text{ as } x > N, \\ f(x), \text{ as } |x| \le N, \\ f(-N) + f'(-N)(x+N), \text{ as } x < -N, \end{cases}$$

and let τ_n as (1.5). Then by 1) $\forall t \ge 0$

$$f_N(x_{t\wedge\tau_n}) - f_N(x_0) - \int_0^{t\wedge\tau_n} f'_{N-}(x_{s-})dx_s$$
$$- \sum_{0<s\le t\wedge\tau_n}(f_N(x_s) - f_N(x_{s-}) - f'_{N-}(x_{s-}) \triangle x_s) = \int_R L^u_{t\wedge\tau_n}(x)\mu_{f_N}(du)$$
$$= \int_{[-N,N]} L^u_{t\wedge\tau_n}(x)\mu_{f_N}(du) = \int_{[-N,N]} L^u_{t\wedge\tau_n}(x)\mu_f(du).$$

As $n = N^2, \forall 0 \le t < \tau_n$ one has that

$$f(x_t) - f(x_0) - \int_0^t f'_-(x_{s-})dx_s$$
$$- \sum_{0<s\le t}(f(x_s) - f(x_{s-}) - f'_-(x_{s-}) \triangle x_s) = \int_R L^u_t(x)\mu_f(du),$$

where we have used the result that by Remark 1.7 as $|u| > N, t < \tau_{N^2}$,

$$L^u_t(x) = \lim_{\varepsilon\downarrow 0} \varepsilon^{-1} \int_0^t I_{[u,u+\varepsilon)}(x_s)ds = 0, \; a, e. \; u \in R.$$

Now since $\tau_n \uparrow \infty$, as $n \uparrow \infty$. Hence (2.1) is obtained. ∎

2.3 Corollary. If f is the difference of two convex functions, then (2.1) is still true.

We can have another useful generalization of Ito's formula.

2.4 Lemma. Assume that $f \in C^1(R)$, and there exists a $h(x)$ such that $\forall \varphi \in C_0^\infty(R)$,

$$\int_R h(x)\varphi(x)dx = -\int_R f'(x)\varphi'(x)dx,$$

(in this sense it is called $h(x)$ the second derivative of $f(x)$ in the distribution sense), and $\forall a < b$

$$\int_a^b |h(x)|\, dx < \infty,$$
$$f\prime(b) - f\prime(a) = \int_a^b h(x)dx.$$

Furthermore, if $\forall a < b$

$$\sup_{a\le x\le b} |h(x)| < \infty,$$

then

$$(2.4)\;\; f(x_t) - f(x_0) = \int_0^t f'(x_{s-})dx_s + \tfrac{1}{2}\int_0^t h(x_s)d\langle M\rangle_s$$
$$+ \sum_{0<s\le t}(f(x_s) - f(x_{s-}) - f'(x_{s-}) \triangle x_s), \forall t \ge 0.$$

Proof. Denote

the right side of (2.4)$= \sum_{i=1}^3 I_i(t)$,

$$\tau_a = \inf\{t \ge 0 : |x_t| + |A|_t + \langle M\rangle_t + \langle N\rangle_t > a\}, \forall a > 0.$$

Obviously, $\tau_a \uparrow \infty$, as $n \uparrow \infty$. Denote the smoothness of h by

$$h_n(x) = \int_R h(x - n^{-1}\overline{x})J(\overline{x})d\overline{x},$$

where

$$J(x)=\begin{cases} c\exp(-(1-|x|^2)^{-1}), & \text{as } |x|<1,\\ 0, & \text{otherwise,}\end{cases}$$

and $\int_R J(x)dx=1$. Then $\forall a\geq 0$

$$\int_{-a}^{a}|h_n(x)-h(x)|\,dx\to 0, \text{ as } n\to\infty.$$

Let

$$f_n'(x)=\int_0^x h_n(y)dy+f'(0),$$
$$f_n(x)=\int_0^x f_n'(y)dy+f(0).$$

By Ito's formula

(2.5) $f_n(x_{t\wedge\tau_a})-f_n(x_0)=\int_0^{t\wedge\tau_a}f_n'(x_{s-})dx_s+\frac{1}{2}\int_0^{t\wedge\tau_a}h_n(x_{s-})d\langle M\rangle_s$
$+\sum_{0<s\leq t\wedge\tau_a}(f_n(x_s)-f_n(x_{s-})-f_n'(x_{s-})\bigtriangleup x_s)=\sum_{i=1}^{3}I_i(t\wedge\tau_a).$

Note that as $n\to\infty,\forall x\in R$

(2.6) $|f_n'(x)-f'(x)|\to 0.$

Indeed, for any given $-\infty<b<a<\infty$ by definition uniformly for $x\in[a,b]$

$$\lim_{n,m\to\infty}|f_n'(x)-f_m'(x)|\leq\lim_{n,m\to\infty}\int_0^x|h_n(y)-h_m(y)|\,dy$$
$$\leq\lim_{n,m\to\infty}\int_a^b|h_n(y)-h_m(y)|\,dy=0.$$

Denote

$$\widetilde{f}(x)=\lim_{n\to\infty}f_n'(x).$$

$\widetilde{f}(x)$ is continuous. On the other hand, $\forall\varphi\in C_0^\infty(R)$ as $n\to\infty$

$$\int_R f_n'(x)\varphi'(x)dx=-\int_R h_n(x)\varphi(x)dx\to-\int_R h(x)\varphi(x)dx$$
$$=\int_R f'(x)\varphi'(x)dx.$$

By the uniqueness of a limit and the continuity of $\widetilde{f}(x)$ and $f'(x)$

$$\widetilde{f}(x)=f'(x),\ \forall x\in[a,b].$$

(2.6) is proved. Note also that as $n\to\infty,\forall x\in R$

$$|f_n(x)-f(x)|\leq\int_0^x|f_n'(y)-f'(y)|\,dy\to 0,$$

where we have used the result that as $a\leq x\leq b$

$$|f_n'(x)|\leq k_{a,b}<\infty,\forall n,$$

where $k_{a,b}\geq 0$ is a constant independent of n. Indeed, for example,

$$|f_n'(x)|\leq\int_0^x|h_n(y)|\,dy\leq\int_0^b dy\int_R\left|h(y-n^{-1}\overline{x})\right|J(\overline{x})d\overline{x}$$
$$\leq\int_{|\overline{x}|\leq 1}J(\overline{x})d\overline{x}\int_{-1}^{b+1}|h(y)|\,dy=k_{a,b}<\infty, \text{ as } 0\leq a\leq x\leq b.$$

Hence, $\forall s\geq 0$,as $n\to\infty$

$$f_n(x_s)\to f(x_s),$$

and

$$\int_0^{t\wedge\tau_a}f_n'(x_s)dA_s\to\int_0^{t\wedge\tau_a}f'(x_s)dA_s.$$

Since as $0\leq s<\tau_a$

(2.7) $|f_n(x_s)-f_n(x_{s-})-f_n'(x_{s-})\bigtriangleup x_s|\leq\sup_{|x|\leq a+1}|h(x)|\,(\Delta x_s)^2.$

This yields that $\forall t\geq 0$,as $n\to\infty$

$$I_3^n(t\wedge\tau_a)=\sum_{0<s<t\wedge\tau_a}(f_n(x_s)-f_n(x_{s-})-f_n'(x_{s-})\bigtriangleup x_s)$$
$$+f_n(x_{t\wedge\tau_a})-f_n(x_{t\wedge\tau_a-})-f_n'(x_{t\wedge\tau_a-})\bigtriangleup x_{t\wedge\tau_a}\to I_3(t\wedge\tau_a).$$

Note also that

$$E\left|\int_0^{t\wedge\tau_a}(h_n(x_s)-h(x_s))d\langle M\rangle_s\right|\leq E\int_0^{t\wedge\tau_a}I_{|x_s|\leq a}|h_n(x_s)-h(x_s)|\,d\langle M\rangle_s$$
$$\leq\sup_{|u|\leq a}EL^u_{t\wedge\tau_a}(x)\int_{-a}^{a}|h_n(u)-h(u)|\,du\to 0, \text{ as } n\to\infty,$$

where we have applied Proposition 2.6 below. Hence as $n\to\infty$

$I_2^n(t \wedge \tau_a) \to \frac{1}{2}\int_0^{t\wedge\tau_a} h(x_s)d\langle M\rangle_s$, in probability; $\forall t \geq 0$.

Similarly, as $n \to \infty$

$E\int_0^{t\wedge\tau_a} |f_n'(x_s) - f'(x_s)|^2 d\langle M\rangle_s \to 0.$

Hence as $n \to \infty$

$I_1^n(t \wedge \tau_a) \to I_1(t \wedge \tau_a)$,in probability; $\forall t \geq 0$.

Therefore in (2.5) letting $n \to \infty$ first and $a \to \infty$ secondly, (2.4) is obtained. ∎

2.5 Proposition. Let

$\frac{1}{2}\text{Ł}_t^a(x) = \frac{1}{2}L_t^a(x) + \sum_{0<s\leq t} I_{(x_{s-}>a)}(x_s - a)^- + \sum_{0<s\leq t} I_{(x_{s-}\leq a)}(x_s - a)^+.$

Then

(2.7) $\frac{1}{2}\sup_{a\leq u\leq b} E\text{Ł}_t^u(x) \leq 2c_0 E(|A|_t + \langle M\rangle_t^{1/2} + \langle N\rangle_t^{1/2}),$

(2.8) $\frac{1}{2}\sup_{a\leq u\leq b} E(\text{Ł}_t^u(x)^2) \leq 4c_0' E(|A|_t^2 + \langle M\rangle_t + \langle N\rangle_t),$

where $c_0, c_0' \geq 0$ are constants.

Proof. By definition

$\frac{1}{2}\text{Ł}_t^a(x) = (x_t - a)^+ - (x_0 - a)^+ - \int_0^t I_{(x_{s-}>a)}dx_s$

$\leq |x_t - x_0| + \left|\int_0^t I_{(x_{s-}>a)}dx_s\right| \leq 2|A|_t + |M_t| + |N_t|$

$+\left|\int_0^t I_{(x_{s-}>a)}dM_s\right| + \left|\int_0^t I_{(x_{s-}>a)}dN_s\right|.$

Hence (2.7) and (2.8) are derived. ∎

2.6 Remark. By the proof of (2.4) it is seen that if x_t is a continuous semi-martingale, then the proof of (2.7) is not necessary, since

$I_3(t) = 0, \forall t \geq 0.$

Hence the condition

$\sup_{a\leq x\leq b} |h(x)| < \infty$

can also be erased. In this case Ito's formula can be found in Rogers and Williams, 1987.

1.3 The Continuity of Local Time

To discuss the continuity of local time we need the following additional assumption besides (1.2): $\forall t \geq 0$

(3.1) $\int_0^t \int_Z |c(t,z,\omega)| p(ds,dz)$ $(= \sum_{0<s\leq t} |\triangle x_s|) < \infty$, $P-a.s.$

3.1 Lemma. (Yor, 1978). Under assumptions (1.2) and (3.1) $L_t^a(x)$ has a version, which is cadlag in $a \in R$, $P-a.s.$

Proof. By (3.1) one can rewrite (1.1) as

$x_t = x_t + \overline{A}_t + M_t + \sum_{0<s\leq t} \triangle x_s,$

where $\overline{A}_t$ is a continuous finite variational process. By the equivalent definition $(1.3)'$ for local time one has that

$\frac{1}{2}L_t^a(x) = (x_t - a)^+ - (x_0 - a)^+ - \int_0^t I_{(x_{s-}>a)}d\overline{A}_s - \int_0^t I_{(x_{s-}>a)}dM_s$

$- \sum_{0<s\leq t} I_{(x_{s-}>a)} \triangle x_s - \sum_{0<s\leq t} I_{(x_{s-}>a)}(x_s - a)^-$

$- \sum_{0<s\leq t} I_{(x_{s-}\leq a)}(x_s - a)^+ = (x_t - a)^+ - (x_0 - a)^+ - \int_0^t I_{(x_{s-}>a)}d\overline{A}_s$

$-\int_0^t I_{(x_{s-}>a)}dM_s - \sum_{0<s\leq t} I_{(x_{s-}>a)} \triangle x_s$
$-\int_0^t \int_Z ((x_{s-} + c(s,z,\omega) - a)^+ - (x_{s-} - a)^+ - I_{(x_{s-}>a)}c(s,z,\omega))p(ds,dz)$
$= \sum_{i=1}^6 I_i(a,t).$

Obviously, $I_1(a,t)$ and $I_2(a,t)$ are continuous in $a \in R$. Moreover, by the Lebesgue dominated convergence theorem and by (3.1) it is easily seen that $I_3(a,t)$, $I_5(a,t)$, and $I_6(a,t)$ are cadlag in $a \in R$, e.g.

$\lim_{b\to a+} I_5(b,t) = I_5(a,t)$, $\lim_{b\to a-} I_5(b,t) = \sum_{0<s\leq t} I_{(x_{s-}\geq a)} \triangle x_s;$

and

$\lim_{b\to a+} I_6(b,t) = I_6(a,t)$, $\lim_{b\to a-} I_6(b,t)$
$= \int_0^t \int_Z ((x_{s-} + c(s,z,\omega) - a)^+ - (x_{s-} - a)^+ - I_{(x_{s-}\geq a)}c(s,z,\omega))p(ds,dz).$

Now let us apply Kolmogorov's theorem to show that $I_4(a,t)$ has a version that is continuous in $a \in R$. In fact, for $a < b \in R$ by (2.7)

$$E\left|\int_0^t (I_{(x_{s-}>b)} - I_{(x_{s-}>a)})dM_s\right|^4 \leq c_0 E\left|\int_0^t I_{(b\geq x_{s-}>a)} d\langle M\rangle_s\right|^2$$
$$= c_0 E\left|\int_R I_{(b\geq u>a)} L_t^u(x)du\right|^2 \leq c_0 \sup_{a\leq u\leq b} E\left|L_t^u(x)\right|^2 |b-a|^2$$
$$\leq 48\,|b-a|^2 c_0 E(|A|_t^2 + \langle M\rangle_t + \int_0^t \int_Z |c(s,z,\omega)|^2 \pi(dz)ds),$$

where $c_0 \geq 0$ is a constant. Since by localization we can suppose that $|A|_t$, $\langle M\rangle_t$, and $\langle N\rangle_t$ are all bounded, where $\langle N\rangle_t$ is defined by (1.5). Hence, Kolmogorov's theorem on the continuous version applies. ■

3.2 Remark. By the proof above it is not difficult to see that

$\lim_{b\to a+} \sup_{0\leq t\leq T} \left|L_t^b(x) - L_t^a(x)\right| = 0,$
$\lim_{b\to a-} \sup_{0\leq t\leq T} \left|L_t^b(x) - L_t^{a-}(x)\right| = 0,$

for a suitable version of $L_t^a(x)$. Furthermore, one can get

(3.2) $\lim_{b\to a+, s\to t} L_s^b(x) = L_t^a(x)$, $\lim_{b\to a-, s\to t} L_s^b(x)$ exists.

1.4 Krylov Estimation

By means of the occupation density formula one easily derives the so-called Krylov estimation with respect to the SDE with jumps as follows:

4.1 Theorem. (Situ, R. 1987c). Assume that x_t satisfies (1.1) with c satisfying (1.2). Let τ_N be defined as (1.5). If $f \in L^1(R)$, then for any $0 \leq T < \infty$, one has that

(4.1) $E\int_0^{T\wedge\tau_N} |f(x_s)|\, d\langle M\rangle_s \leq k_{N,T} \int_R |f(y)|\, dy \widehat{=} k_{N,T} \|f\|_{L^1(R)}$,

where $k_{N,T} \geq 0$ is a constant independent of f such that

$k_{N,T} = k_0 E(|A|_{T\wedge\tau_N} + \langle M\rangle_{T\wedge\tau_N}^{1/2} + \langle N\rangle_{T\wedge\tau_N}^{1/2}).$

Furthermore, assume that

(4.2) M_t is a square integrable martingale, c satisfies

$E\int_0^T \int_Z |c(s,z,\omega)|^2 \pi(dz)ds < \infty;$

and

(4.3) $E|A|_T < \infty.$

Then (4.1) can be strengthened to

(4.4) $E\int_0^T |f(x_s)|\, d\langle M\rangle_s \leq k_T \|f\|_{L^1(R)}$,
where $0 \leq k_T = k_{\infty,T}$.
Proof. Let $f(x) \geq 0$ and
$f_n(x) = f(x) \wedge n$.
Then by the formula (1.7)
$\int_0^t |f_n(x_s)|\, d\langle M\rangle_s = \int_R |f_n(a)|\, L_t^a(x)da, \forall t \geq 0.$
By the proof of (2.7) one sees that
$0 \leq \frac{1}{4}EL^a_{T\wedge\tau_N}(x) \leq k_0 E(|A|_{T\wedge\tau_N} + \langle M\rangle^{1/2}_{T\wedge\tau_N} + \langle N\rangle^{1/2}_{T\wedge\tau_N}) \widehat{=} k_{N,T}.$
Hence,
$E\int_0^{T\wedge\tau_N} |f_n(x_s)|\, d\langle M\rangle_s \leq k_{N,T}\int_R |f_n(y)|\, dy \leq k_{N,T}\int_R |f(y)|\, dy.$
Letting $n \to \infty$ and applying Fatou's lemma, one derives (4.1). Now under the further assumptions (4.2) and (4.3) one has that
$k_{N,T} \leq k_{\infty,T} < \infty.$
Hence (4.4) is also derived. ■

4.2 Corollary. If
$M_t = \int_0^t \sigma(s,\omega)dw_s,\ A_t = \int_0^t b(s,\omega)ds,$
and
$|b(t,\omega)|^2 + |b(t,\omega)|^2 + \int_Z |c(t,z,\omega)|^2\, \pi(dz) \leq k_0,$
where $0 \leq k_0$ is a constant, and $w_t, t \geq 0$, s a standard Brownian Motion process, then for any f, which is a Borel measurable function valued in R,
(4.5) $E\int_0^T |f(x_s)|\, ds \leq k_T \|f\|_{L^1(R)}$,
where $0 \leq k_T$ is a constant depending on k_0 and T only.
Proof. If
$\|f\|_{L^1(R)} = \infty,$
then (4.5) is obviously true. In case
$\|f\|_{L^1(R)} < \infty,$
then (4.5) is derived by (4.4). ■

1.5 Tanaka Formula

In this section we shall discuss a Tanaka formula with respect to the difference of two semi-martingales. Assume that semi-martingales like (1.1) satisfy
(5.1) $x_t^i = x_0^i + A_t^i + \widetilde{M}_t^i + \int_0^t \int_Z c(s, x_s^i, z, \omega)q(ds, dz),\ i = 1, 2,$
where $A_t^i, \overline{M}_t^i$ and $q(dt, dz)$ have the same meaning as that in (1.1). If $P-a.s.$
(5.2) $\int_0^t \int_Z |c(s, x_S^i, z, \omega)|^2\, \pi(dz)ds < \infty, i = 1, 2,$
then by (1.1) and (1.10) we have
(5.3) $|x_t^1 - x_t^2| = |x_0^1 - x_0^2| + \int_0^t sgn(x_{s-}^1 - x_{s-}^2)d(x_s^1 - x_s^2) + L_t^0(x^1 - x^2)$
$+J_t,$
where
(5.4) $J_t = \int_0^t \int_Z (|x_{s-}^1 - x_{s-}^2 + c(s, x_{s-}^1, z, \omega) - c(s, x_{s-}^2, z, \omega)| - |x_{s-}^1 - x_{s-}^2|$
$-sgn(x_{s-}^1 - x_{s-}^2)(c(s, x_s^1, z, \omega) - c(s, x_s^2, z, \omega))p(ds, dz).$

By (5.3) and (5.4) it is easily seen that the following Lemma is true.

5.1 Lemma. (Situ, R. 1985a). Assume that (5.2) holds, and
(5.5) $x \geq y \Rightarrow x + c(t, x, z, \omega) \geq y + c(t, y, z, \omega)$.
Then
(5.6) $J_t = 0, \forall t \geq 0$.
Now assume that in (5.1) there exists a local square integrable martingale M_t such that
(5.7) $\widetilde{M_t}^i = \int_0^t \sigma^i(s, \omega) dM_s$,
then $\widetilde{M_t}^i$ is a local square integrable martingale as is M_t, $i = 1, 2$, if for any given $x \in R$ $\sigma^i(s, \omega)$ is $\Im_t$-adapted and jointly measurable such that
(5.8) $\int_0^t |\sigma^i(s, \omega)|^2 d\langle M \rangle_s < \infty, \forall t \geq 0, P - a.s.\ i = 1, 2$.
Now introduce some assumptions as follows: It is said that $\overline{\sigma}$ with σ satisfies condition (5.9), if for each $N = 1, 2, \cdots$, and $T > 0$ as $t \in [0, T]$; $|x|, |y| \leq N$
(5.9) $|\overline{\sigma}(t, x, \omega) - \overline{\sigma}(t, y, \omega)| \leq (k_{N,T}(t, \omega)$
$+\varphi_{N,T}(|x - y|) \, |\sigma(t, x, \omega) - \sigma(t, y, \omega)|) \cdot \rho_{N,T}(|x - y|)$,
where $k_{N,T}(t, \omega) \geq 0$ and $\varphi_{N,T} \geq 0$ is non-random such that
$E \int_0^T |k_{N,T}(t, \omega)|^2 d\langle M \rangle_t < \infty$, $\int_{|u| \leq 2N} |\varphi_{N,T}(u)|^2 du < \infty$,
and $0 \leq \rho_{N,T}$ is strictly increasing, $\rho_{N,T}(0) = 0$ such that
$\int_{0+} du / \rho_{N,T}(u)^2 = +\infty$.
We have

5.2 Theorem. (Situ, R. 1985a, 1987c). If (5.2) and (5.8) hold, and one of the following conditions is fulfilled:

(i) σ with σ satisfies (5.9);

(ii) $\sigma(t, x, \omega) = a\sigma^H(t, x, \omega) + b\sigma^V(t, x, \omega)$,
where $a, b > 0$ are constants, σ^H with σ satisfies (5.9), and
$\sigma^H(t, x, \omega) \geq 0$,
moreover, $\sigma^V(t, x, \omega) \geq r > 0$, r is a constant, and there exists a non-random locally finite variational function $v(x)$ such that
$|\sigma^V(t, x, \omega) - \sigma^V(t, y, \omega)|^2 \leq k_N |v(x) - v(y)|$, as $|x|, |y| \leq N$,
where $0 \leq k_N$ is a constant depending on N only, and
(5.10) $|\sigma^V(t, x, \omega)| + |\sigma^H(t, x, \omega)| \leq g(x)$,
where $g(x)$ is a continuous function, which is non-random;

(iii) $\sigma(t, x, \omega) = \sigma^H(t, x, \omega) \cdot \sigma^V(t, x, \omega)$,
where $\sigma^H(t, x, \omega)$ and $\sigma^V(t, x, \omega)$ satisfy conditions in (ii) respectively, but without assumption that $\sigma^H \geq 0$;
then under the additional condition (3.1) one has that (5.3) holds with
(5.11) $L_t^0(x_1 - x_2) = 0, \forall t \geq 0$.
Furthermore, if (5.5) holds, then
$J_t = 0, \forall t \geq 0$.
In this case we have a Tanaka formula
(5.12) $|x_t^1 - x_t^2| = |x_0^1 - x_0^2| + \int_0^t sgn(x_{s-}^1 - x_{s-}^2) d(x_s^1 - x_s^2)$.

5.3 Remark. Conditions (i), (ii), and (iii) mean that σ can be degenerate, σ is uniformly non-degenerate and σ can be degenerate but is locally bounded (satisfying (5.10)), respectively. The interesting thing is that in (ii) and (iii) σ can be discontinuous.
To prove Theorem 5.2 we need a lemma as follows:

5.4 Lemma. If $f(x)$ and $g(x)$ are any two real functions, then

$$|f(x)g(x) - f(y)g(y)| \leq (|f(x)| \wedge |f(y)|)\, |g(x) - g(y)| + (|g(x)| \vee |g(y)|)\, |f(x) - f(y)|,$$

$$|f(x)| \wedge |f(y)| \leq |\lambda f(x) + (1-\lambda) f(y)| + |f(x) - f(y)|, \forall \lambda \in [0,1].$$

Furthermore, if $g(x), g(y) \geq r > 0$, then $\forall \lambda \in [0,1]$

$$|f(x)| \wedge |f(y)| \leq r^{-1} |\lambda f(x)g(x) + (1-\lambda) f(y)g(y)| + |f(x) - f(y)|.$$

Proof. Obviously, the first inequality is true. We now discuss the second one. If $f(x) \cdot f(y) \geq 0$, then

$$|f(x)| \wedge |f(y)| \leq |\lambda f(x) + (1-\lambda) f(y)|, \forall \lambda \in [0,1].$$

If $f(x) \cdot f(y) < 0$, assume that

$$0 < |f(y)| \leq |f(x)|,$$

then

$$|f(x)| \wedge |f(y)| = |f(y)| \leq |f(x) - f(y)|.$$

Assume now that $g(x)$ and $g(y)$ are greater than $r > 0$. If $f(x) \cdot f(y) \geq 0$, then $\forall \lambda \in [0,1]$

$$|f(x)| \wedge |f(y)| \leq r^{-1} |\lambda f(x)g(x) + (1-\lambda) f(y)g(y)|.$$

Hence the second and third inequalities are also true. ■

Now we are in a position to establish Theorem 5.2.

Proof. Assume that N and $T > 0$ are given arbitrarily. For simplicity we just denote all the functions $k_{N,T}(t)$, $\varphi_{N,T}$, $\rho_{N,T}, \cdots$, which depend on N and T in the condition (5.9) above and in the following, simply by $k(t)$, φ, $\rho, \cdots$ etc. As the method of proving Lemma 1.3 we take $a_n \downarrow\downarrow 0$ such that

$$\int_{a_n}^{a_{n-1}} (\rho(u)^2 \vee u)^{-1} du = n,$$

and take continuous functions $g_n(x) \geq 0$ such that

$$g_n(x) = 0, \text{ as } x \notin (a_n, a_{n-1});\ g_n(x) \leq 2(n(\rho(x)^2 \vee x))^{-1}, \forall x;$$

and

$$\int_{a_n}^{a_{n-1}} g_n(x)dx = 1.$$

Let

$$h_n(x) = \int_0^{|x|} \int_0^y g_n(u)du dy.$$

Then by Ito's formula

$$\begin{aligned}(5.13)\ h_n(x_t^1 - x_t^2) &= h_n(x_0^1 - x_0^2) + \int_0^t h_n'(x_{s-}^1 - x_{s-}^2)d(x_s^1 - x_s^2) \\ &+ \tfrac{1}{2}\int_0^t g_n(|x_s^1 - x_s^2|)\, \left|\sigma(s, x_s^1, \omega) - \sigma(s, x_s^2, \omega)\right|^2 d\langle M \rangle_s \\ &+ \int_0^t \int_Z (h_n(x_{s-}^1 - x_{s-}^2 + c(s, x_{s-}^1, \omega) - c(s, x_{s-}^2, \omega)) - h_n(x_{s-}^1 - x_{s-}^2) \\ &- h_n'(x_{s-}^1 - x_{s-}^2)(c(s, x_{s-}^1, \omega) - c(s, x_{s-}^2, \omega)))p(ds, dz) = \textstyle\sum_{i=1}^4 I_i^n(t).\end{aligned}$$

Case (i): Let τ_N^i as (1,5) for x_t^i, and $\tau_N = \tau_N^1 \wedge \tau_N^2$. Then

$$\begin{aligned}(5.14)\ E\left|I_3^n(t \wedge \tau_N)\right| &\leq 2E\int_0^{t \wedge \tau_N} g_n(|x_s^1 - x_s^2|)(k(s)^2 \\ &+ \varphi(|x_s^1 - x_s^2|)^2 \left|\sigma(t, x_s^1, \omega) - \sigma(t, x_s^2, \omega)\right|^2) \cdot \rho(|x_s^1 - x_s^2|)^2 I_{|x_s^1 - x_s^2| \leq 2N} d\langle M \rangle_s\end{aligned}$$

$\leq 2n^{-1}(E\int_0^t k(s)^2 d\langle M\rangle_s + \int_R E\ L^a_{t\wedge\tau_N}(x^1 - x^2)\ I_{|a|\leq 2N}\varphi(a)^2 da)$

$\leq 2n^{-1}(k'_N + k_N) \to 0$, as $n \to \infty$,

where we have applied that

$0 \leq E\ L^a_{t\wedge\tau_N}(x^1 - x^2) \leq k_0 N$, k_0- constant.

(See (2.7)). Now by assumption (3.1), applying the Lebesgue dominated convergence theorem one easily sees that

$\lim_{n\to\infty} I_4^n(t) = J_t$.

Hence

$L^0_t(x^1 - x^2) = 0, \forall t \geq 0$;

i.e. (5.3) holds with (5.11).

Case (iii): Let

$\sigma_m(x) = \int_R v(x - m^{-1}y)\overline{J}(y)dy$,

where

$$\overline{J}(x) = \begin{cases} k_0 \exp(-(1-x^2)^{-1}), & \text{as } |x| < 1, \\ 0, & \text{otherwise,} \end{cases}$$

$k_0 = (\int_{-1}^1 \exp(-(1-x^2)^{-1})dx)^{-1}$.

Since $v(x)$ is locally finite variational, $\sigma_m(x)$ is uniformly bounded (with respect to m) in any finite interval of x, and $\sigma_m(x)$ is continuously differentiable such that

$\lim_{m\to\infty}\sigma_m(x) = v(x), \forall x \in R$, *a.e.*

and for each $N > 0$ there exits a constant k_N depending on N only such that

$\sup_{m\geq 1}\int_{-N}^N |\sigma'_m(x)|\,dx \leq k_N < \infty$.

Note that as $|x|, |y| \leq N$

$|\sigma(t,x,\omega) - \sigma(t,y,\omega)|^2 \leq (|\sigma^H(t,x,\omega)| \wedge |\sigma^H(t,y,\omega)|)^2 |v(x) - v(y)| \cdot k_N$

$+\widetilde{k_N}\,|\sigma^H(t,x,\omega) - \sigma^H(t,y,\omega)|^2 = I_1(x,y,\omega) + I_2(x,y,\omega)$,

$E\int_0^{t\wedge\tau_N} g_n(|x^1_s - x^2_s|)\,|\sigma^H(s,x^i_s,\omega)|^2\,|v(x^i_s) - \sigma_m(x^i_s)|\,d\langle M\rangle_s$

$\leq 2(n\widetilde{a}_n)^{-1}\int_R EL^u_{t\wedge\tau_N}(x^i)r^{-2}\,|v(u) - \sigma_m(u)|\,I_{|u|\leq N}du \to 0$, as $m \to \infty$,

for each n and N; $i = 1, 2$; where $\widetilde{a}_n = \rho(a_n)^2 \wedge a_n$. Therefore by Fatou's lemma from (5.13)

$EI_3^n(t\wedge\tau_N) \leq \lim_{m\to\infty} E\int_0^{t\wedge\tau_N} g_n(|x^1_s - x^2_s|)\,|\sigma^H(s,x^1_s,\omega)\wedge\sigma^H(s,x^2_s,\omega)|^2$

$\cdot|\sigma_m(x^1_s) - \sigma_m(x^2_s)|\,d\langle M\rangle_s \cdot k_N + \int_0^{t\wedge\tau_N} g_n(|x^1_s - x^2_s|)I_2(x,y,\omega)d\langle M\rangle_s$

$= \sum_{i=1}^2 J^i_{n,N}$.

Moreover, by Lemma 5.4

$J^1_{n,N} \leq \lim_{m\to\infty} E\int_0^{t\wedge\tau_N} g_n(|x^1_s - x^2_s|)r^{-2}k_N$

$\cdot\int_0^1 |\alpha\sigma(s,x^1_s,\omega) + (1-\alpha)\sigma(s,x^2_s,\omega)|^2\,|\sigma'_m(\alpha x^1_s + (1-\alpha)x^2_s)|\,d\alpha$

$\cdot|x^1_s - x^2_s|\,I_{|x^1_s|\leq N}I_{|x^2_s|\leq N}d\langle M\rangle_s + \overline{k}_N J^2_{n,N} = \widetilde{J}^1_{n,N} + \overline{k}_N J^2_{n,N}$,

and

$\widetilde{J}^1_{n,N}$

$\leq \lim_{m\to\infty} 2n^{-1}r^{-2}\int_0^1\int_R I_{|u|\leq 2N}EL^u_{t\wedge\tau_N}(\alpha x^1 + (1-\alpha)x^2)\,|\sigma'_m(u)|\,dud\alpha k_N$

$\leq n^{-1}\widetilde{k}_N\sup_{m\geq 1}\int_{-2N}^{2N}|\sigma'_m(u)|\,du \leq n^{-1}\overline{k}_N \to 0$, as $n \to \infty, \forall N$.

By (5.14) we similarly get

$J^2_{n,N} \to 0$, as $n \to \infty, \forall N$.
Hence we also get
$L^0_{t\wedge\tau_N}(x^1 - x^2) = 0.$
The rest of the proof is similar to case (i). Case (ii) can be shown in the same manner as (iii). ■

5.5 Remark. Le Gall (1983) obtained (5.11) for continuous systems ($c = 0$) under condition that (5.9) is true for all x, y (i.e. it is a global condition) with $\varphi_{N,T} = 0$ and $k_{N,T} = 1$, and under assumption that $\sigma^H = 1$ or $\sigma^V = 1$ in (iii). However, he used the right continuity of $L^a_t(x^1 - x^2)$ in a at $a = 0$ to establish this result. Here we do not make assumption (3.1). So for our systems with jumps this technique can not be applied. However, our result (Theorem 5.2) implies his.

Furthermore, we can have another theorem for Tanaka type formula as follows:

5.6 Theorem. If we modify conditions (ii) and (iii) in Theorem 5.2 to the following conditions (ii)' and (iii)', respectively:
(ii)' $\sigma(t, x, \omega) = a\sigma^H(t, x, \omega) + b\widetilde{\sigma}^H(t, \sigma^V(t, x, \omega))$,
where σ^H satisfies the same condition in (ii), but for $\widetilde{\sigma}^H$ we make the following assumption
$\widetilde{\sigma}^H(t, \sigma^V(t, x, \omega)) \geq r > 0$, r is a constant,
and
$\left|\widetilde{\sigma}^H(t, x) - \widetilde{\sigma}^H(t, y)\right| \leq \widetilde{\rho}_N(|x - y|)$, as $|x|, |y| \leq k_N$,
where $k_N \geq N$ is such a constant dependent on N that as $|x| \leq N + 1$
$|V(x)| \leq k_N$,
and $0 \leq \widetilde{\rho}_N(u)$, as $u \geq 0; \widetilde{\rho}_N(0) = 0$, and $\widetilde{\rho}_N(u)$ is increasing such that
$\int_{0+} du/\widetilde{\rho}_N(u)^2 = \infty$,
$\widetilde{\rho}_N(ku) \leq c_N k^{1/2}\widetilde{\rho}_N(u)$, $\forall u, k \geq 0$, as $0 \leq u \leq 2k_N$,
where c_N – a non-negative constant depending on N only and, moreover, there exists a non-random locally finite variational function $V(x)$ such that
$\left|\sigma^V(t, x, \omega) - \sigma^V(t, y, \omega)\right| \leq |V(x) - V(y)|$,
furthermore,
$\left|\sigma^H(t, x, \omega)\right| + \left|\widetilde{\sigma}^H(t, \sigma^V(t, x, \omega))\right| \leq g(x)$,
where $g(x)$ is a non-random continuous function;
(iii)' $\sigma(t, x, \omega) = \sigma^H(t, x, \omega) \cdot \widetilde{\sigma}^H(t, \sigma^V(t, x, \omega))$,
where $\sigma^H(t, x, \omega), \widetilde{\sigma}^H(t, x), \sigma^V(t, x, \omega)$, and $\widetilde{\sigma}^H(t, \sigma^V(t, x, \omega))$ satisfy conditions in (ii)' respectively, but without assumption $\sigma^H(t, x, \omega) \geq 0$; then all conclusions in Theorem 5.2 hold provided that (3.1) is satisfied.

5.7 Remark. Condition
$\widetilde{\rho}_N(ku) \leq c_N k^{1/2}\widetilde{\rho}_N(u)$, $\forall u, k \geq 0$, as $0 \leq u \leq 4k_N$,
implies that
$\widetilde{\rho}_N(u + v) \leq \widetilde{c}_N\left(\widetilde{\rho}_N(u) + \widetilde{\rho}_N(v)\right)$, as $0 \leq u, v \leq 2k_N$,

where $\widetilde{c}_N = c_N\sqrt{2}$. Indeed, since $\widetilde{\rho}_N(u)$ is increasing in u, if $0 \leq v \leq u \leq N$, then

$\widetilde{\rho}_N(u+v) \leq c_N 2^{1/2}\widetilde{\rho}_N(u) \leq \widetilde{c}_N\left(\widetilde{\rho}_N(u) + \widetilde{\rho}_N(v)\right).$

Now let us show Theorem 5.6.

Proof. We only need to make some mild modification in the proof of Theorem 5.2. For example, now let $a_n \downarrow\downarrow 0$ such that

$\int_{a_n}^{a_{n-1}}(\rho(u)^2 \vee \widetilde{\rho}(u)^2)^{-1}du = n,$

where like the proof of Theorem 5.2 for simplicity we just write ρ and $\widetilde{\rho}$ for ρ_N and $\widetilde{\rho}_N$, etc. Take continuous functions $g_n(x) \geq 0$ such that

$g_n(x) \leq 2(n(\rho(u)^2 \vee \widetilde{\rho}(u)^2))^{-1},$

$g_n(x) = 0$, as $x \notin (a_n, a_{n-1})$,

and

$\int_{a_n}^{a_{n-1}} g_n(x)dx = 1.$

Then denote $h_n(x)$ as in proof of Theorem 5.2 by $g_n(x)$. Discuss step by step as there, but notice that

(5.15) $E\int_0^{t\wedge\tau_N} g_n(|x_s^1 - x_s^2|)\left|\sigma^H(s,x_s^i,\omega)\right|^2 \widetilde{\rho}(|v(x_s^i) - \sigma_m(x_s^i)|)^2 d\langle M\rangle_s$

$\leq 2(n\rho(a_n)^2)^{-1}r^{-2}\int_R EL^u_{t\wedge\tau_N}(x^i)\widetilde{\rho}(|v(u) - \sigma_m(u)|)^2 I_{|u|\leq N}du.$

Since $\widetilde{\rho}(u)$ is continuous at $u = 0$ and as $m \to \infty$

$\sigma_m(u) \to v(u)$, in Lebesgue measure.

Hence it is not difficult to show that as $m \to \infty$

(5.15)$\to 0$.

Note also that

$\lim_{m\to\infty} E\int_0^{t\wedge\tau_N} g_n(|x_s^1 - x_s^2|)r^{-2}\int_0^1 \left|\alpha\sigma(s,x_s^1,\omega) + (1-\alpha)\sigma(s,x_s^2,\omega)\right|^2 \cdot$

$\cdot\widetilde{\rho}\left(\left|\sigma'_m(\alpha x_s^1 + (1-\alpha)x_s^2)\right| \cdot \left|x_s^1 - x_s^2\right|\right)^2 d\alpha \cdot I_{|x_s^1|\leq N,|x_s^2|\leq N} d\langle M\rangle_s$

$\leq \lim_{m\to\infty} 2n^{-1}r^{-2}2Nc_N^2\int_0^1\int_R EL^u_{t\wedge\tau_N}(\alpha x^1 + (1-\alpha)x^2)\left|\sigma'_m(u)\right| I_{|u|\leq 2N}dud\alpha$

$\to 0$, as $n \to \infty$.

Hence, all conclusions in Theorem 5.2 still hold. ■

5.8 Example. Let

$\widetilde{\sigma}^H(t,x) = |x|^{1/2},$

$\sigma^V(t,x,\omega) = V(x) \geq r > 0,$

$\widetilde{\rho}_N(u)^2 = u$, as $u \geq 0$,

where $V(x)$ is a non-random local finite variational function. Then all conditions on $\widetilde{\sigma}^H$ in Theorem 5.2 are fulfilled.

Tanaka formulas (5.12) and (5.3) are very powerful for discussing the properties of solutions to SDE with jumps on uniqueness, comparison, convergence, and so on. (For cases without jumps such discussion can be seen in Le Gall, 1983, 1984). In the next section we will consider such problems.

The Tanaka formula for SDE with different jump coefficients is also useful. In this case one still has the formula (5.3) with (5.4) but for $c^1 - c^2$. More precisely, assume that semi-martingales $x_t^i, i = 1,2$, satisfy

(5.16) $x_t^i = x_0^i + A_t^i + \widetilde{M}_t^i + \int_0^t\int_Z c^i(s, x_{s-}^i, z, \omega)q(ds,dz), i = 1,2,$

where $A_t^i, \widetilde{M}_t^i$ and $q(dt,dz)$ have the same meaning as (5.1). If $P - a.s.$

(5.17) $\int_0^t \int_Z \left|c^i(s,x^i_{s-},z,\omega)\right|^2 \pi(dz) < \infty, i = 1,2,$
then we have the following

5.9 Theorem. The following Tanaka formula is true:
(5.18) $\left|x^1_t - x^2_t\right| = \left|x^1_0 - x^2_0\right| + \int_0^t sgn(x^1_{s-} - x^2_{s-})d(x^1_s - x^2_s) + L^0_t(x^1 - x^2) + J_t,$
where
(5.19) $J_t = \int_0^t \int_Z (\left|x^1_{s-} - x^2_{s-} + c^1(s,x^1_{s-},z,\omega) - c^2(s,x^2_{s-},z,\omega)\right|$
$- \left|x^1_{s-} - x^2_{s-}\right| - sgn(x^1_{s-} - x^2_{s-})\left(c^1(s,x^1_{s-},z,\omega) - c^2(s,x^2_{s-},z,\omega)\right))p(ds,dz).$
Furthermore, if $c^i, i = 1,2$, satisfy condition (3.1), i.e.
(5.20) $\int_0^t \int_Z \left|c^i(s,x^i_{s-},z,\omega)\right| p(ds,dz) < \infty,\ i = 1,2,\ P - a.s.$
and conditions in Theorem 5.2 or Theorem 5.6 are satisfied except (5.5), then
(5.21) $L^0_t(x^1 - x^2) = 0, \forall t \geq 0.$

The following corollary is obviously true.

5.10 Corollary. Under assumption of Theorem 5.9 and the additional condition

$\int_0^t \int_Z \left|c^i(s,x^i_{s-},z,\omega)\right| \pi(dz)ds < \infty,\ i = 1,2,\ P - a.s.$

one has that
(5.22) $\left|x^1_t - x^2_t\right| = \left|x^1_0 - x^2_0\right| + \int_0^t sgn(x^1_{s-} - x^2_{s-})(d(A^1_s - A^2_s)$
$+d(\widetilde{M}^1_s - \widetilde{M}^2_s)) + \int_0^t \int_Z (\left|x^1_{s-} - x^2_{s-} + c^1(s,x^1_{s-},z,\omega) - c^2(s,x^2_{s-},z,\omega)\right|$
$- \left|x^1_{s-} - x^2_{s-}\right|)q(ds,dz) + \int_0^t \int_Z (\left|x^1_s - x^2_s + c^1(s,x^1_s,z,\omega) - c^2(s,x^2_s,z,\omega)\right|$
$- \left|x^1_s - x^2_s\right| - sgn(x^1_s - x^2_s)\left(c^1(s,x^1_s,z,\omega) - c^2(s,x^2_s,z,\omega)\right))\pi(dz)ds.$

1.6 Uniqueness of Solutions to Stochastic Differential Equations

Now consider 1-dimensional SDE with Poisson jumps as follows:
(6.1) $x^i_t = x^i_0 + \int_0^t b(s,x^i_s,\omega)dA_s + \int_0^t \sigma(s,x^i_s,\omega)dM_s$
$+ \int_0^t \int_Z c(s,x^i_{s-},z,\omega)q(ds,dz), i = 1,2, \forall t \geq 0,$
where A_t, N_t and $q(dt,dz)$ have the same meaning as that of (1.1), and we always assume that $\forall t \geq 0,\ \int_0^t \left|b(s,x^i_s,\omega)\right| dA_s < \infty, P - a.s., i = 1,2.$

6.1 Theorem. (Pathwise uniqueness. Situ, R. 1985a, 1986a). If coefficients σ and c satisfy all conditions in Theorem 5.2 except (5.5), and c also satisfies one of the following conditions: for any given $T, N > 0$
$(6.2)_1$ (i) $\int_Z \left|c(s,x,z,\omega) - c(s,y,z,\omega)\right| \pi(dz) \leq k_{N,T}(t)\rho_{N,T}\left(|x-y|\right),$
as $|x|, |y| \leq N, \forall t \in [0,T]$,
where $k_{N,T}(t) \geq 0$ is non-random such that $\forall t \in [0,T]$

$\int_0^t k_{N,T}(s)ds < \infty,$

and $\rho_{N,T}(u) \geq 0$ is non-random, concave and strictly increasing on $u \geq 0$ such that

$\int_{0+} du/\rho_{N,T}(u) = \infty, \rho_{N,T}(0) = 0,$

moreover,

$(6.2)_2$ $\int_Z |c(s,x,z,\omega)| \pi(dz) < \infty, \forall x \in R, s \geq 0, \omega \in \Omega;$

(ii) (5.5) holds;

furthermore, if b satisfies

(6.3) $\operatorname{sgn}(x-y) \cdot (b(s,x,\omega) - b(s,y,\omega)) \leq \overline{k}_{N,T}(t)\overline{\rho}_{N,T}(|x-y|),$

as $|x|, |y| \leq N, \forall t \in [0,T]$,

where $\overline{\rho}_{N,T}$ has the same property as that of $\rho_{N,T}$ in (i) above, and $\overline{k}_{N,T}(t) \geq 0$ is non-random such that $\forall t \in [0,T]$

$\int_0^t \overline{k}_{N,T}(s)dA_s < \infty, \forall t \in [0,T],$

then $x_0^1 = x_0^2$, $P-a.s.$ implies that

(6.4) $P(\omega : \sup_{t\geq 0} |x_t^1 - x_t^2| = 0) = 1.$

6.2 Definition. If for any two solutions of (6.1) with $x_0^1 = x_0^2$ defined on the same probability space with the same A_t, M_t and $q(dt,dz)$ (6.5) is always true, then it is said that the pathwise uniqueness of solutions for (6.1) holds.

Now let us establish Theorem 6.1.

Proof. We may suppose M_t is a square integrable martingale without loss of any generality. Denote

(6.5) $\tau_N = \inf\{ t \geq 0 : \sum_{i=1}^2 [|x_t^i| + \int_0^t |b(s,x_s^i,\omega)| dA_s$

$+(\int_0^t |\sigma(s,x_s^i,\omega)|^2 d\langle M\rangle_s)^{1/2} + (\int_0^t \int_Z |c(s,x_{s-}^i,z,\omega)|^{1/2} \pi(dz))^{1/2}] > N \},$

$X_t = x_t^1 - x_t^2.$

Then for any given $T, N > 0$ by Theorem 5.6 in case (i) we have

$E|X_{t\wedge\tau_N}| \leq E\int_0^{t\wedge\tau_N} \widetilde{k}_{N,T}(s)\widetilde{\rho}_{N,T}(|X_s|)\, d(|A|_s + s), \forall t \in [0,T],$

where we write

$\widetilde{k}_{N,T}(s) = 2k_{N,T}(s) + \overline{k}_{N,T}(s),\ \widetilde{\rho}_{N,T} = \rho_{N,T} + \overline{\rho}_{N,T}.$

Obviously, $\widetilde{\rho}_{N,T}$ is still increasing, concave with $\widetilde{\rho}_{N,T}(0) = 0$, and by Lemma 6.3 below

$\int_{0+} du/\, \widetilde{\rho}_{N,T}(u) = \infty.$

Write

$\overline{A}_t = \int_0^t \widetilde{k}_{N,T}(s)d(|A|_s + s), \forall t \geq 0,$

$T_t = \overline{A}_t^{-1}.$

Then

$t = \overline{A}(T_t),$

and T_t is strictly increasing and continuous. Moreover, T_s is a $\Im_t$-stopping time, for each $s \geq 0$. Indeed,

$\{T_s \leq t\} = \left\{\overline{A}_s^{-1} \leq t\right\} = \left\{s \leq \overline{A}_t\right\} \in \Im_t.$

Therefore,

$E|X(T_t \wedge T \wedge \tau_N)| \leq E\int_0^{T_t\wedge T\wedge\tau_N} \widetilde{\rho}_{N,T}(|X_s|)\, d\overline{A}_s, \forall t \in [0,T]$

Set

$$Y_t = \begin{cases} X_t, & \text{as } t \in [0,T), \\ 0, & \text{otherwise.} \end{cases}$$

Then, obviously, Y_t is still a cadlag process, and $\forall t \geq 0$

$E|Y(T_t \wedge \tau_N)| \leq E\int_0^{T_t\wedge\tau_N} \widetilde{\rho}_{N,T}(|Y_s|)\, d\overline{A}_s \leq E\int_0^{t\wedge\overline{A}(\tau_N)} \widetilde{\rho}_{N,T}(|Y(T_s)|)\, ds$

$\leq E\int_0^t \widetilde{\rho}_{N,T}(|Y(T_s \wedge \tau_N)|)\, ds \leq \int_0^t \widetilde{\rho}_{N,T}(E|Y(T_s \wedge \tau_N)|)\, ds,$

where we have used the fact that T_t and $\overline{A}_t$ are strictly increasing. Hence, $\forall t \geq 0$

$$E|Y(T_t \wedge \tau_N)| = 0.$$

(See Remark 6.8 below). This shows that $P - a.s. \forall t \geq 0$

$$Y(T_t \wedge \tau_N) = 0,$$

where we have used the fact that $Y(t)$ is right continuous and T_t is continuous. Note that T_t strictly and continuously increases to ∞,as t does. Hence, $P - a.s.$ $Y(t \wedge \tau_N) = 0, \forall t \geq 0. \forall N = 1, 2, \cdots$.Therefore, $P - a.s.$

$$Y(t) = 0, \forall t \geq 0,$$

i.e., $P - a.s.$

$$X(t) = 0, \forall t \in [0, T).$$

Since $T > 0$ is arbitrary given, (6.4) is valid. ■

Now let us show

6.3 Lemma. If $\rho_i(u)$ is strictly increasing, concave in $u \geq 0$ with $\rho_i(0) = 0$ such that

$$\int_{0+} du/\rho_i(u) = \infty, \; i = 1, 2,$$

then $\rho_1(u)+\rho_2(u)$ is still strictly increasing, concave in $u \geq 0$ with $\rho_1(0)+\rho_2(0) = 0$ and satisfies

(6.6) $\int_{0+} du/ (\rho_1(u) + \rho_2(u)) = \infty.$

Proof. One only needs to prove (6.6). The rest of the conclusions in Lemma 6.3, are obviously true. By the proof of proposition 2.2 $(\rho_i)'_+ (0)$ exists, $i = 1, 2.$

1) If $(\rho_1)'_+ (0) > (\rho_2)'_+ (0)$, then it is easily seen that there exists a $\delta > 0$ such that for $0 < u < \delta$

$$\rho_1(u) > \rho_2(u).$$

Hence (6.6) is true.

2) If $(\rho_1)'_+ (0) = (\rho_2)'_+ (0)$, then it is also easily seen that there exists a $\delta > 0$ such that for $0 < u < \delta$

$$\rho_1(u) < 2u + \rho_2(u).$$

Hence we only need to show that

(6.7) $\int_{0+} du/ (u + \rho_2(u)) = \infty.$

If

$$u \geq \rho_2(u),$$

then (6.7) is true. If there exists a $u_0 > 0$ such that

$$u_0 < \rho_2(u_0),$$

then by the concavity of ρ_2 with $\rho_2(0) = 0$ it yields that as $0 < u \leq u_0$

$$u < \rho_2(u).$$

Hence (6.7) is still true. ■

Tanaka formula (5.12) can also be applied to show that for (6.1) the weak law uniqueness of solutions implies the pathwise uniqueness of solutions. First of all let us show the following

6.4 Theorem. (Situ, R 1985a). Suppose that all assumptions in Theorem 5.2 hold. Assume that $x^i_t, i = 1, 2,$ satisfy (6.1). Let

$$y^1_t = \max(x^1_t, x^2_t), \; y^2_t = \min(x^1_t, x^2_t).$$

Then y_t^i satisfies

(6.8) $y_t^i = y_0^i + \int_0^t b(s, y_s^i, \omega) dA_s + \int_0^t \sigma(s, y_s^i, \omega) dM_s$
$+ \int_0^t \int_Z c(s, y_{s-}^i, z, \omega) q(ds, dz), i = 1, 2, \forall t \geq 0.$

Proof. $y_t^1 = \frac{1}{2}(|x_t^1 - x_t^2| + x_t^1 + x_t^2) = \frac{1}{2}(x_t^1 + x_t^2$
$+ \int_0^t sgn\,(x_{s-}^1 - x_{s-}^2)\, d\,(x_s^1 - x_s^2) + |x_0^1 - x_0^2|)$
$= y_0^1 + \int_0^t b(s, y_s^1, \omega) dA_s + \int_0^t \sigma(s, y_s^1, \omega) dM_s + \int_0^t \int_Z c(s, y_{s-}^1, z, \omega) q(ds, dz).$
Hence (6.8) holds for $i = 1$. The proof is the same for $i = 2$. ■

6.5 Definition. We say that the weak law uniqueness of solutions for (6.1) holds, if whenever $\{x_t^1, t \geq 0\}$ and $\{x_t^2, t \geq 0\}$ are two solutions (can be defined on different probability spaces) whose initial laws of $x_0^i, i = 1, 2$, coincide, then the laws of $\{x_t^1, t \geq 0\}$ and $\{x_t^2, t \geq 0\}$ coincide.

Now let us show that the weak law uniqueness of solutions implies the pathwise uniqueness of solutions for (6.1) under some mild conditions.

6.6 Theorem. (Situ, R. 1985a). Under assumption of Theorem 5.2 the weak law uniqueness of solutions implies the pathwise uniqueness of solutions for (6.1).

Proof. Assume that $\{x_t^i, t \geq 0\}, i = 1, 2$, satisfy (6.1). If the pathwise uniqueness for these two solutions is not true, then it is not difficult to see that there exist a rational number r, a positive number ε and a $t > 0$ such that

$P(y_t^1 > r \geq y_t^2) \geq \varepsilon > 0.$

Since $y_t^1 \geq y_t^2$, and

$\{y_t^1 > r\} = \{y_t^2 > r\} \cup \{y_t^1 > r \geq y_t^2\}.$

Hence,

$P\{y_t^1 > r\} \geq P\{y_t^2 > r\} + \varepsilon > P\{y_t^2 > r\}.$

However, by Theorem 6.4 and the weak law uniqueness, the laws of $\{y_t^i, t \geq 0\}$, $i = 1, 2$, should coincide. This is a contradiction. ■

6.7. Remark. Obviously, Theorems 6.1, 6.4 and 6.6 still hold under assumption of Theorem 5.6.

6.8. Remark. The following fact is true: (Situ, R. 1985a).
If $\forall t \geq 0$ a real non-random function y_t satisfies

$0 \leq y_t \leq \int_0^t \rho(y_s) ds < \infty,$

where $0 \leq \rho(u)$, as $u \geq 0$, is increasing such that $\rho(0) = 0, \rho(u) > 0$, as $u > 0$; and

$\int_{0+} du/\rho(u) = \infty,$

then

$y_t = 0, \forall t \geq 0.$

Proof. Let

$z_t = \int_0^t \rho(y_s) ds (\geq y_t \geq 0).$

Obviously, one only needs to show that $\forall t \geq 0$

$z_t = 0.$

Indeed, z_t is absolutely continuous, increasing and *a.e.*

(6.9) $\dot{z}_t = \rho(y_t) \leq \rho(z_t)$.
Set

$$t_0 = \sup\{t \geq 0 : z_s = 0, \forall s \in [0,t]\}.$$

If $t_0 < \infty$, then $z_t > 0$, as $t > t_0$. Hence by assumption and (6.9) for any $\delta > 0$

$$\infty = \int_{(0,z(t_0+\delta))} du/\rho(u) = \int_{(t_0,t_0+\delta)} dz_t/\rho(z_t) \leq \int_{(t_0,t_0+\delta)} dt \leq \delta.$$

This is a contradiction. Therefore $t_0 = \infty$. ■

1.7 Comparison for Solutions of Stochastic Differential Equations

The Tanaka formula is also useful for discussing the comparison of solutions to SDE. Assume that $x_t^i, i = 1, 2$, satisfy

(7.1) $x_t^i = x_0^i + \int_0^t \beta^i(s,\omega)dA_s + \int_0^t \sigma(s, x_s^i, \omega)dM_s$
$\quad + \int_0^t \int_Z c(s, x_{s-}^i, z, \omega)q(ds, dz), i = 1, 2, \forall t \geq 0,$

where M_t and $q(dt, dz)$ have the same meaning as that in (6.1), and assume that A_t is a continuous increasing process. We also assume that

(7.2) $\int_0^t |\beta^i(s,\omega)| \, dA_s < \infty, i = 1, 2, \forall t \geq 0, P-a.s.$
(7.3) σ and c satisfy all conditions in Theorem 5.2;
(7.4) $b^1(t, x, \omega) \geq b^2(t, x, \omega)$,
(7.5) $\beta^1(t,\omega) \geq b^1(t, x_t^1, \omega), b^2(t, x_t^2, \omega) \geq \beta^2(t,\omega)$;
(7.6) $\operatorname{sgn}(x-y) \cdot (\, b^2(t,x,\omega) - b^2(t,y,\omega))$
$\quad \leq k_{N,T}(t)\rho_{N,T}(|x-y|)$, as $|x|, |y| \leq N, \forall t \in [0,T]$,

where $N, T > 0$ are arbitrarily given, and $k_{N,T}$ and $\rho_{N,T}$ satisfy the same condition as in 1° of Theorem 6.1. (Or b^1 satisfied (7.6)). We have

7.1 Theorem. (Comparison Theorem. Situ R. 1985a, 1987c). Under the assumption (7.1)-(7.6) if

(7.7) $x_0^1 \geq x_0^2$,

then $P-a.s.$

$$x_t^1 \geq x_t^2, \forall t \geq 0.$$

Proof. We can assume that M_t is a square integrable martingale without loss of any generality. Define τ_N as (6.5). By Tanaka formula (5.12) for any given $N, T > 0$ we have

(7.8) $E(x_{t\wedge\tau_N}^2 - x_{t\wedge\tau_N}^1)^+ \leq E\int_0^{t\wedge\tau_N} I_{x_s^2 > x_s^1}(\beta^2(s,\omega) - \beta^1(s,\omega))dA_s$
$\quad \leq E\int_0^{t\wedge\tau_N} I_{x_s^2 > x_s^1}(b^2(s, x_s^2, \omega) - b^1(s, x_s^1, \omega))dA_s$
$\quad \leq E\int_0^{t\wedge\tau_N} k_{N,T}(s)\rho_{N,T}\left((x_s^2 - x_s^1)^+\right) dA_s, \forall t \in [0,T].$

Now let us use the method of proof in Theorem 6.1 .
Set

$$\overline{A}_t = \int_0^t k_{N,T}(s)d(A_s + s), \forall t \geq 0,$$

$$T_t = \overline{A}_t^{-1},$$

$$Y_t = \begin{cases} (x_t^2 - x_t^1)^+, & \text{as } t \in 0, T), \\ 0, & \text{otherwise.} \end{cases}$$

Then by (7.8) we have that $\forall t \geq 0$

$E|Y(T_{t\wedge\tau_N})| \leq \int_0^t \rho_{N,T}(E|Y(T_{s\wedge\tau_N})|)ds.$

Hence, as in the proof of Theorem 6.1, this yields that $P-a.s.$

$Y(t) = 0, \forall t \geq 0,$

i.e. $P-a.s.$

$x_t^1 \geq x_t^2, \forall t \in [0,T).$

Since T is given arbitrarily , the desired result is obtained. ■

7.2 Theorem. In Theorem 7.1 if we get rid of the condition (7.6), then we need to strengthen the condition (7.4) to the requirement that $b^1(t,x)$ and $b^2(t,x)$ are non-random and jointly continuous such that

$b^1(t,x) > b^2(t,x).$

Furthermore, for simplicity, assume that σ and c satisfy the Lipschitzian condition as follows:

$|\sigma(t,x,\omega)-\sigma(t,y,\omega)|^2 + \int_Z |c(t,x,z,\omega)-c(t,y,z,\omega)|^2 \pi(dz)$

$\leq k_0 |x-y|^2,$

and assume that all coefficients are bounded:

$|b^1(t,x)|^2 + |b^2(t,x)|^2 + |\sigma(t,x,\omega)|^2 + \int_Z |c(t,x,z,\omega)|^2 \pi(dz) \leq k_0.$

In this case, the conclusion of Theorem 7.1 is still true.

Proof. Indeed, there exists a jointly Lipschitzian continuous function $\bar{b}(t,x)$ such that

$b^1(t,x) > \bar{b}(t,x) > b^2(t,x).$

It is obvious that $b^1$and $\bar{b}$ satisfy (7.4) - (7.7). Hence by Theorem 7.1 one obtains that $P-a.s.$

$x_t^1 \geq x_t, \forall t \geq 0,$

where x_t is the pathwise unique strong (strong means that x_t is $\Im_t^{M,q}-$ adapted) solution of the following SDE

$x_t = x_0 + \int_0^t \bar{b}(s,x_s)dA_s + \int_0^t \sigma(s,x_s,\omega)dM_s$

$+ \int_0^t \int_Z c(s,x_{s-},z,\omega)q(ds,dz), \forall t \geq 0,$

such a solution exists that can be obtained by the Picard iteration technique. Similarly, one also obtains that $P-a.s.$

$x_t \geq x_t^2, \forall t \geq 0.$

Hence, the conclusion is derived. ■

In the following we discuss the strong comparison theorem for solutions to (7.1). For this we first need to prepare a lemma.

7.3 Lemma. If V_t is a continuous semi-martingale and N_t is a real cadlag semi-martingale, then SDE

(7.9) $Z_t = V_t + \int_0^t Z_{s-}dN_s,\ t \geq 0,$

has a unique solution

$Z_t = \epsilon(N)_t \left\{ V_0 + \int_0^t \epsilon(N)_s^{-1}(dV_s - d\langle V^c, N^c\rangle_s) \right\}, \forall t \geq 0,$

where V^c and N^c are the continuous local martingale component of V and N, respectively, (indeed, $V^c = V$ here), and

(7.10) $\epsilon(N)_t = \exp(N_t - N_0 - \frac{1}{2}\langle N^c\rangle_t)\prod_{0<s\leq t}(1+\triangle N_s)e^{-\triangle N_s}$. $(\epsilon(N)_0 = 1)$.

Lemma 7.3 is a generalization of result on the continuous semi-martingale obtained by Yoeurp and Yor, 1977, for the semi-martingale with jumps.

Proof. First we show that $\prod_{0<s\leq t}(1+\triangle N_s)e^{-\triangle N_s}$ is a finite variational process. Note that

$$(1+\triangle N_s)e^{-\triangle N_s} = (1+\triangle N_s I_{|\triangle N_s|>2^{-1}})\exp(-\triangle N_s I_{|\triangle N_s|>2^{-1}})$$
$$\cdot(1+\triangle N_s I_{|\triangle N_s|\leq 2^{-1}})\exp(-\triangle N_s I_{|\triangle N_s|\leq 2^{-1}}) = I_s^1 \cdot I_s^2.$$

By the cadlag property of N_t for all $t \geq 0$ one has that $\{s \in (0,t] : |\triangle N_s| > \frac{1}{2}\}$ is a finite set. Hence,

$$\prod_{0<s\leq t} I_s^1 < \infty.$$

Moreover, $\prod_{0<s\leq t} I_s^1$ is a finite variational function of t, since it is a finite product of finite variational functions. On the other hand,

$$x - \log(1+x) = x^2(1/2 - x/3 + x^2/4 - \cdots), \text{ as } |x| < 1.$$

Hence, as $|x| \leq 1/2$

$$89x^2/240 \leq x - \log(1+x) \leq 5x^2/6.$$

Since N_t is a semi-martingale, $P - a.s.$

$$\sum_{0<s\leq t} |\triangle N_s|^2 < \infty.$$

Hence,

$$U_t = \log\prod_{0<s\leq t} I_s^2 = \sum_{0<s\leq t}\log I_s^2$$

is absolutely convergent. (Note that $\{s \in (0,t] : \triangle N_s > 0\}$ is at most a countable set). Hence,

$$\sum_{0<s\leq t} \left|\log I_s^2\right| \leq k_t(\omega) < \infty,$$

where $k_t(\omega)$ is a constant depending on t and ω only. Note also that

$$e^b - e^a = e^{a+\theta(b-a)}(b-a),$$

where $0 < \theta < 1$. Hence, for $0 = t_0 < t_1 < \cdots < t_n = t$

$$\sum_{k=0}^n \left|\exp(U_{t_k}) - \exp(U_{t_{k-1}})\right| \leq e^{1+k_t(\omega)}\sum_{k=0}^n \left|U_{t_k} - U_{t_{k-1}}\right|$$
$$\leq e^{1+k_t(\omega)}\sum_{0<s\leq t}\left|\log I_s^2\right| \leq e^{1+k_t(\omega)}k_t(\omega) < \infty.$$

Therefore

$$\prod_{0<s\leq t} I_s^2 = e^{U_t}$$

is also finite variational. This yields that $\prod_{0<s\leq t}(1+\triangle N_s)e^{-\triangle N_s}$ is a finite variational process. Now let

$$Z_t = e^{x_t}y_t^1 y_t^2,$$

where

$$x_t = N_t - N_0 - \tfrac{1}{2}\langle N^c\rangle_t,$$
$$y_t^1 = V_0 + \int_0^t \epsilon(N)_s^{-1}(dV_s - d\langle V^c, N^c\rangle_s),$$
$$y_t^2 = \prod_{0<s\leq t}(1+\triangle N_s)e^{-\triangle N_s}.$$

By Ito's formula

$$Z_t = Z_0 + \int_0^t e^{x_{s-}}y_{s-}^1 y_{s-}^2 dx_s + \int_0^t e^{x_{s-}}y_{s-}^1 dy_s^2 + \int_0^t e^{x_{s-}}y_{s-}^2 dy_s^1$$
$$+\sum_{0<s\leq t}(Z_s - Z_{s-} - e^{x_{s-}}y_{s-}^1 y_{s-}^2 \triangle x_s - e^{x_{s-}}y_{s-}^1 \triangle y_s^2)$$
$$+\int_0^t e^{x_{s-}}y_{s-}^2 d\langle x^c, y^{1c}\rangle_s + \tfrac{1}{2}\int_0^t e^{x_{s-}}y_{s-}^1 y_{s-}^2 d\langle x^c\rangle_s$$
$$= V_0 + \int_0^t Z_{s-}dN_s + \int_0^t \epsilon(N)_s\epsilon(N)_s^{-1}(dV_s - d\langle V^c, N^c\rangle_s)$$
$$+\int_0^t \epsilon(N)_s\epsilon(N)_s^{-1}d\langle V^c, N^c\rangle_s$$

$+\sum_{0<s\leq t} Z_{s-}(e^{\triangle N_s}(1+\triangle N_s)e^{-\triangle N_s} - 1 - \triangle N_s) = V_t + \int_0^t Z_{s-}dN_s.$

Therefore Z_t satisfies (7.9). The uniqueness is easily derived by the linearity of equation (7.9). Indeed, if $Z_t^i, i = 1, 2$, both satisfy (7.9), then

$$\left|Z_t^1 - Z_t^2\right|^2 \leq \left|\int_0^t (Z_{s-}^1 - Z_{s-}^2)dN_s\right|^2.$$

Now one can discuss similarly to the proof of Theorem 6.1 to obtain the conclusion. ■

Now we are in a position to give a strong comparison theorem for solutions of (7.1).

7.4 Theorem. (Strong comparison. Situ, R. 1987c). Assume that $x_t^i, i = 1, 2$, satisfy (7.1), A_t in (7.1) is a continuous strictly increasing process, $0 < \langle M\rangle_t, \forall t \geq 0$, and

(7.11) $\beta^i(t,\omega) = b^i(t, x_t^i, \omega),\ i = 1, 2,$

$(7.12)_1$ $|\sigma(t,x,\omega) - \sigma(t,y,\omega)|^2 + \int_Z |c(t,x,z,\omega) - c(t,y,z,\omega)|^2 \pi(dz)$
$\leq k_0 |x-y|^2,$

$x \geq y \Rightarrow x + c(t,x,z,\omega) \geq y + c(t,y,z,\omega),$

$\int_Z |c(t,x,z,\omega)| p(ds,dz) < \infty,$

$\int_Z |c(t,x,z,\omega) - c(t,y,z,\omega)| p(ds,dz) < |x-y|,$

$(7.12)_2$ $|b^1(t,x,\omega)|^2 + |b^2(t,x,\omega)|^2$
$+ |\sigma(t,x,\omega)|^2 + \int_Z |c(t,x,z,\omega)|^2 \pi(dz) \leq g(x),$

where $g(x)$ is a non-random continuous function, moreover, assume that

(7.13) $b^1(t,x,\omega) \geq b^2(t,x,\omega),$

(7.14) $x_0^1 \geq x_0^2,$

furthermore, assume that one of the following conditions is fulfilled:

(i) $b^1(t,x,\omega)$ satisfies

(7.15) $|b^1(t,x,\omega) - b^1(t,y,\omega)|^2 \leq k_0 |x-y|^2,$

(or b^2 satisfies this condition), and

(7.16) $x_0^1 > x_0^2;$

(ii) (7.15) holds, and

(7.17) $b^1(t,x,\omega) > b^2(t,x,\omega);$

(iii) $b^1(t,x)$ and $b^2(t,x)$ are non-random and jointly continuous such that (7.17) holds, and in $(7.12)_2$ $g(x)$ satisfies

$g(x) \leq k_0(1+|x|);$

(iv) $b^1(x)$, $b^2(x)$ and $\sigma(x)$ are non-random functions depending on x only, and $A_t = \langle M\rangle_t > 0$, as $t > 0$; $x_0^2 = \overline{x}_0 \in R$, $\overline{x}_0$ is a non-random constant, and there exist constants $\varepsilon, \delta > 0$ such that

$|\sigma(x)| \geq \varepsilon > 0,$

$\int_{|a-x_0|<\delta} I_{(b^1(a)=b^2(a))} da = 0,$

moreover,

$|c(t,x,z,\omega)| \leq k_0,$

and (7.15) holds.

Then $P - a.s.$

(7.18). $x_t^1 > x_t^2, \forall t > 0.$

Proof. By assumption as $t \geq 0$

$x_t^1 - x_t^2 = V_t + \int_0^t (x_{s-}^1 - x_{s-}^2) dN_s,$

where

$V_t = x_0^1 - x_0^2 + \int_0^t (b^1(t, x_s^2, \omega) - b^2(t, x_s^2, \omega)) dA_s,$

$N_t = \int_0^t I_{(x_{s-}^1 = x_{s-}^2)} (x_{s-}^1 - x_{s-}^2)^{-1} [(b^1(t, x_s^1, \omega) - b^1(t, x_s^2, \omega)) dA_s$

$+ (\sigma(t, x_s^1, \omega) - \sigma(t, x_s^2, \omega)) dM_s + \int_Z (c(t, x_{s-}^1, z, \omega) - c(t, x, x_{s-}^2, \omega)) q(ds, dz)].$

Note that by assumption $(7.12)_1$ $\triangle N_t > -1$. Now by Lemma 7.3

(7.19) $x_t^1 - x_t^2 = \epsilon(N)_t \left\{ x_0^1 - x_0^2 + \int_0^t \epsilon(N)_s^{-1} dV_s \right\},$

where $\epsilon(N)_t$ is defined by (7.10), and $\epsilon(N)_t > 0$. In the following we will use formula (7.19) to prove the conclusion.

(i) : Since $x_0^1 > x_0^2$. By (7.19) one finds that (7.18) is true.

(ii) : By (7.17) V_t is a strictly increasing process. Hence (7.18) is also true.

(iii) By (7.17) there exists a jointly Lipschitzian continuous function $\bar{b}(t, x)$ such that

$b^1(t, x) > \bar{b}(t, x) > b^2(t, x).$

Since now coefficients $\bar{b}, \sigma$ and c are less than linear growth and satisfy the Lipschitzian condition, hence, by the Picard iteration technique it is easily seen that there exists a unique solution $\bar{x}_t$ satisfying

$\bar{x}_t = x_0^1 + \int_0^t \bar{b}(s, \bar{x}_s) dA_s + \int_0^t \sigma(s, \bar{x}_s, \omega) dM_s + \int_0^t \int_Z c(s, \bar{x}_{s-}, z, \omega) q(ds, dz),$

$\forall t \geq 0.$

Applying (ii), one gets that $P - a.s.$

$x_t^1 > \bar{x}_t > .x_t^2, \forall t \geq 0.$

(iV) : Without loss of generality we can assume that N_t is a bounded semimartingale, otherwise one can use the localization technique. Note that $\forall t \geq 0$

$V_t = \int_R \sigma^{-2}(a)(b^1(a) - b^2(a)) L_t^a(x^2) da + x_0^1 - \bar{x}_0$

$\geq x_0^1 - \bar{x}_0 + \int_R I_{(b^1(a) > b^2(a)) \cap (|a - \bar{x}_0| < \delta)} \sigma^{-2}(a)(b^1(a) - b^2(a)) L_t^a(x^2) da > 0,$

where we have applied the Lemma 7.5 below. Note also that since N_t is bounded, there exists a constant $k_t(\omega)$ depending on ω and t only such that

$0 < \epsilon(N)_s \leq k_t(\omega)$, as $s \in [0, t]$.

Hence,

$\int_0^t \epsilon(N)_s^{-1} dV_s \geq k_t(\omega)^{-1} V_t > 0, \forall t \geq 0.$

Hence, by (7.19) (7.18) still follows. ■

7.5 Lemma. Assume that x_t satisfies (7.1) for $i = 2$ (or $i = 1$). If

$\langle M \rangle_t > 0, \forall t \geq 0;$

$\sigma(t, x, \omega) \geq r_0 > 0$, $x_0 = x_0 \in R$ (x_0 is a constant),

then for any $\varepsilon > 0$ there exists a $t_\varepsilon > 0$, and for each $t \in (0, t_\varepsilon]$ there exists a set

$\Lambda_{t,\varepsilon} \subset \{u : |u - x_0| < \varepsilon\}$

such that its Lebesgue measure $m(\Lambda_{t,\varepsilon}) > 0$, and one has

$L_t^u(x) > 0$, as $u \in \Lambda_{t,\varepsilon}$, $t \in (0, t_\varepsilon]$.

Proof. Since x_t is right continuous at $t = 0$, for any $\varepsilon > 0$ there exists a $t_\varepsilon > 0$, and for each $t \in [0, t_\varepsilon]$

$|x_t - x_0| < \varepsilon$.
Therefore as $t \in (0, t_\varepsilon]$

$$\int_R I_{|u-x_0|<\varepsilon} L_t^u(x)du = \int_0^t I_{|x_s-x_0|<\varepsilon}\sigma^2(s,x_s,\omega)d\langle M\rangle_s \geq r_0 \langle M\rangle_t > 0.$$

This shows that for each $t \in (0, t_\varepsilon]$ the conclusion is true. ■

To get rid of the condition

(7.20) $\int_Z |c(t,x,z,\omega) - c(t,y,z,\omega)|\, p(ds,dz) < |x-y|$

in $(7.12)_1$ we have the following strong comparison theorem.

7.6 Theorem. If condition (7.20) in $(7.12)_1$ of Theorem 7.4 is replaced by the following condition:

(i)' $x \geq y \Rightarrow c(t,x,z,\omega) \geq c(t,y,z,\omega)$,

then the conclusion of Theorem 7.4 still holds.

Proof. Since by Theorem 7.1 $P-a.s.$

(7.22) $x_t^1 \geq x_t^2$, $\forall t \geq 0$.

Hence by (i)'

$$c(t,x_t^1,z,\omega) \geq c(t,x_t^2,z,\omega).$$

Therefore in (7.19) we still have

$$\epsilon(N)_t > 0.$$

Then the proof in Theorem 7.4 still goes through. ■

1.8 Convergence of Solutions to Stochastic Differential Equations

In this paragraph we are going to use the Tanaka formula to discuss the convergence and stability of solutions to SDE with jumps. First we will discuss SDEs with the same diffusion coefficients.

8.1 Theorem. (Situ, R 1985a). Assume that $x_t^n, n = 1,2,\cdots$, and $x_t = x_t^0$ satisfy

(8.1) $x_t^n = x_0^n + \int_0^t b^n(s,x_s^n,\omega)dA_s + \int_0^t \sigma(s,x_s^n,\omega)dM_s$
$\quad + \int_0^t \int_Z c^n(s,x_{s-}^n,z,\omega)q(ds,dz)$, $\forall t \geq 0, n = 0,1,2,\cdots$,

where A_t is a continuous increasing process, M_t is a continuous local square integrable martingale, and

$$q(dt,dz) = p(dt,dz) - \pi(dz)dt,$$

where $p(dt,dz)$ is a Poisson random point measure with compensator $\pi(dz)dt$, $\pi(\cdot)$ is a $\sigma-$finite measure in Z. Moreover, assume that $\forall t,x,y$ $P-a.s.$

1° $|\sigma(t,x,\omega) - \sigma(t,x,\omega)| \leq \overline{k}(s,\omega)\rho(|x-y|)$,

where $\overline{k}(s,\omega) \geq 0$ satisfies

$$\int_0^t |\overline{k}(s,\omega)|^2 d\langle M\rangle_s < \infty,$$

$\rho(u) > 0$, as $u > 0$; $\rho(0) = 0$, it satisfies

$$\int_{0+} du/\rho(u)^2 = \infty;$$

2° $|b^n(t,x,\omega)| + |\sigma(t,x,\omega)| + \sum_{i=1}^2 \int_Z |\sigma(t,x,\omega)|^i \pi(dz) \leq g(s,\omega)$, $n = 0,1,2,\cdots$, where $g(s,\omega)$ is continuous, and $b^n(t,x,\omega), \sigma(t,x,\omega)$ are jointly measurable;

3° $\lim_{n\to\infty} E\left|x_0^n - x_0\right| = 0$;

4° $sgn(x-y)(b(t,x,\omega)-b(t,y,\omega)) \leq F(t,\omega)\left|x-y\right|,\ \int_0^t F(s,\omega)dA_s < \infty,$

$\left|b^n(t,x,\omega)-b(t,y,\omega)\right| \leq F^n(t,\omega),\ \lim_{n\to\infty} E\int_0^t F^n(s,\omega)dA_s = 0, F \geq 0;$

5° $\left|c(t,x,z,\omega)-c(t,y,z,\omega)\right| \leq G(t,z,\omega)\left|x-y\right|,$

$\left|c^n(t,x,z,\omega)-c(t,y,z,\omega)\right| \leq G^n(t,z,\omega),,$

$\int_0^t\int_Z G(s,z,\omega)\pi(dz)ds < \infty,\ \lim_{n\to\infty} E\int_0^t\int_Z G^n(s,z,\omega)\pi(dz)ds = 0.$

Then we have that $\forall t \geq 0$

(8.2) $\lim_{n\to\infty} E[\exp(-\int_0^t F(s,\omega)dA_s - 2\int_0^t\int_Z G(s,z,\omega)\pi(dz)ds)$
$\cdot\left|x_t^n - x_t\right|] = 0$

Proof. For arbitrary given $0 \leq T < \infty$ let us show (8.2) holds as $t \in [0,T]$. Applying (5.22) we get

(8.3) $\left|x_{t\wedge\tau_r}^n - x_{t\wedge\tau_r}\right| = \left|x_0^n - x_0\right|$
$+\int_0^{t\wedge\tau_r} sgn(x_s^n - x_s)(b^n(s,x_s^n,\omega) - b(s,x_s,\omega))dA_s$
$+\int_0^{t\wedge\tau_r} sgn(x_s^n - x_s)(\sigma(s,x_s^n,\omega) - \sigma(s,x_s,\omega))dM_s$
$+\int_0^{t\wedge\tau_r}\int_Z[\left|x_{s-}^n - x_{s-} + c^n(s,x_{s-}^n,z,\omega) - c(s,x_{s-},z,\omega)\right|$
$-\left|x_{s-}^n - x_{s-}\right|]q(ds,dz)$
$+\int_0^{t\wedge\tau_r}\int_Z[\left|x_s^n - x_s + c^n(s,x_s^n,z,\omega) - c(s,x_s,z,\omega)\right|$
$-\left|x_s^n - x_s\right| - sgn(x_s^n - x_s)(c^n(s,x_s^n,z,\omega) - c(s,x_s,z,\omega))]\pi(dz)ds,$

where

$\tau_r = \inf\left\{t \in [0,T] : \int_0^t g(s,\omega)^2 d(\langle M\rangle_s + s) > .r\right\}.$

Denote

$X_t^n = x_t^n - x_t,\ \sigma_n^*(t,\omega) = \sigma(t,x_t^n,\omega) - \sigma(t,x_t,\omega),$
$c_n^*(t,z,\omega) = c(t,x_{t-}^n,z,\omega) - c(t,x_{t-},z,\omega).$

Then by (8.3)

$\left|X_{t\wedge\tau_r}^n\right| \leq \left|X_0^n\right| + \int_0^{t\wedge\tau_r} F(s,\omega)\left|X_s^n\right|dA_s + \int_0^{t\wedge\tau_r} F^n(s,\omega)dA_s$
$+2\int_0^{t\wedge\tau_r}\int_Z G^n(s,z,\omega)\pi(dz)ds + 2\int_0^{t\wedge\tau_r}\int_Z G(s,z,\omega)\left|X_s^n\right|\pi(dz)ds$
$+N_{t\wedge\tau},$

where

$N_t = \int_0^t sgn(X_s^n)\cdot\sigma_n^*(s,\omega)dM_s + \int_0^t\int_Z[\left|X_{s-}^n + c_n^*(s,z,\omega)\right| - \left|X_{s-}^n\right|]q(ds,dz).$

Applying Lemma 8.2 below we get

$\exp(-\int_0^{t\wedge\tau_r} F(s,\omega)dA_s - 2\int_0^{t\wedge\tau_r}\int_Z G(s,z,\omega)\pi(dz)ds))\cdot\left|X_{t\wedge\tau_r}^n\right|$
$\leq \left|X_0^n\right| + \int_0^{t\wedge\tau_r}\exp(-\int_0^s F(u,\omega)dA_u - 2\int_0^s\int_Z G(u,z,\omega)\pi(dz)du))$
$\cdot(F^n(s,\omega)dA_s + 2\int_Z G^n(s,z,\omega)\pi(dz)ds + dN_s).$

Hence

$EH_{t\wedge\tau_r}\left|X_{t\wedge\tau_r}^n\right| \leq E\left|X_0^n\right| + E\int_0^{t\wedge\tau_r} H_s(F^n(s,\omega)dA_s + 2\int_Z G^n(s,z,\omega)\pi(dz)ds)$
$\leq E\left|X_0^n\right| + E\int_0^t H_s(F^n(s,\omega)dA_s + 2\int_Z G^n(s,z,\omega)\pi(dz)ds),$

where

$H_s = \exp(-\int_0^s F(u,\omega)dA_u - 2\int_0^s\int_Z G(u,z,\omega)\pi(dz)du)) \leq 1.$

Note that by assumption for each ω, when r is large enough

$\tau_r(\omega) = \infty.$

Therefore by the above result letting $r \to \infty$, by Fatou's lemma we have

$EH_t |X_t^n| \leq E|X_0^n| + E\int_0^t F^n(s,\omega)dA_s + 2\int_Z G^n(s,z,\omega)\pi(dz)ds.$

Now let $n \to \infty$ we find that

$\lim_{n\to\infty} EH_t |X_t^n| = 0.$ ■

Theorem 8.1 implies theorem 3 in Mel'nikov (1985), if we let $c, c^n = 0, \forall n$. Moreover, condition 4° here is weaker than the usual Lipschitzian condition there. Because now $b(t,x,\omega)$ can be discontinuous . For example, $b(t,x) = A_0 - A_1 sgn(x)$, where A_0, A_1 are constants.

8.2 Lemma. (Stochastic Gronwall's inequality). Assume that V_t, N_t are cadlag processes, where N_t is a semi-martingale, and assume that B_t is a continuous increasing process. If

$V_t \leq N_t + \int_0^t V_s dB_s, \forall t \geq 0; V_0 = 0,$

then

$e^{-B_t}V_t \leq N_0 + \int_0^t e^{-B_s}dN_s, \forall t \geq 0.$

Proof. Note that by Ito's formula

$d(N_tB_t) = B_t dN_t + N_t dB_t,$

$d_s(N_s(B_t - B_s)^2/2) = [(B_t - B_s)^2/2]dN_s - (B_t - B_s)N_s dB_s$, etc.

Hence, by assumption it can be seen that

$V_t \leq N_t + \int_0^t V_s dB_s \leq N_t + \int_0^t N_s dB_s + \int_0^t dB_s \int_0^s V_u dB_u$

$= N_0 + N_0 B_t + \int_0^t (B_t - B_s)dN_s + \int_0^t dN_s + \int_0^t V_s(B_t - B_s)dB_s$

$\leq \cdots \leq N_0(\sum_{n=0}^k (B_t)^n/n!) + \int_0^t (\sum_{n=0}^k (B_t - B_s)^n/n!)dN_s$

$+ \int_0^t V_s(B_t - B_s)^k/k! dB_s.$

Since V_t is cadlag, it must be bounded locally. (Its bound can depend on ω). Hence

$\left|\int_0^t V_s(B_t - B_s)^k/k! dB_s\right| \leq k_0(\omega) B_t^{k+1}/(k+1)! \to 0$, as $n \to \infty$.

Therefore

$V_t \leq N_0 e^{B_t} + \int_0^t e^{B_t - B_s}dN_s, \forall t \geq 0.$ ■

Now let us discuss the convergence theorems for solutions of SDE with different diffusion coefficients. Assume that $x_t^n, n = 0, 1, 2, \cdots$, satisfy

(8.5) $x_t^n = x_0^n + \int_0^t b^n(s, x_s^n, \omega)dA_s + \int_0^t \sigma^n(s, x_s^n, \omega)dM_s$

$+ \int_0^t \int_Z c^n(s, x_{s-}^n, z, \omega)q(ds, dz),\ n = 0, 1, 2, \cdots,$

where A_t, M_t and $q(dt, dz)$ have the same meaning as that in Theorem 8.1. We have

8.3 Theorem. Assume that

1° $|b^n(t,x,\omega)|^2 + |\sigma^n(t,x,\omega)|^2 + \int_Z |c^n(t,x,z,\omega)|^2 \pi(dz) \leq g(t,\omega),\ n = 0, 1, 2, \cdots$ where $g(t,\omega)$ is a continuous process, $b^n(t,x,\omega), \sigma^n(t,x,\omega)$, and $c^n(t,x,z,\omega)$ are all jointly measurable,

2° $(x - y)(b^0(t,x,\omega) - b^0(t,y,\omega)) \leq F_1(t,\omega)|x - y|^2,$

$|\sigma^0(t,x,\omega) - \sigma^0(t,y,\omega)|^2 \leq F_2(t,\omega)|x - y|^2,$

$+ \int_Z |c^0(t,x,z,\omega) - c^0(t,y,z,\omega)|^2 \pi(dz) \leq F_3(t,\omega)|x - y|^2,$

where $F_i(t,\omega) \geq 0, i = 1, 2, 3$, are such that

$\int_0^t F_1(s,\omega)dA_s + \int_0^t F_2(s,\omega)d\langle M\rangle_s + \int_0^t F_3(s,\omega)ds < \infty,$

3° $\left|b^n(t,x,\omega)-b^0(t,x,\omega)\right|^2 \le F_1^n(t,\omega)$,

$\left|\sigma^n(t,x,\omega)-\sigma^0(t,x,\omega)\right|^2 \le F_2^n(t,\omega)$,

$\int_Z \left|c^n(t,x,z,\omega)-c^0(t,x,z,\omega)\right|^2 \pi(dz) \le F_3^n(t,\omega)$,

where

$\lim_{n\to\infty} E(\int_0^t F_1^n(s,\omega)dA_s + \int_0^t F_2^n(s,\omega)d\langle M\rangle_s + \int_0^t F_3^n(s,\omega)ds) = 0$,

4° $\lim_{n\to\infty} E\left|x_0^n - x_0^0\right|^2 = 0$.

Then

(8.6) $\lim_{n\to\infty} E(H_t |X_t^n|^2) = 0, \forall t \ge 0$,

where

$X_t^n = x_t^n - x_t^0$,

$H_t = \exp(-2\int_0^t((F_1(s,\omega)+\frac{1}{2})dA_s + F_2(s,\omega)d\langle M\rangle_s + F_3(s,\omega)ds))$.

Proof. By Ito's formula

(8.7) $|X_t^n|^2 = |X_0^n|^2 + 2\int_0^t X_s^n(b^n(t,x_s^n,\omega)-b^0(t,x_s^0,\omega))dA_s$

$+\int_0^t \left|\sigma^n(t,x_s^n,\omega)-\sigma^0(t,x_s^0,\omega)\right|^2 d\langle M\rangle_s$

$+\int_0^t\int_Z \left|c^n(t,x_s^n,z,\omega)-c^0(t,x_s^0,z,\omega)\right|^2 \pi(dz)ds + N_t$

$\le |X_0^n|^2 + 2\int_0^t |X_s^n|^2 (F_1(s,\omega)dA_s + F_2(s,\omega)d\langle M\rangle_s + F_3(s,\omega)ds)$

$+\int_0^t |X_s^n|^2 dA_s + \int_0^t (F_1^n(s,\omega)dA_s + 2F_2^n(s,\omega)d\langle M\rangle_s + 2F_3^n(s,\omega)ds) + N_t$,

where

$N_t = 2\int_0^t X_s^n(\sigma^n(t,x_s^n,\omega)-\sigma^0(t,x_s^0,\omega))dM_s$

$+\int_0^t\int_Z(\left|X_{s-}^n + c^n(t,x_{s-}^n,z,\omega)-c^0(t,x_{s-}^0,z,\omega)\right|^2 - \left|X_{s-}^n\right|^2)q(ds,dz)$.

By Lemma 8.2 and note that $0\le H_t\le 1$, one has

$H_t|X_t^n|^2 \le |X_0^n|^2 + \int_0^t 2H_s(F_1^n(s,\omega)dA_s + F_2^n(s,\omega)d\langle M\rangle_s$

$+F_3^n(s,\omega)ds + \frac{1}{2}dN_s)$.

Hence as in the proof of Theorem 8.1 we find that

$\lim_{n\to\infty} EH_t|X_t^n|^2 = 0$. ■

We also have some other conditions on σ^n and σ^0 for the convergence of solutions.

8.4 Theorem. Assume that A_t and M_t in Theorem 8.3 are integrable and square integrable, respectively, and

$g(s,\omega) = g(s)$

is non-random as given in condition 1° in Theorem 8.3, and g is a continuous function; and assume that the condition on σ^n and σ^0 in 3° of Theorem 8.3 is modified to

$\sigma^n(t,x,\omega) = \sigma^n(x), \forall n = 0,1,2,\cdots$

which do not depend on t and ω, and $\exists\delta_0 > 0$,

$0 < \delta_0 \le \sigma^n(x), \forall n = 0,1,2,\cdots$

$\int_R \left|\sigma^n(a)-\sigma^0(a)\right|^2 da \to 0$, as $n\to\infty$.

1) Besides, assume that all other conditions in Theorem 8.3 remain true. Then the conclusion of Theorem 8.3 still holds.

2) If $A_t = \langle M\rangle_t$, and the condition on b^n and b in 3° of Theorem 8.3 is modified to that

$b^n(t,x,\omega) = b^n(x), \forall n = 0,1,2,\cdots$
which do not depend on t and ω, and as $n \to \infty$

$\int_R |b^n(a) - b^0(a)|^2 da \to 0,$

and suppose that all other conditions in Theorem 8.3 remain true. Then the conclusion of Theorem 8.3 still holds.

Proof. 1): Similar to (8.7), one gets that

$|X_t^n|^2 \leq |X_0^n|^2 - \int_0^t |X_s^n|^2 d(\log H_s) + \int_0^t F_1^n(s,\omega)dA_s + \int_0^t |X_s^n|^2 dA_s$
$+2\int_0^t |\sigma^n(x_s^n) - \sigma^0(x_s^n)|^2 d\langle M\rangle_s + 2\int_0^t F_3^n(s,\omega)ds + N_t.$

By Lemma 8.2 and the Krylov type estimate (4.4) $\forall 0 \leq T < \infty$

$EH_t |X_t^n|^2 \leq E|X_0^n|^2 + E\int_0^t (F_1^n(s,\omega)dA_s + 2F_3^n(s,\omega)ds)$
$+k_T\delta_0^{-2}\int_R |\sigma^n(a) - \sigma^0(a)|^2 da$, as $t \in [0,T]$.

Hence, as $n \to \infty$

$EH_t |X_t^n|^2 \to 0.$

2): Since now $A_t = \langle M\rangle_t$ by the Krylov type estimate (4.4) from (8.7) one gets that

$EH_t |X_t^n|^2 \leq E|X_0^n|^2 + E\int_0^t 2F_3^n(s,\omega)ds$
$+k_T\delta_0^{-2}(\int_R |\sigma^n(a) - \sigma^0(a)|^2 + |b^n(a) - b^0(a)|^2)da$, as $t \in [0,T]$.

Hence the conclusion again follows. ■

Now let us discuss the convergence of solutions to (8.5) under some other conditions by using the local time technique. Suppose that in (8.5) M_t is a continuous square integrable martingale, and A_t is a continuous integrable increasing process. In the following we always assume that

(8.8) $\sigma^n(t,x,\omega) = \sigma^n(x), \forall n = 0,1,2,\cdots$

which do not depend on t and ω, and assume that

$$(8.9)\begin{cases} |b^n(t,x,\omega)|^2 + |\sigma^n(x)|^2 + \sum_{i=1}^2 \int_Z |c(t,x,z,\omega)|^2 \pi(dz) \leq g(t), \\ c^n = c^0 = c(y,x,z,\omega), \forall n = 0,1,2,\cdots \end{cases}$$

where $g(t)$ is a non-random real continuous function depending on t only.

We have

8.5 Lemma. If

(8.10) $b^n = 0, \forall n = 0,1,2,\cdots$

(8.11) $\int_Z |c(t,x,z,\omega) - c(t,y,z,\omega)| \pi(dz) \leq k_0 |x-y|,$

(8.12) where $k_0 \geq 0$ is a constant, and

$x \geq y \Rightarrow c(t,x,z,\omega) \geq c(t,y,z,\omega),$

then

1)$|EL_t^a(x^n - x^0) - EL_t^0(x^n - x^0)| \leq (4 + 2k_0T)|a|, \forall n = 0,1,2,\cdots; a \in R;$ $t \in [0,T]$;

2) $E|X_t^n| \to 0$, as $n \to \infty$; iff (if and only if)

$E|X_0^n| \to 0,\ EL_t^0(X^n) \to 0$,and as $n \to \infty$;

where $X_t^n = x_t^n - x_t^0$.

Proof. By (5.3) and (5.6)

$E|X_t^n| = E|X_0^n| + EL_t^0(X^n).$

2) is derived.

Let us show 1): By (1.3)

$(X_t^n - a)^+ = (X_0^n - a)^+ + \int_0^t I_{(X_{s-}^n > a)} dX_s^n$

$+ \int_0^t \int_Z (X_{s-}^n - a + c(s, x_{s-}^n, z, \omega) - c(s, x_{s-}^0, z, \omega))^+$

$-(X_{s-}^n - a)^+ - I_{(X_{s-}^n > a)}(c(s, x_{s-}^n, z, \omega) - c(s, x_{s-}^0, z, \omega)))p(ds, dz) + \frac{1}{2} L_t^a(X^n)$

$= \sum_{i=1}^3 I_i^n(a) + \frac{1}{2} L_t^a(X^n).$

Note that as $0 \leq t \leq T$

$EI_3^n(a) \leq E \int_0^t \int_Z I_{(0 < X_{s-}^n \leq a)}(c(s, x_{s-}^n, z, \omega) - c(s, x_{s-}^0, z, \omega))^+ \pi(dz) ds$

$\leq ak_0 T$, for $a > 0$;

and

$EI_3^n(a) \leq E \int_0^t \int_Z I_{(0 \geq X_{s-}^n \geq a)}(c(s, x_{s-}^n, z, \omega) - c(s, x_{s-}^0, z, \omega))^- \pi(dz) ds$

$\leq (-a) k_0 T$, for $a \leq 0$;

where we have applied (8.12) and used the fact that

$a^+ - b^+ \leq (a - b)^+, \forall a, b \in R.$

Again, by (8.12)

$I_3^n(0) = 0.$

Note also that by assumption (8.10) X_t^n is a martingale. Hence it is easily seen that

$\left| EL_t^a(x^n - x^0) - EL_t^0(x^n - x^0) \right| \leq (4 + 2k_0 T) \left| a \right|.$ ■

8.6 Lemma. Under the assumption of Lemma 8.5, if there exists a $\rho_1(a)$ such that

$\sup_{\varepsilon > 0} \overline{\lim}_{n \to \infty} \int_\varepsilon^\infty (da/\rho_1(a)) EL_t^a(X^n) < \infty,$

where $0 < \rho_1(a)$, as $a > 0$; and

$\int_{0+} da/\rho_1(a) = \infty.$

then as $n \to \infty$

$EL_t^0(X^n) \to 0.$

Proof. If for a fixed $t > 0$

$\overline{\lim}_{n \to \infty} EL_t^0(X^n) = r > 0,$

then by 1) of Lemma 8.5

$\overline{\lim}_{n \to \infty} \int_\varepsilon^\infty (da/\rho_1(a)) EL_t^a(X^n) \geq \overline{\lim}_{n \to \infty} \int_\varepsilon^\infty I_{(r \geq (4 + 2k_0 T)a)}(a)(EL_t^0(X^n)$

$+ EL_t^a(X^n) - EL_t^0(X^n)) da/\rho_1(a) \geq \int_\varepsilon^\infty (r - (4 + 2k_0 T)a)^+ da/\rho_1(a).$

Hence

$\sup_{\varepsilon > 0} \overline{\lim}_{n \to \infty} \int_\varepsilon^\infty (da/\rho_1(a)) EL_t^a(X^n) = \infty.$

This is a contradiction. ■

8.7 Lemma. Under the assumption of Lemma 8.6 if $\sigma = \sigma^0$ satisfies all conditions in Theorem 5.2 globally, i.e. all dominant functions and constants are independent of N, $k_{N,T} = k_T$, $\varphi_{N,T} = \varphi_T$, $\rho_{N,T} = \rho_T$, $k_N = k, etc.$, and

$\int_R \left| \varphi_T(u) \right|^2 du < \infty;$

and if as $n \to \infty$

$\int_R \left| \sigma^n(y) - \sigma^0(y) \right|^2 dy \to 0,$

$E \left| x_0^n \right| \to 0$, (recall that $X_t^n = x_t^n - x_t^0$),

and there exists a $\delta > 0$ such that

$\delta \leq \sigma^n(x), \forall n = 0, 1, 2, \cdots,$
then $\forall t \geq 0$, as $n \to \infty$
$E|X_t^n| \to 0.$

Proof. Note that
$\int_\varepsilon^\infty EL_t^a(X^n)da/(\rho(a)^2 + a) = E\int_0^t I_{(X_s^n > \varepsilon)} \left|\sigma^n(x_s^n) - \sigma^0(x_s^0)\right|^2$
$\cdot d\langle M\rangle_s / (\rho(|X_s^n|)^2 + |X_s^n|) \leq 2E\int_0^t I_{(X_s^n > \varepsilon)} \left|\sigma^n(x_s^n) - \sigma^0(x_s^n)\right|^2$
$\cdot d\langle M\rangle_s / (\rho(|X_s^n|)^2 + |X_s^n|) + 2E\int_0^t I_{(X_s^n > \varepsilon)} \left|\sigma^0(x_s^n) - \sigma^0(x_s^0)\right|^2$
$\cdot d\langle M\rangle_s / (\rho(|X_s^n|)^2 + |X_s^n|) = A(n) + B(n),$
where we denote $\rho = \rho_T$ for $t \in [0, T]$ for simplicity. However, by assumption and Proposition 2.6 as $t \in [0, T]$
$A(n) \leq 2(\rho(\varepsilon)^2 + \varepsilon)^{-1} E\int_0^t \left|\sigma^n(x_s^n) - \sigma^0(x_s^n)\right|^2 d\langle M\rangle_s$
$= 2(\rho(\varepsilon)^2 + \varepsilon)^{-1} E\int_R L_t^a(x^n) \left|\sigma^n(a) - \sigma^0(a)\right|^2 / \left|\sigma^n(a)\right|^2 da$
$\leq 2[(\rho(\varepsilon)^2 + \varepsilon)\delta^2]^{-1} k'_T \int_R \left|\sigma^n(a) - \sigma^0(a)\right|^2 da \to 0$, as $n \to \infty$.
Moreover, by assumption 5.9
$B(n) \leq k''_T < \infty.$
Hence conditions in Lemma 8.6 are verified. Applying Lemma 8.6 and 8.5 the conclusion follows. ■

Now we are in a position to give a theorem on the convergence of solutions to SDEs with different diffusion and non-zero drift coefficients.

8.8 Theorem. Under assumption of Lemma 8.7 except (8.10) if
(8.13) $b^n(t, x, \omega) = b^0(t, x, \omega), \forall n = 1, 2, \cdots,$
and
$\text{sgn}(x - y) \cdot (b^0(t, x, \omega) - b^0(t, y, \omega)) \leq k_0 |x - y|,$
then $\forall t \geq 0$ as $n \to \infty$
$Ee^{-k_0 A_t} |x_t| \to 0.$

Proof. Denote
$\overline{I}_2^n(a) = \int_0^t I_{(X_s^n > a)}(b^0(s, x_s^n, \omega) - b^0(s, x_s^0, \omega))dA_s.$
Then for $a \geq 0$
$E\left|\overline{I}_2^n(a) - \overline{I}_2^n(0)\right| \leq E\left|\int_0^t I_{(0 < X_s^n \leq a)} X_s^n dA_s\right| k_0 \leq k_0 a E(A_T) = a k_T.$
Hence as $0 \leq t \leq T$, for $a \geq 0$
$\left|EL_t^a(X^n) - EL_t^0(X^n)\right| \leq (4 + 2k_0 T + 2k_T)a.$
A similar idea shows that as $0 \leq t \leq T$
$\left|EL_t^a(X^n) - EL_t^0(X^n)\right| \leq (4 + 2k_0 T + 2k_T)|a|.$
Therefore Lemma 8.6 is still true, i.e. as $n \to \infty$
$EL_t^0(X^n) \to 0.$
Note that
$|X_t^n| = |X_0^n| + \int_0^t sgn(X_s^n) \cdot (b^0(s, x_s^n, \omega) - b^0(s, x_s^0, \omega))dA_s$
$+ N_t + L_t^0(X^n) \leq |X_0^n| + \int_0^t k_0 |X_s^n| dA_s + N_t + L_t^0(X^n),$
where
$N_t = \int_0^t sgn(X_s^n) \cdot (\sigma^n(x_s^n) - \sigma^0(x_s^0))dM_s$
$+ \int_0^t \int_Z (\left|X_{s-}^n + c(s, x_{s-}^n, z, \omega) - c(s, x_{s-}^0, z, \omega)\right| - \left|X_{s-}^n\right|) q(ds, dz).$

By Lemma 8.2

$e^{-k_0 A_t}|X_t^n| \leq |X_0^n| + \int_0^t e^{-k_0 A_s} d(N_{st} + L_s^0(X^n)) \leq |X_0^n| + N_t + L_t^0(X^n).$

Hence as $n \to \infty$

$Ee^{-k_0 A_t}|X_t^n| \leq E[|X_0^n| + L_t^0(X^n)] \to 0.$ ∎

A similar proof shows the following theorem is true.

8.9 Theorem. (Situ, R 1984). The conclusion of Theorem 8.8 is still true, provided that (8.13) is replaced by the following conditions:

$b^n(t, x, \omega) = b^n(x), \forall n = 0, 1, 2, \cdots,$

$\int_R |b^n(a) - b^0(a)| da \to 0$, as $n \to \infty$, and $A_t = \langle M \rangle_t$.

Proof. Similarly to the proof of Theorem 8.8 one finds that for arbitrary given $\varepsilon > 0$ there exists an N such that as $n \geq N$

$|EL_t^a(X^n) - EL_t^0(X^n)| \leq (4 + 2k_0T + 2k_T)|a| + \varepsilon.$

Hence the proof follows in a similar fashion. ∎

Furthermore, we can also derive in a same way a convergence theorem on solutions of SDE's with different jump coefficients.

8.10 Theorem. The conclusion is still true, provided that (8.11) in the assumption of Theorem 8.9 is replaced by the following conditions:

$c^n(t, x, z, \omega) = c^n(x, z), \forall n = 0, 1, 2, \cdots,$

$\int_R \int_Z |c^n(a, z) - c^n(a, z)| \pi(dz) da \to 0$, as $n \to \infty$.

Since the proof is similar, we omit it here.

1.9 Existence of Solutions to Stochastic Differential Equations

Consider

(9.1) $x_t = x_0 + \int_0^t b(s, x_s, \omega) dA_s + \int_0^t \sigma(s, x_s, \omega) dM_s$

$+ \int_0^t \int_Z c(s, x_{s-}, z, \omega) q(ds, dz), \forall t \geq 0,$

where A_t, M_t and $q(dt, dz)$ have the same meaning as that in (7.1). Then by using the standard iterative method and stopping time technique (see Ikeda & Watanabe, 1989, and the proof of Theorem 9.6 below) one easily derives the following

9.1 Lemma. Assume that

1° (bounded condition) b, σ and c are jointly measurable such that

$|b(s, x, \omega)|^2 + |\sigma(s, x, \omega)|^2 + \int_Z |c(s, x, z, \omega)|^2 \pi(dz) \leq k_0,$

2° (Lipschitzian condition) $|b(s, x, \omega) - b(s, y, \omega)|^2 + |\sigma(s, x, \omega) - \sigma(s, y, \omega)|^2$

$+ \int_Z |c(s, x, z, \omega) - c(s, y, z, \omega)|^2 \pi(dz) \leq k_0 |x - y|^2.$

The (9.1) has a unique solution. Actually, it is pathwise unique. (For the definition of pathwise uniqueness to solutions of SDE see Definition 6.2).

In the following we shall use the comparison theorem for solutions to derive the existence of solution to (9.1) with non-Lipschitzian coefficients.

9.2 Theorem. Assume that

1° b, σ and c satisfy condition 1° in Lemma 9.1, (3.1) and (5.5) in Theorem 5.2, and $b(t,x,\omega) = b(t,x)$, (i.e. it is independent of ω);

2° $b(t,x)$ is jointly continuous, and σ, c satisfy condition 2° in Lemma 9.1.

Then (9.1) has a solution.

Proof. Take jointly Lipschitzian continuous functions $\overline{b}^n(t,x)$ and $\underline{b}^n(t,x)$ such that $\forall n = 1, 2, \cdots$

$$b(t,x) - 1/n < \underline{b}^n(t,x) < b(t,x) - 1/(n+1) < b(t,x)$$
$$< b(t,x) + 1/(n+1) < \overline{b}^n(t,x) < b(t,x) + 1/n.$$

Then by Lemma 9.1 there exists a unique $\overline{x}_t^n$ and $\underline{x}_t^n$ such that $\forall t \geq 0$

(9.2) $\overline{x}_t^n = x_0 + \int_0^t \overline{b}^n(s, \overline{x}_s^n, \omega) dA_s + \int_0^t \sigma(s, \overline{x}_s^n, \omega) dM_s$
$+ \int_0^t \int_Z c(s, \overline{x}_{s-}^n, z, \omega) q(ds, dz),$

and

$$\underline{x}_t^n = x_0 + \int_0^t \underline{b}^n(s, \underline{x}_s^n, \omega) dA_s + \int_0^t \sigma(s, \underline{x}_s^n, \omega) dM_s$$
$$+ \int_0^t \int_Z c(s, \underline{x}_{s-}^n, z, \omega) q(ds, dz).$$

By comparison Theorem 7.1 $P - a.s.$

$$\underline{x}_t^1 \leq \underline{x}_t^n \leq \underline{x}_t^{n+1} \leq \overline{x}_t^{n+1} \leq \overline{x}_t^n \leq \overline{x}_t^1, \forall t \geq 0, \forall n.$$

Hence there exist $P - a.s.$

$$\overline{x}_t^0 = \lim_{n \to \infty} \overline{x}_t^n.$$

By the boundedness of σ and its continuity with respect to x it is easily seen that

$$\lim_{n \to \infty} E \left| \int_0^t \sigma(s, \overline{x}_s^n, \omega) dM_s - \int_0^t \sigma(s, \overline{x}_s^0, \omega) dM_s \right|^2 = 0.$$

Similarly,

$$\lim_{n \to \infty} E \left| \int_0^t \int_Z c(s, \overline{x}_{s-}^n, z, \omega) q(ds, dz) - \int_0^t \int_Z c(s, \overline{x}_{s-}^0, z, \omega) q(ds, dz) \right|^2 = 0.$$

If one can show that as $n \to \infty$

(9.3) $\left| \int_0^t b^n(s, \overline{x}_s^n) dA_s - \int_0^t b(s, \overline{x}_s^0) dA_s \right| \to 0, \ P - a.s.$

then one can take the limit in probability in (9.2) and find that $\overline{x}_t^0$ satisfies (9.1). Let us show (9.3).

Indeed,

$$\text{the left hand side of (9.3)} \leq \left| \int_0^t b^n(s, \overline{x}_s^n) dA_s - \int_0^t b(s, \overline{x}_s^n) dA_s \right|$$
$$+ \left| \int_0^t b(s, \overline{x}_s^n) dA_s - \int_0^t b(s, \overline{x}_s^0) dA_s \right| = I_1^n + I_2^n.$$

By the continuity and boundedness of b it yields that as $n \to \infty$

$$I_2^n \to 0.$$

However, as $n \to \infty$

$$I_1^n \leq t/n \to 0.$$

Hence (9.3) is proved. ■

9.3 Corollary. If $b(t,x)$ also satisfies that for any $0 < T < \infty, \forall N = 1, 2, \cdots$ as $t \in [0,T]$

3° $(x - y) \cdot (b(t,x) - b(t,y)) \leq k_{N,T}(t) \rho_{N,T}(|x - y|^2)$, as $|x|, |y| \leq N$, where $k_{N,T}(t) \geq 0, \rho_{N,T}(u) > 0$, as $u > 0$; $\rho_{N,T}(0) = 0, \rho_{N,T}(u)$ is concave, and both $k_{N,T}(t)$ and $\rho_{N,T}(u)$ are non-random such that

$$\int_0^T k_{N,T}(t)dt < \infty, \int_{0+} du/\rho_{N,T}(u) = \infty,$$

then the solution of (9.1) is pathwise unique.

Proof. If $x_t^i, i = 1, 2$, both satisfy (9.1), then by Ito's formula as $t \in [0, T]$

$$E\left|x_{t\wedge\tau_N}^1 - x_{t\wedge\tau_N}^2\right|^2 \leq E(\int_0^{t\wedge\tau_N} k_{N,T}(s)\rho_{N,T}(\left|x_s^1 - x_s^2\right|^2)dA_s$$
$$+k_0 \int_0^{t\wedge\tau_N} \left|x_s^1 - x_s^2\right|^2 d\langle M\rangle_s + k_0 \int_0^{t\wedge\tau_N} \left|x_s^1 - x_s^2\right|^2 ds)$$
$$\leq E\int_0^{t\wedge\tau_N} \widetilde{k}_{N,T}(s)\widetilde{\rho}_{N,T}(\left|x_s^1 - x_s^2\right|^2)d\widetilde{A}_s,$$

where

$$\tau_N = \inf(t \geq 0 : \left|x_t^1\right| + \left|x_t^2\right| > N),$$
$$\widetilde{k}_{N,T}(t) = k_{N,T}(t) + k_0,\ \widetilde{\rho}_{N,T}(u) = \rho_{N,T}(u) + u,$$
$$\widetilde{A}_t = A_t + \langle M\rangle_t + t,$$

and for simplicity, without loss of generality we may assume that M_t and A_t is integrable and square integrable process, respectively. Thus applying the same technique as the proof of Theorem 6.1, one derives without difficulty that $P - a.s.$

$$x_t^1 = x_t^2, \forall t \geq 0. \blacksquare$$

9.4 Remark. Theorem 9.2 is a generalization of Theorem 1 in Mel'nikov, 1985, to the case of SDE with jumps.

Now let us relax the bounded condition in 1° of Lemma 9.1 and Theorem 9.2 to the less than linear growth condition. Since the SDE here is with an increasing process A_t, a local martingale M_t, (both are continuous), and with jumps, considerable discussion is necessary. We have

9.5 Lemma. Assume that

1° jointly measurable functions b, σ, c satisfy

$$xb(t, x, \omega) \leq k_0(1 + |x|^2),$$
$$|\sigma(t, x, \omega)|^2 + \int_Z |c(t, x, z, \omega)|^2 \pi(dz) \leq k_0(1 + |x|^2),$$

2° $E|x_0|^2 < \infty$.

If $x_t, t \geq 0$, is a solution of (9.1), then

$$E\sup_{s\leq\widetilde{T}(t)} |x_s|^2 \leq \overline{k}_T e^{\overline{k}_T} E|x_0|^2 = k_T < \infty, \text{ as } t \in [0, T],$$

where $\overline{k}_T \geq 0$ is a constant depending on $T, E|x_0|^2, E\langle M\rangle_T$ and k_0 only, $\overline{k}_T$ is independent of $x_t, t > 0$, and $\widetilde{T}(t)$ is the inverse of $\widetilde{A}_t$,

$$\widetilde{A}_t = A_t + \langle M\rangle_t + t.$$

Proof. By Ito's formula one has that

$$|x_t|^2 \leq |x_0|^2 + k_0 \int_0^t (1 + |x_s|^2)dA_s + 2\int_0^t x_s\sigma(t, x_s, \omega)dM_s$$
$$+ \int_0^t |\sigma(t, x_s, \omega)|^2 d\langle M\rangle_s + 2\int_0^t \int_Z x_s c(t, x_{s-}, z, \omega)q(ds, dz)$$
$$+ \int_0^t \int_Z |c(t, x_{s-}, z, \omega)|^2 p(ds, dz).$$

Denote

$$Y_t = \sup_{s\leq t} |x_s|^2.$$

By the martingale inequality

$EY_t \leq E|x_0|^2 + k_0 E\int_0^t (1+|x_s|^2)dA_s$
$+k_0^{1/2} 8E(\int_0^t (|x_s|(1+|x_s|^2))^2 d\langle M\rangle_s)^{1/2}) + k_0 E\int_0^t (1+|x_s|^2)d(s+\langle M\rangle_s)$
$+8E((\int_0^t \int_Z |x_s|^2 |c|^2 \pi(dz)ds)^{1/2}),$
where for simplicity we may assume that A_t and M_t are integrable and square integrable process, respectively. Otherwise, we can use the localization technique.
Again, for simplicity we may assume that
$EY_{\widetilde{T}(t)} = E\sup_{s\leq \widetilde{T}(t)} |x_s|^2 < \infty.$
(Otherwise, we can use the localization technique. For example, introduce stopping times
$\tau_N = \inf(t \geq 0 : |x_t| > N),\ N = 1, 2, \cdots$
Then
$EY_{\widetilde{T}(t)\wedge\tau_N} < \infty,$
and as $N \uparrow \infty, \tau_N \uparrow \infty$. Hence, if one can prove
$EY_{\widetilde{T}(t)\wedge\tau_N} \leq k_T, \forall 0 \leq t \leq T, \forall N,$
then applying Fatou's lemma, one derives the conclusion of Lemma 9.5). However, for any $\varepsilon > 0$
$EY_{\widetilde{T}(t)} \leq E|x_0|^2 + k'_T\varepsilon^{-1}\int_0^t (1 + EY_{\widetilde{T}(s)})ds + \varepsilon EY_{\widetilde{T}(t)}$, as $t \in [0,T]$.
By Gronwall's inequality we have
$EY_{\widetilde{T}(t)} \leq E|x_0|^2 \overline{k}_T e^{\overline{k}_T}$, as $t \in [0,T]$. ■

Now we are in a position to prove the following theorem.

9.6 Theorem. Assume that all assumptions in Lemma 9.5 hold, and b, σ, c satisfy the condition 2° in Lemma 9.1, and, moreover,
$|b(t,x,\omega)|^2 \leq k_0(1+|x|^2).$
Then (9.1) has a unique solution.

Proof. Let us use the Picard iteration technique. Denote
$x_t^1 = x_0,$
(9.4) $x_t^{n+1} = x_0 + \int_0^t b(s,x_s^n,\omega)dA_s + \int_0^t \sigma(s,x_s^n,\omega)dM_s$
$+ \int_0^t \int_Z c(s,x_{s-}^n,z,\omega)q(ds,dz),\ \forall t \geq 0, \forall n = 1,2,\cdots$
Then
$|x_t^{n+1} - x_t^n|^2 \leq 3(\left|\int_0^t (b(s,x_s^n,\omega) - b(s,x_s^{n-1},\omega))dA_s\right|^2$
$+\left|\int_0^t (\sigma(s,x_s^n,\omega) - \sigma(s,x_s^{n-1},\omega))dM_s\right|^2$
$+\left|\int_0^t \int_Z (c(s,x_{s-}^n,z,\omega) - c(s,x_{s-}^{n-1},z,\omega))q(ds,dz)\right|^2).$
Let
$Y_t^{n+1} = \sup_{s\leq t}|x_s^{n+1} - x_s^n|^2.$
By the martingale inequality
$EY_t^{n+1} \leq 3(k_0 E\left|\int_0^t (Y_s^n)^{1/2} dA_s\right|^2 + 4k_0 E\int_0^t Y_s^n d\langle M\rangle_s$
$+4k_0 E\int_0^t Y_s^n ds) \leq 3k_0 E\left|\int_0^t (Y_s^n)^{1/2} d\widetilde{A}_s\right|^2 + 24k_0 E\int_0^t Y_s^n d\widetilde{A}_s,$

where $\widetilde{A}_s$ is defined in Lemma 9.5. Hence

(9.5) $EY^{n+1}_{\widetilde{T}(t)} \leq 3k_0 E\left|\int_0^t \left(Y^n_{\widetilde{T}(s)}\right)^{1/2} ds\right|^2 + 24k_0 E\int_0^t Y^n_{\widetilde{T}(s)} ds$

$\leq (3k_0T + 24k_0)\int_0^t EY^n_{\widetilde{T}(s)} ds = \widetilde{k}_T \int_0^t EY^n_{\widetilde{T}(s)} ds$, as $t \in [0,T]$.

After iteration one obtains that

(9.6) $EY^{n+1}_{\widetilde{T}(t)} \leq \overline{k}_T(\widetilde{k}_T t)^n/n!$,

where

$\overline{k}_T = 2(E|x_0|^2 + k_T)T$,

and k_T comes from Lemma 9.5. Hence,

$\sum_{n=1}^\infty P(\left(Y^{n+1}_{\widetilde{T}(T)}\right)^{1/2} > n^{-2}) \leq \overline{k}_T \sum_{n=1}^\infty (\widetilde{k}_T T)^n n^4/n! < \infty.$

By Borel-Cantelli lemma one has that $P - a.s.$

$x_0 + \sum_{n=1}^\infty \left(Y^{n+1}_{\widetilde{T}(T)}\right)^{1/2} < \infty,$

i.e. $P - a.s.$

$x_0 + \sum_{n=0}^\infty \left|x_t^{n+1} - x_t^n\right| < \infty$, uniformly for $t \in [0, \widetilde{T}(T)]$.

Hence, there exists a limit: $P - a.s.$

$x_t^0 = x_0 + \sum_{n=1}^\infty (x_t^{n+1} - x_t^n) = \lim_{n\to\infty} x_t^n$, uniformly for $t \in [0, \widetilde{T}(T)]$.

By the uniform convergence x_t^0 is cadlag. Moreover, by Lemma 9.5 and Fatou's lemma

$\sup_{t\leq T} E\left|x^0_{\widetilde{T}(t)}\right|^2 \leq k_T < \infty.$

Let us show that as $t \in [0,T]$

(9.6) $\lim_{n\to\infty} E\left|x^n_{\widetilde{T}(t)} - x^0_{\widetilde{T}(t)}\right|^2 = 0.$

Indeed, let

$Y_t^{n,p} = \sup_{s\leq t} |x_s^{n+p} - x_s^n|^2.$

Then by (9.6) one gets that

$EY^{n,p}_{\widetilde{T}(t)} \leq \sum_{j=n+1}^{n+p} EY^j_{\widetilde{T}(t)} \leq \sum_{j=n+1}^{n+p} \overline{k}_T(\widetilde{k}_T t)^j/j!.$

Let $p \to \infty$ by Fatou's lemma

$E\left|x^n_{\widetilde{T}(t)} - x^0_{\widetilde{T}(t)}\right|^2 \leq \sum_{j=n+1}^\infty \overline{k}_T(\widetilde{k}_T t)^j/j!.$

Hence (9.6) is obtained. From this as $t \in [0,T]$ by Lemma 9.5 and Lebesgue's dominated convergence theorem

$E\left|\int_0^{t\wedge\widetilde{T}(t)} \left|b(s,x_s^n,\omega) - b(s,x_s^0,\omega)\right| dA_s\right|^2$

$\leq k_0 T\int_0^t E\left|x^n_{\widetilde{T}(s)} - x^0_{\widetilde{T}(s)}\right|^2 ds \to 0$, as $n \to \infty$.

Similarly, for $t \in [0,T]$ as $n \to \infty$

$E\left|\int_0^{t\wedge\widetilde{T}(t)} (\sigma(s,x_s^n,\omega) - \sigma(s,x_s^0,\omega))dM_s\right|^2$

$\leq k_0 \int_0^t E\left|x^n_{\widetilde{T}(s)} - x^0_{\widetilde{T}(s)}\right|^2 ds \to 0,$

and

$E\left|\int_0^{t\wedge\widetilde{T}(t)} \int_Z (c(s,x_{s-}^n,z,\omega) - \sigma(s,x_{s-}^0,z,\omega))q(ds,dz)\right|^2 \to 0.$

Therefore in (9.4) let $n \to \infty$, one obtains that $P-a.s.$

$$x^0_{t\wedge\widetilde{T}(t)} = x_0 + \int_0^{t\wedge\widetilde{T}(t)} b(s,x^0_s,\omega)dA_s + \int_0^{t\wedge\widetilde{T}(t)} \sigma(s,x^0_s,\omega)dM_s$$
$$+\int_0^{t\wedge\widetilde{T}(t)}\int_Z c(s,x^0_{s-},z,\omega)q(ds,dz),\ \forall t\geq 0.$$

Note that as T tends continuously to infinity, so does $\widetilde{T}(T)$. Therefore, this yields that $P-a.s.$

$$x^0_t = x_0 + \int_0^t b(s,x^0_s,\omega)dA_s + \int_0^t \sigma(s,x^0_s,\omega)dM_s$$
$$+\int_0^t\int_Z c(s,x^0_{s-},z,\omega)q(ds,dz),\ \forall t\geq 0.$$

The uniqueness is derived by Corollary 9.3. ■

Now we can derive

9.7 Theorem. Under assumption 1° and 2° of Lemma 9.5 if

1°′ $b(t,x,\omega)=b(t,x)$, $|b(t,x)|^2 \leq k_0(1+|x|^2)$,

2°′ $b(t,x)$ is a continuous function in (t,x), and σ,c satisfy condition 2° (Lipschitzian condition) in Lemma 9.1,

then (9.1) has a solution.

Proof. The proof of this Theorem is similar to that of Theorem 9.2. The only difference is that now we cannot use the bounded convergence theorem to show that as $n\to\infty$

$(9.7)_1$ $E\int_0^{t\wedge\widetilde{T}(t)} \left|\overline{b}^n(s,\overline{x}^n_s) - b(s,\overline{x}^0_s)\right| dA_s \to 0.$

However, since

$$\underline{x}^1_t \leq \overline{x}^n_t \leq \overline{x}^1_t, \forall n,$$

and by Lemma 9.5 as $t\in[0,T]$

$$E\sup_{s\leq\widetilde{T}(t)}\left|\overline{x}^1_s\right|^2 < \infty,\ E\sup_{s\leq\widetilde{T}(t)}\left|\underline{x}^1_s\right|^2<\infty.$$

Hence by the dominated convergence theorem as $n\to\infty$ $\forall t\in[0,T]$

$$E\left|\overline{x}^n_{\widetilde{T}(t)} - \overline{x}^0_{\widetilde{T}(t)}\right|^2 \to 0,$$

$(9.7)_2$ $E\left|\int_0^{t\wedge\widetilde{T}(t)}(\sigma(s,\overline{x}^n_s,\omega)-\sigma(s,\overline{x}^0_s,\omega))dM_s\right|^2$

$$\leq k_0\int_0^{t\wedge\widetilde{T}(s)} E\left|x^n_s - x^0_s\right|^2 d\langle M\rangle_s \leq k_0\int_0^t E\left|x^n_{\widetilde{T}(s)} - x^0_{\widetilde{T}(s)}\right|^2 ds \to 0.$$

Similarly,

$(9.7)_3$ $E\left|\int_0^{t\wedge\widetilde{T}(t)}\int_Z(c(s,x^n_{s-},z,\omega)-c(s,x^0_{s-},z,\omega))q(ds,dz)\right|^2 \to 0.$

Let us show $(9.7)_1$. Note that

$$\text{the left hand side of } (9.7)_1 \leq E\int_0^{t\wedge\widetilde{T}(t)}\left|\overline{b}^n(s,\overline{x}^n_s) - b(s,\overline{x}^n_s)\right| dA_s$$
$$+E\int_0^{t\wedge\widetilde{T}(t)}\left|b(s,\overline{x}^n_s)-b(s,\overline{x}^0_s)\right| dA_s = I^n_1 + I^n_2.$$

Obviously, we still have that $n\to\infty$

$$I^n_1 \leq t/n \to 0.$$

On the other hand,

$$I^n_2 \leq \int_0^T E\left|b(\widetilde{T}(s),\overline{x}^n_{\widetilde{T}(s)}) - b(\widetilde{T}(s),\overline{x}^0_{\widetilde{T}(s)})\right| ds = \int_0^T EZ^n_s ds.$$

Note that by Lemma 9.5 as $s\in[0,T]$

$$E\left|b(\widetilde{T}(s),\overline{x}^n_{\widetilde{T}(s)})-b(\widetilde{T}(s),\overline{x}^0_{\widetilde{T}(s)})\right|^2\leq 2k_0(1+k_T)<\infty.$$

Hence $\left\{b(\widetilde{T}(s),\overline{x}^n_{\widetilde{T}(s)})-b(\widetilde{T}(s),\overline{x}^0_{\widetilde{T}(s)})\right\}_{n=1}^{\infty}$ are uniformly integrable with respect to measure $dP\times ds$. However, by the jointly continuity of b this shows that

$\lim_{n\to\infty} Z^n_s=0,\ \forall s\in[0,T]$.

Hence

$\lim_{n\to\infty} I^n_2=0$.

Therefore the conclusion can still be obtained in similar fashion. ■

9.8 Corollary. If $b(t,x)$ also satisfies the condition 3° in Corollary 9.3, then the solution of (9.1) is also pathwise unique.

The proof is completely the same as that of Corollary 9.3.

9.9 Remark. Theorem 9.7 and Corollary 9.8 seem to be new.

1.10 Tanaka Formula for SDE with Poisson Jumps in n-Dimensional Space

In the above we have already seen that the Tanaka type formula is a powerful tool in the study of 1-dimensional SDE with Poisson jumps. It is natural to discuss such a formula for the n-dimensional space. In this case it needs a little more discussion. Consider

(10.1) $x_t=x_0+\int_0^t b(s,x_s,\omega)dA_s+\int_0^t\sigma(s,x_s,\omega)dM_s$
$\quad+\int_0^t\int_Z c(s,x_{s-},z,\omega)q(ds,dz),\ \forall t\geq 0,$

where $A_t=(A_{1t},\cdots,A_{mt}),M_t=(M_{1t},\cdots,M_{rt})$; A_{it} is a continuous increasing process, $1\leq i\leq m$; M_{it} is a continuous local square integrable martingale, $1\leq i\leq r$; such that $\langle M_i,M_j\rangle_t=0$, as $i=j$; $q(dt,dz)$ is a Poisson martingale measure with the compensator $\pi(dz)dt$ such that

$q(dt,dz)=p(dt,dz)-\pi(dz)dt,$

where $p(dt,dz)$ is the Poisson counting measure; c is a R^d-valued vector, and b and σ is a $d\times m$ and $d\times r$ matrix, respectively. We have

10.1 Theorem. Assume that

1° $|b(t,x,\omega)|^2+\|\sigma(t,x,\omega)\|^2+\int_Z\|c(t,x,z,\omega)\|^2\pi(dz)\leq g(|x|)$,
where $g(u)$ is a continuous function on $u\geq 0$;

2° $\|\sigma(t,x,\omega)-\sigma(t,y,\omega)\|^2\leq k_N(t)\rho_N(|x-y|)\,|x-y|$, as $|x|,|y|\leq N$;
where $0\leq k_N(t)$, $\int_0^T k_N(t)dt<\infty$, for each $T<\infty;0\leq\rho_N(u)$ is strictly increasing,in $u\geq 0$ with $\rho_N(0)=0$, and

$\int_{0+}du/\rho_N(u)=\infty.$

If x^i_t satisfies (10.1) with coefficients (b^i,σ,c) and initial value $x^i_0,i=1,2$, respectively, but defined on the same probability space and with the same M_t,A_t and $q(dt,dz)$, then $\forall t\geq 0$

(10.2) $\left|x_t^1 - x_t^2\right| = \left|x_0^1 - x_0^2\right|$
$+ \int_0^t sgn(x_s^1 - x_s^2) \cdot (b^1(s, x_s^1, \omega) - b^2(s, x_s^2, \omega)) dA_s$
$+ \int_0^t sgn(x_s^1 - x_s^2) \cdot (\sigma(s, x_s^1, \omega) - \sigma(s, x_s^2, \omega)) dM_s$
$+ \frac{1}{2} \sum_{i,j,k=1}^{d,d,r} \int_0^t I_{(x_s^1 = x_s^2)} \frac{\left|x_s^1 - x_s^2\right|^2 \delta_{ij} - (x_{is}^1 - x_{is}^2)(x_{js}^1 - x_{js}^2)}{\left|x_s^1 - x_s^2\right|^3}$
$\cdot (\sigma_{ik}(s, x_s^1, \omega) - \sigma_{ik}(s, x_s^2, \omega))(\sigma_{jk}(s, x_s^1, \omega) - \sigma_{jk}(s, x_s^2, \omega)) d\langle M_k \rangle_s$
$+ \int_0^t \int_Z sgn(x_{s-}^1 - x_{s-}^2) \cdot (c(s, x_{s-}^1, z, \omega) - c(s, x_{s-}^2, z, \omega)) q(ds, dz)$
$+ \int_0^t \int_Z [\left|x_{s-}^1 - x_{s-}^2 + c(s, x_{s-}^1, z, \omega) - c(s, x_{s-}^2, z, \omega)\right| - \left|x_{s-}^1 - x_{s-}^2\right|$
$- sgn(x_{s-}^1 - x_{s-}^2) \cdot (c(s, x_{s-}^1, z, \omega) - c(s, x_{s-}^2, z, \omega))] p(ds, dz),$

where

$$sgn x = \begin{cases} (\frac{x_1}{|x|}, \cdots, \frac{x_d}{|x|}), & \text{as } x = 0, \\ 0, & \text{as } x = 0. \end{cases}$$

Proof. Let

$\tau_N = \inf\left\{t \geq 0 : \left|x_t^1\right| + \left|x_t^2\right| > N\right\}.$

Now for any fixed N take $1 \geq a_n \downarrow 0$ such that

$\int_{a_n}^{a_{n-1}} \rho_N^{-1}(u) du = n,$

and for each n take a continuous function $f_n(u)$ such that

$$f_n(u) = \begin{cases} 0, & \text{as } 0 \leq u \leq a_n, \text{ or } u \geq a_{n-1}, \\ \text{any number less than } 2(n\rho_N(u))^{-1}, & \text{otherwise;} \end{cases}$$

with $\int_{a_n}^{a_{n-1}} f_n(u) du = 1$. Set

$\varphi_n(x) = \int_0^{|x|} dy \int_0^y f_n(u) du, n = 1, 2, \cdots.$

Then $\varphi_n \in C^2(R^d)$, and

$(\partial/\partial x_i)\varphi_n(x) = I_{(a_n < |x|)} \frac{x_i}{|x|} \int_0^{|x|} f_n(u) du,$

$(\partial^2/\partial x_i \partial x_j)\varphi_n(x) = I_{(a_n < |x|)} (\frac{|x|^2 \delta_{ij} - x_i x_j}{|x|^3} \int_0^{|x|} f_n(u) du + f_n(|x|) \frac{x_i x_j}{|x|^2}).$

Obviously,

$|(\partial/\partial x_i)\varphi_n(x)| \leq 1,$

and as $n \to \infty$

$\varphi_n(x) \to |x|, (\partial/\partial x_i)\varphi_n(x) \to I_{(x=0)} \frac{x_i}{|x|}.$

By Ito's formula

$\varphi_n(x_{t\wedge\tau_N}^1 - x_{t\wedge\tau_N}^2) = \varphi_n(x_0^1 - x_0^2)$
$+ \sum_{i,k=1}^{d,m} \int_0^{t\wedge\tau_N} (\partial/\partial x_i)\varphi_n(x_s^1 - x_s^2) \cdot (b_{i,k}^1(s, x_s^1, \omega) - b_{i,k}^2(s, x_s^2, \omega)) dA_{ks}$
$+ \sum_{i,k=1}^{d,r} \int_0^{t\wedge\tau_N} (\partial/\partial x_i)\varphi_n(x_s^1 - x_s^2) \cdot (\sigma_{ik}(s, x_s^1, \omega) - \sigma_{ik}(s, x_s^2, \omega)) dM_{ks}$
$+ \frac{1}{2} \sum_{i,j,k=1}^{d,d,r} \int_0^{t\wedge\tau_N} (\partial^2/\partial x_i \partial x_j)\varphi_n(x_s^1 - x_s^2)$
$\cdot (\sigma_{ik}(s, x_s^1, \omega) - \sigma_{ik}(s, x_s^2, \omega))(\sigma_{jk}(s, x_s^1, \omega) - \sigma_{jk}(s, x_s^2, \omega)) d\langle M_k \rangle_s$
$+ \sum_{i=1}^d \int_0^{t\wedge\tau_N} \int_Z (\partial/\partial x_i)\varphi_n(x_{s-}^1 - x_{s-}^2) \cdot (c_i(s, x_{s-}^1, z, \omega)$
$- c_i(s, x_{s-}^2, z, \omega)) q(ds, dz)$
$+ \int_0^{t\wedge\tau_N} \int_Z [\varphi_n(x_{s-}^1 - x_{s-}^2 + c(s, x_{s-}^1, z, \omega) - c(s, x_{s-}^2, z, \omega))$
$- \varphi_n(x_{s-}^1 - x_{s-}^2) - \sum_{i=1}^d (\partial/\partial x_i)\varphi_n(x_{s-}^1 - x_{s-}^2)$
$\cdot (c_i(s, x_{s-}^1, z, \omega) - c_i(s, x_{s-}^2, z, \omega))] p(ds, dz) = \sum_{i=1}^6 I_i^n.$

By Lebesgue's dominated convergence theorem it is easily seen that as $n \to \infty$

$I_1^n \to \int_0^{t\wedge\tau_N} sgn(x_s^1 - x_s^2)\cdot(b(s,x_s^1,\omega) - b(s,x_s^2,\omega))dA_s,$
$E|\sum_{i,k=1}^{d,r}\int_0^{t\wedge\tau_N}(\partial/\partial x_i)\varphi_n(x_s^1 - x_s^2)\cdot(\sigma_{ik}(s,x_s^1,\omega) - \sigma_{ik}(s,x_s^2,\omega))dM_{ks}$
$-\int_0^{t\wedge\tau_N} sgn(x_s^1 - x_s^2)\cdot(\sigma(s,x_s^1,\omega) - \sigma(s,x_s^2,\omega))dM_s\,|^2 \to 0.$

Note that as $n\to\infty, \forall t\in[0,T], \forall N$

$\sum_{i,j,k=1}^{d,d,r}\int_0^{t\wedge\tau_N} f_n\left(\left|x_s^1 - x_s^2\right|\right)\frac{(x_{is}^1 - x_{is}^2)(x_{js}^1 - x_{js}^2)}{|x_s^1 - x_s^2|^2}\cdot(\sigma_{ik}(s,x_s^1,\omega) - \sigma_{ik}(s,x_s^2,\omega))$
$\cdot(\sigma_{jk}(s,x_s^1,\omega) - \sigma_{jk}(s,x_s^2,\omega))d\left\langle M_k\right\rangle_s \le 2Nk_0'T/n \to 0.$

Moreover, as $0\le s\le t\wedge\tau_N$

$\sum_{i,j,k=1}^{d,d,r}\left|I_{(a_n<|x_s^1 - x_s^2|)}\left(\frac{|x_s^1 - x_s^2|^2\delta_{ij} - (x_{is}^1 - x_{is}^2)(x_{js}^1 - x_{js}^2)}{|x_s^1 - x_s^2|^3}\int_0^{|x_s^1 - x_s^2|} f_n(u)du\right|\right.$
$\cdot\left|(\sigma_{ik}(s,x_s^1,\omega) - \sigma_{ik}(s,x_s^2,\omega))(\sigma_{jk}(s,x_s^1,\omega) - \sigma_{jk}(s,x_s^2,\omega))\right|$
$\le \sum_{i,j,k=1}^{d,d,r}\rho(2N)k_N(s).$

Hence by Lebesgue's dominated convergence theorem it is seen that as $n\to\infty$

$I_4^n \to$ the 4th term in the right hand side of (10.2)
(with $t\wedge\tau_N$ substituting t).

Similarly, it is seen that as $n\to\infty$

$I_5^n \to$ the 5th term in the right hand side of (10.2)
(with $t\wedge\tau_N$ substituting t).

Moreover, by Lebesgue's dominated convergence theorem as $n\to\infty$

$E|\sum_{i=1}^{d}\int_0^{t\wedge\tau_N}\int_Z(\partial/\partial x_i)\varphi_n(x_{s-}^1 - x_{s-}^2)\cdot(c_i(s,x_{s-}^1,z,\omega)$
$-c_i(s,x_{s-}^2,z,\omega))q(ds,dz) - \int_0^t\int_Z sgn(x_{s-}^1 - x_{s-}^2)$
$\cdot(c(s,x_{s-}^1,z,\omega) - c(s,x_{s-}^2,z,\omega))q(ds,dz)\,|^2 \to 0.$

Finally, note that

$\lim_{N\to\infty}\tau_N = \infty.$

The desired conclusion can now be derived. ■

As an application of formula (10.2) we give the following

10.2 Theorem. Assume that conditions 1° and 2° in Theorem 10.1 hold, moreover,

3° $\rho_N(u)$ in 2° is concave, $\forall N = 1,2,\cdots$;

4° $\sum_{i=1}^d(sgn(x-y))_i\cdot(b(s,x,\omega) - b(s,y,\omega))_{ik} \le \overline{G}_N(|x-y|)$, as $|x|, |y|\le N$, $\forall k = 1,\cdots,m$;

where $\overline{G}_N(u)$ has the same property as that of $\rho_N(u)$,

5° $\int_Z |c(t,x,z,\omega)|\,\pi(dz) < \infty,$
$\int_Z |c(t,x,z,\omega) - c(t,x,z,\omega)|\,\pi(dz) \le G_N'(|x-y|)$, as $|x|, |y|\le N$;

where $G_N'(u)$ has the same property as that of $\rho_N(u)$.

Then the pathwise uniqueness of solutions for (10.1) holds.

Proof. Assume that $x_t^i, i=1,2,$ are two solutions of (10.1) on the same probability space with the same $A_t, M_t, q(dt,dz)$. Then by formula (10.2)

$E\left|x_{t\wedge\tau_N}^1 - x_{t\wedge\tau_N}^2\right| \le \sum_{k=1}^m E\int_0^{t\wedge\tau_N}\overline{G}_N(\left|x_s^1 - x_s^2\right|)dA_{ks}$
$+k_0'\sum_{k=1}^m E\int_0^{t\wedge\tau_N}k_N(s)\rho_N(\left|x_s^1 - x_s^2\right|)d\left\langle M\right\rangle_{ks}$
$+2\sum_{k=1}^m E\int_0^{t\wedge\tau_N}G_N'(\left|x_s^1 - x_s^2\right|)ds \le E\int_0^{t\wedge\tau_N}\widetilde{G}_N(\left|x_s^1 - x_s^2\right|)d\widetilde{A}_s,$

where $\tau_N = \inf\left\{t\ge 0: \left|x_t^1\right| + \left|x_t^2\right| > N\right\},$

$\widetilde{A}_s = \sum_{k=1}^m A_{ks} + \sum_{k=1}^m \int_0^{s\wedge\tau_N} k_N(r)d\langle M\rangle_{kr} + \sum_{k=1}^m s,$

$\widetilde{G}_N(u) = \overline{G}_N(u) + \rho_N(u) + G'_N(u),$

and $\widetilde{G}_N(u)$ still has the property as that of $\rho_N(u)$ by Lemma 6.3. Let

$\widetilde{T}_t = \widetilde{A}_t^{-1}.$

Note that $\widetilde{A}_t > t$. Hence $\widetilde{T}_t = \widetilde{A}_t^{-1} < t$. Thus

$E\left|x^1_{\widetilde{T}_{t\wedge\tau_N}} - x^2_{\widetilde{T}_{t\wedge\tau_N}}\right| \leq E\int_0^{t\wedge\tau_N} \widetilde{G}_N(\left|x^1_{\widetilde{T}_s} - x^2_{\widetilde{T}_s}\right|)ds$

$\leq \int_0^t \widetilde{G}_N(E\left|x^1_{\widetilde{T}_{s\wedge\tau_N}} - x^2_{\widetilde{T}_{s\wedge\tau_N}}\right|)ds.$

From this it easily found that $P - a.s.$ $\forall t \geq 0$

$x_t^1 = x_t^2$. ■

We can also use Theorem 10.1 to discuss the convergence of solutions to (10.1). Suppose that $x_t^n, n = 0,1,2,\cdots$ satisfy the following SDEs, respectively,

(10.3) $x_t^n = x_0 + \int_0^t b^n(s, x_s^n, \omega)dA_s + \int_0^t \sigma(s, x_s^n, \omega)dM_s$

$+\int_0^t\int_Z c(s, x_{s-}^n, z, \omega)q(ds, dz), \forall t \geq 0, n = 0,1,2,\cdots$

We also denote $x_t^0 = x_t, b^0 = b$.

10.3 Theorem. Assume that

1° $\sum_{i=1}^d (sgn(x-y))_i \cdot (b(s,x,\omega) - b(s,y,\omega))_{ik} \leq F_k(s,\omega)\,|x-y|,$

where $0 \leq F_k(s,\omega), k = 1,\cdots,m$, satisfy $\forall t \geq 0$

$\int_0^t F_k(s,\omega)dA_{ks} < \infty,$

2° $\|\sigma(t,x,\omega) - \sigma(t,y,\omega)\|^2 \leq K(s,\omega)\,|x-y|^2,$

where $0 \leq K(s,\omega)$ satisfies $\forall t \geq 0$

$\sum_{k=1}^r \int_0^t K(s,\omega)^2 d\langle M_k\rangle_s < \infty,$

3° conditions 3°–5° in Theorem 8.1 hold here for b^n, b being matrices, and $x_{0,}^n x_0; c^n, c$ being vectors, e.g.

$\|b^n(t,x,\omega) - b(t,x,\omega)\| \leq F^n(t,\omega)$, $\lim_{n\to\infty}\sum_{k=1}^m E\int_0^t F^n(s,\omega)dA_{ks} = 0,$

$|c^n(t,x,z,\omega) - c(t,x,z,\omega)| \leq G^n(t,z,\omega),$

$\lim_{n\to\infty} E\int_0^t\int_Z G^n(s,z,\omega)\pi(dz)ds = 0,$

$|c(t,x,z,\omega) - c(t,y,z,\omega)| \leq G(t,z,\omega)\,|x-y|,$

$E\int_0^t\int_Z G(s,z,\omega)\pi(dz)ds < \infty$, etc.

Then we have that $\forall t \geq 0$

(10.4) $\lim_{n\to\infty} E[\exp(-\sum_{k=1}^m \int_0^t F_k(s,\omega)dA_{ks}$

$-2\int_0^t\int_Z G(s,z,\omega)\pi(dz)ds) \cdot |x_t^n - x_t|] = 0.$

The proof is similar to that of Theorem 8.1 if here we use formula (10.2), so we omit it.

Conditions 3°, 4° in Theorem 10.2 can be some non-Lipschitzian condition. Let us give an example as follows.

10.4 Example. Let

$b_i(t,x) = -k_0\frac{x_i}{|x|}I_{x=0} + k_1 x_i + k_2|x_i| + k_3|x|.$

Then it is easily seen that as $x, y = 0$,

$\sum_{i=1}^d \frac{x_i - y_i}{|x-y|}\left(-\frac{x_i}{|x|} + \frac{y_i}{|y|}\right) = \sum_{i=1}^d \frac{1}{|x-y|}\left(-\frac{x_i^2}{|x|} - \frac{y_i^2}{|y|} + \frac{x_i y_i}{|x|} + \frac{x_i y_i}{|y|}\right)$

$\leq \frac{1}{|x-y|}(-|x|-|y|+\frac{|x||y|}{|x|}+\frac{|x||y|}{|y|})=0,$
where we have applied the Schwarz's inequality to get that
$\sum_{i=1}^{d} x_i y_i \leq |x|\,|y|$.
Note also that as $x=0, y=0$,
$\sum_{i=1}^{d} \frac{x_i}{|x|}\left(-\frac{x_i}{|x|}\right) \leq 0.$
Hence $b(t,x)$ satisfies the condition 3°.

In the following we are going to derive a Tanaka formula for the component $(x_{it}^1 - x_{it}^2)$ of $(x_t^1 - x_t^2)$, which is useful for the discussion of the comparison of solutions and even for the reflecting SDE in multi-dimensional space to be considered later.

Suppose that $x_t^i, i=1,2$, satisfy the following SDEs similar to (10.1), respectively,

(10.5) $x_t^j = x_0 + \int_0^t b^j(s,x_s^j,\omega)dA_s + \int_0^t \sigma(s,x_s^j,\omega)dM_s$
$+\int_0^t \int_Z c(s,x_{s-}^j,z,\omega)q(ds,dz), \forall t \geq 0, j=1,2.$

Then the component x_{it}^j of x_t^j satisfies

(10.6) $x_{it}^j = x_0 + \sum_{k=1}^m \int_0^t b_{ik}^j(s,x_s^j,\omega)dA_{ks} + \sum_{k=1}^r \int_0^t \sigma_{ik}(s,x_s^j,\omega)dM_{ks}$
$+\int_0^t \int_Z c_i(s,x_{s-}^j,z,\omega)q(ds,dz), \forall t \geq 0, i=1,2,\cdots,d.$

By (1.3)

(10.7) $(x_{it}^j - a)^+ = (x_{i0}^j - a)^+ + \int_0^t I_{(x_{is-}^j > a)} dx_{is}^j + \frac{1}{2}L_t^0(x_i^j)$
$+\int_0^t \int_Z [(x_{is-}^j - a + c_i(s,x_{s-}^j,z,\omega))^+ - (x_{is-}^j - a)^+$
$-I_{(x_{is-}^1 > a)} c_i(s,x_{s-}^j,z,\omega)]p(ds,dz),\ j=1,2;$

(10.8) $(x_{it}^1 - x_{it}^2)^+ = (x_{i0}^1 - x_{i0}^2)^+ + \int_0^t I_{(x_{is-}^1 > x_{is-}^2)} d(x_{is}^1 - x_{is}^2) + \frac{1}{2}L_t^0(x_i^1 - x_i^2)$
$+\int_0^t \int_Z [(x_{is-}^1 - x_{is-}^2 + c_i(s,x_{s-}^1,z,\omega) - c_i(s,x_{s-}^2,z,\omega))^+$
$-(x_{is-}^1 - x_{is-}^2)^+ - I_{(x_{is-}^1 > x_{is-}^2)}(c_i(s,x_{s-}^1,z,\omega) - c_i(s,x_{s-}^2,z,\omega)]p(ds,dz).$

By (1.7) for any non-negative Borel measurable function f

(10.9) $\sum_{k=1}^r \int_0^t f(x_{is}^j)\left|\sigma_{ik}(s,x_s^j,\omega)\right|^2 d\langle M_k\rangle_s = \int_{R^1} f(a)L_t^a(x_i^j)da.$

Checking the proof of Theorem 10.1, we can derive the following

10.5 Theorem. Assume that
1° $|b(t,x,\omega)|^2 + \|\sigma(t,x,\omega)\|^2 + \int_Z \|c(t,x,z,\omega)\|^2 \pi(dz) \leq g(|x|)$,
where $g(u)$ is a continuous function on $u \geq 0$.
Then

(10.10) $L_{t\wedge\tau_N}^0(x_i^1 - x_i^2) = \lim_{n\to\infty} \sum_{k=1}^r \int_0^{t\wedge\tau_N} I_{(x_{is}^1 > x_{is}^2)} \varphi_n''(x_{is}^1 - x_{is}^2)$
$\cdot(\sigma_{ik}(s,x_s^1,\omega) - \sigma_{ik}(s,x_s^2,\omega))^2 d\langle M_k\rangle_s,$

where
$\varphi_n(x) = \int_0^{x^+} dy \int_0^y f_n(u)du, n=1,2,\cdots.$
and $f_n(u)$ is defined in the proof of Theorem 10.1.

Proof. Note that for $x \in R^1$
$\varphi_n'(x) = I_{(x>a_n)} \int_0^{x^+} f_n(u)du,\ \varphi_n''(x) = I_{(x>a_n)} f_n(x^+),$
$|\varphi_n'(x)| \leq 1,\ \varphi_n(x) \to x^+,\ \varphi_n'(x) \to I_{(x>a_n)}$, as $n \to \infty$.
Hence by the same approach as that in the proof of Theorem 10.1 one finds that

$(x_{it}^1 - x_{it}^2)^+ = (x_{i0}^1 - x_{i0}^2)^+ + \int_0^t I_{(x_{is-}^1 > x_{is-}^2)} d(x_{is}^1 - x_{is}^2)$
$+ \int_0^t \int_Z [(x_{is-}^1 - x_{is-}^2 + c_i(s, x_{s-}^1, z, \omega) - c_i(s, x_{s-}^2, z, \omega))^+$
$-(x_{is-}^1 - x_{is-}^2)^+ - I_{(x_{is-}^1 > x_{is-}^2)}(c_i(s, x_{s-}^1, z, \omega) - c_i(s, x_{s-}^2, z, \omega)] p(ds, dz)$
$+\frac{1}{2} \lim_{n\to\infty} \sum_{k=1}^r \int_0^{t\wedge\tau_N} I_{(x_{is}^1 > x_{is}^2)} \varphi_n''(x_{is}^1 - x_{is}^2)$
$\cdot(\sigma_{ik}(s, x_s^1, \omega) - \sigma_{ik}(s, x_s^2, \omega))^2 d\langle M_k \rangle_s .$
Comparing this with (10.8), one obtains the conclusion. ■

Now we are in a position to give the following

10.6 Theorem. Assume that condition 1° in Theorem 10.5 holds, and
2° $|\sigma_{ik}(t, x + y, \omega) - \sigma_{ik}(t, x, \omega)|^2 \leq \rho_N(|y_i|)^2$
$\cdot(k_N(t, \omega) + g_N(x_i) |\sigma_{ik}(t, x, \omega)|^2)$, as $|x|, |y| \leq N$, $N = 1, 2, \cdots$
$\forall\, x, y \in R^d, x = (x_1, \cdots, x_i, \cdots, x_d), y = (y_1, \cdots, y_i, \cdots, y_d)$,
where $0 \leq k_N(t, \omega)$ and $0 \leq g_N(u)$ satisfy conditions: $\forall 0 \leq T < \infty$
$\sum_{k=1}^r \int_0^T k_N(t, \omega) d\langle M_k \rangle_t < \infty$, $\int_{|u| \leq N'} g_N(u) du < \infty, \forall N'$;
and $\rho_N(u) \geq 0$ is a strictly increasing function on $u \geq 0$ with $\rho_N(0) = 0$ such that
$\int_{0+} du/\rho_N(u)^2 = \infty.$
Then
$L_t^0(x_i^1 - x_i^2) = 0, \forall t \geq 0, \forall i = 1, 2, \cdots, d;$
i.e. $\forall i = 1, 2, \cdots, d$,
(10.11) $(x_{it}^1 - x_{it}^2)^+ = (x_{i0}^1 - x_{i0}^2)^+ + \int_0^t I_{(x_{is-}^1 > x_{is-}^2)} d(x_{is}^1 - x_{is}^2)$
$+ \int_0^t \int_Z [(x_{is-}^1 - x_{is-}^2 + c_i(s, x_{s-}^1, z, \omega) - c_i(s, x_{s-}^2, z, \omega))^+$
$-(x_{is-}^1 - x_{is-}^2)^+ - I_{(x_{is-}^1 > x_{is-}^2)}(c_i(s, x_{s-}^1, z, \omega) - c_i(s, x_{s-}^2, z, \omega)] p(ds, dz).$
Furthermore, if $x, y \in R^d$
$x_i \geq y_i \Rightarrow x_i + c_i(s, x, z, \omega) \geq y_i + c_i(s, y, z, \omega),$
then $\forall i = 1, 2, \cdots, d$,
(10.12) $(x_{it}^1 - x_{it}^2)^+ = (x_{i0}^1 - x_{i0}^2)^+ + \int_0^t I_{(x_{is-}^1 > x_{is-}^2)} d(x_{is}^1 - x_{is}^2).$

Proof. By (10.10)
$L_{t\wedge\tau_N}^0(x_i^1 - x_i^2) = \lim_{n\to\infty} \sum_{k=1}^r \int_0^{t\wedge\tau_N} I_{(x_{is}^1 > x_{is}^2)} \varphi_n''(x_{is}^1 - x_{is}^2)$
$\cdot(\sigma_{ik}(s, x_s^1, \omega) - \sigma_{ik}(s, x_s^2, \omega))^2 d\langle M_k \rangle_s = \lim_{n\to\infty} I_n.$
However, by assumption
$I_n \leq n^{-1} \sum_{k=1}^r \int_0^{t\wedge\tau_N} (k_N(s, \omega) + g_N(x_{is}^2) |\sigma_{ik}(t, x_s^2, \omega)|^2) d\langle M_k \rangle_s$
$= n^{-1} \sum_{k=1}^r \int_0^{t\wedge\tau_N} k_N(s, \omega) d\langle M_k \rangle_s + n^{-1} \int_{[-N,N]} g_N(a) L_{t\wedge\tau_N}^a(x_i^2) da$
$\to 0$, as $n \to \infty$.
Therefore $L_{t\wedge\tau_N}^0(x_i^1 - x_i^2) = 0$. (10.12) is obviously true by virtue of our assumptions. ■

Now let us consider some more general SDE. Suppose $x_t^i, i = 1, 2$, solve the SDEs as follows:
(10.13) $x_t^i = x_0^i + A_t^i + \int_0^t \sigma(s, x_s^i, \omega) dM_s$
$+ \int_0^t \int_Z c(s, x_{s-}^i, z, \omega) q(ds, dz), \forall t \geq 0, i = 1, 2;$

where A_t^i is a real continuous finite variational process, $i = 1, 2$, M_t and $q(dt, dz)$ have the same meaning as that in (10.1). Then by the same approach we can get the similar theorems as Theorem 10.1, 10.5 and 10.6. For example, we give the similar first one and last one as follows:

10.7 Theorem. Assume that condition 2° in Theorem 10.1 holds, and
$1^{\circ\prime}$ $\|\sigma(t,x,\omega)\|^2 + \int_Z \|c(t,x,z,\omega)\|^2 \pi(dz) \leq g(|x|)$,
where $g(u)$ is a continuous function on $u \geq 0$.
Then $\forall t \geq 0$

$$(10.14)\ |x_t^1 - x_t^2| = |x_0^1 - x_0^2| + \int_0^t sgn(x_{s-}^1 - x_{s-}^2) \cdot d(x_s^1 - x_s^2)$$
$$+\frac{1}{2}\sum_{i,j,k=1}^{d,d,r} \int_0^t I_{(x_s^1 = x_s^2)} \frac{|x_s^1 - x_s^2|^2 \delta_{ij} - (x_{is}^1 - x_{is}^2)(x_{js}^1 - x_{js}^2)}{|x_s^1 - x_s^2|^3}$$
$$\cdot(\sigma_{ik}(s,x_s^1,\omega) - \sigma_{ik}(s,x_s^2,\omega))(\sigma_{jk}(s,x_s^1,\omega) - \sigma_{jk}(s,x_s^2,\omega))d\langle M_k\rangle_s$$
$$+\int_0^t \int_Z [|x_{s-}^1 - x_{s-}^2 + c(s,x_{s-}^1,z,\omega) - c(s,x_{s-}^2,z,\omega)| - |x_{s-}^1 - x_{s-}^2|$$
$$-sgn(x_{s-}^1 - x_{s-}^2) \cdot (c(s,x_{s-}^1,z,\omega) - c(s,x_{s-}^2,z,\omega))]p(ds,dz).$$

10.8 Theorem. Assume that condition 2° in Theorem 10.6 and condition $1^{\circ\prime}$ in Theorem 10.7 hold. Then conclusions of Theorem 10.6 still hold.

Chapter 2

Skorohod Problems with Given Cadlag Functions

In this chapter we will discuss the Skorohod problems with given cadlag (i.e. right continuous and with left-hand limit) functions in a convex domain, and in a half space that has a special jump reflection. The first case is basically adopted from Liptser and Shiryayev, 1989. The second one, which is a generalization from Chaleyal-Maurel et al., 1980, is new.

2.1 The Space D and Skorohod's Topology

The content of this section is adopted from Liptser and Shiryayev, 1986.

The space $D = D([0,\infty), R)$ is a space of cadlag functions $X = (X_t)_{t\geq 0}$,with values in R. Similarly, define $D([0,\infty), R^d)$. For space $D([0,1]) = D([0,1]), R)$ we introduce the following metric: for $x, y \in D([0,1])$

$$d_0(X,Y) = \inf_{\lambda\in\Lambda}\left\{\sup_{0\leq t\leq 1}\left|X_t - Y_{\lambda(t)}\right| + \sup_{0\leq s\leq t\leq 1}\left|\log\frac{\lambda(t)-\lambda(s)}{t-s}\right|\right\},$$

where

$$\Lambda = \left\{\begin{array}{c}\lambda = \lambda(t): \ \lambda \text{ is strictly increasing, continuous} \\ \text{on } t\in[0,1], \text{ such that } \lambda(0)=0, \lambda(1)=1\end{array}\right\}.$$

1.1 Lemma. The space $(D([0,1]), d_0)$ is a complete separable metric space. The metric d_0 is equivalent to Skorohod's metric

$d(X,Y) = \inf_{\lambda\in\Lambda}\left\{\sup_{0\leq t\leq 1}\left|X_t - Y_{\lambda(t)}\right| + \sup_{0\leq t\leq 1}|\lambda(t) - t|\right\}.$

Let

$D^0([0,1]) = \{X \in D([0,1]) : X_1 = X_{1-}\}.$

Obviously, $(D^0([0,1]), d_0)$ is a closed subspace of $(D([0,1]), d_0)$. Similarly, define

$D^0 = \{X \in D : \lim_{t\to\infty} X_t \text{ exists and finite}\}.$

Let

$\rho^0(X,Y) = d_0(X',Y')$,

where $X'_t = X_{\psi^{-1}(t)}$, $Y'_t = Y_{\psi^{-1}(t)}$,

$\psi(t) = -\log(1-t)$, $0 \le t < 1$; and $\psi(t) = \infty$, as $t = 1$.

Then obviously, (D^0, ρ^0) is a complete separable metric space. Now let

$X_t^k = X_t \cdot g_k(t)$,

where

$g_k(t) = I_{[0,k]}(t) + (k+1-t)I_{(k,k+1]}(t)$.

Set

$\rho(X,Y) = \sum_{k=1}^{\infty} 2^{-k} \frac{\rho^0(X^k,Y^k)}{1+\rho^0(X^k,Y^k)}$.

Then we have the following

1.2 Lemma.(Lindvall, 1973). (D,ρ) is a complete separable metric space. Moreover, the convergence in the metric ρ is equivalent to the existence of functions $\lambda_n \in \Lambda_\infty, n \ge 1$, such that for each $L > 0$

$\sup_{0\le t\le L} \left| X_t - X^n_{\lambda_n(t)} \right| \to 0$, $\sup_{0\le t\le L} |\lambda_n(t) - t| \to 0$, as $n \to \infty$,

where Λ_∞ is the set of continuous strictly increasing functions $\lambda(t)$ with $\lambda(0) = 0$.

1.3 Lemma. (Billingsley, 1968). For each $X \in D$ there exist step functions $X^n \in D$ such that

$\sup_{s\le t} |X_s - X_s^n| \to 0$, as $n \to \infty$, for any $0 \le t < \infty$.

Lemma 1.3 can be derived from Lemma 1 in §14 of chapter 3 in Billingsley, 1968.

2.2 Skorohod's Problem in a General Domain. Solution with Jumps

The content of this section is basically adopted from Liptser and Shiryayev, 1986. But Lemma 2.3 here is different, hence the results derived from it are also different. Moreover, some misprints there are corrected here.

From now on we always denote

$D = D([0,\infty), R^d)$.

Suppose we are given a convex domain Θ in R^d; denote its boundary by $\partial\Theta$, and its closure by $\overline{\Theta} = \Theta^{cl}$. We introduce the following

2.1 Definition. For a given $Y \in D$ and $Y_0 \in \Theta^{cl}$, if there exists an $X \in D(\Theta) = \{X \in D : X_t \in \Theta^{cl}, t \ge 0\}$ and a $\phi \in D$ with $\phi_0 = 0$ such that (X,ϕ) satisfies (2.1): 1) $X_t = Y_t + \phi_t$,

2)$|\phi|_t = Var(\phi)_t < \infty$, for all $t > 0$,

where $Var(\phi)_t$ is the total variation of ϕ on $[0,t]$;

3) $|\phi|_t = \int_0^t I_{\partial\Theta}(X_s) d|\phi|_s$;

4) $\phi_t = \int_0^t n_s d|\phi|_s$,

where $n_t \in \aleph_{x_t}$, as $x_t \in \partial\Theta$,

$\aleph_x = \{n \in R^d : |n| = 1, B(x-n,1) \cap \Theta = \}$,

$B(x,r) = \{y \in R^d : |y-x| < r\}$;

5) ϕ is ${}_t$-adapted,

where ${}_t = \bigcap_{n=1}^{\infty} \sigma(X_s; s \le t + n^{-1})$;

then (X, ϕ) is called a solution of Skorohod problem (2.1) with given (Y, Y_0).

Obviously, conditions 3) and 4) indicate that $|\phi|_t$ increases only when x_t takes values in the boundary, and $d\phi_t$ has the same direction as that of n_t. Besides, when $\aleph_{x(t)}$ has only one element, then n_t is the inner normal vector at $x_t \in \partial\Theta$; when $\aleph_{x(t)}$ has more than one element, then n_t is the one belong to $\aleph_{x(t)}$ and it exists. Furthermore, it should be noticed that the reflection ϕ_t here can have jumps, but if its jump increment is not equal to zero, then it only happens at such times t, that x_t takes values at the boundary. We have the following

2.2 Lemma. For (2.1) we have that conditions 3) and 4) are equivalent to 3') and 4'). where

3'): for all f: $\Theta^{cl} \to R^d$, bounded and continuous such that $f\mid_{\partial\Theta} = 0$, and for all $t > 0$

$\int_0^t f(X_s) \cdot d\phi_s = 0$;

4') for all $Y \in D(\Theta)$

$(Y_t - X_t) \cdot d\phi_t \ge 0$, as $t \ge 0$.

(Or equivalently,

$\int_0^t (Y_s - X_s) \cdot d\phi_s$ is increasing, as t does.)

Lemma 2.2 is easily derived by the convexity of Θ and the normal reflection of $d\phi_t$. Now let us derive some estimates for solutions of (2.1), which is different from Lemma 2 in Chapter 10 of Liptser and Shiryayev, 1986.

2.3 Lemma. If (X, ϕ) and (X', ϕ') are solutions of (2.1) with (Y, Y_0) and (Y', Y_0'),respectively, then

$|X_t - X_t'|^2 \le 2(\,|Y_t - Y_t'|^2 - 2\int_0^t (Y_s - Y_s') \cdot d(\phi_s - \phi_s'))$,

$|X_t - X_s|^2 \le 2(\,|Y_t - Y_s|^2 - 2\int_s^t (Y_u - Y_s) \cdot d\phi_u)$, as $s \le t$.

Proof. Note that

$|X_t - X_t'|^2 \le 2(\,|Y_t - Y_t'|^2 + |\phi_t - \phi_t'|^2)$.

By Ito's formula

$$|\phi_t - \phi_t'|^2 = 2\int_0^t (\phi_{s-} - \phi_{s-}') \cdot d(\phi_s - \phi_s') + \sum_{0<s\le t} |\Delta\phi_s - \Delta\phi_s'|^2$$
$$= -\sum_{0<s\le t} |\Delta\phi_s - \Delta\phi_s'|^2 - 2\int_0^t (\phi_s - \phi_{s-} - (\phi_s' - \phi_{s-}')) \cdot d(\phi_s^c - \phi_s'^c)$$
$$+2\int_0^t (\phi_s - \phi_s') \cdot d(\phi_s - \phi_s') = \sum_{i=1}^3 I_i \le \sum_{i=2}^3 I_i,$$

where

$\phi_t^c = \phi_t - \sum_{0<s\le t} \Delta\phi_s$.

Since ϕ and ϕ' have only countable discontinuous points, it follows that

$I_2 = 0$.

On the other hand, by 4')

$I_3 = 2\int_0^t (X_s - X_s' - (Y_s - Y_s')) \cdot d(\phi_s - \phi_s') \leq -2\int_0^t (Y_s - Y_s') \cdot d(\phi_s - \phi_s')$.
Hence we have the first conclusion. Similarly, one has that
$|X_t - X_s|^2 \leq 2(\ |Y_t - Y_s|^2 + |\phi_t - \phi_s|^2)$.
By Ito's formula again
$|\phi_t - \phi_s|^2 = 2\int_s^t (\phi_{u-} - \phi_s) \cdot d\phi_u + \sum_{s<u\leq t} |\triangle \phi_u|^2 \leq 2\int_s^t (\phi_u - \phi_s) \cdot d\phi_u$
$\leq -2\int_s^t (Y_u - Y_s) \cdot d\phi_u$
Hence the second conclusion is also derived. ■

2.4 Corollary. 1) For given $(Y, Y_0) \in D \times \Theta^{cl}$ if the solution of (2.1) exists, then it must be unique.
2) Moreover, if $Y \in C(R^d)$, then the solution $(X, \phi) \in C(\Theta) \times C(R^d)$, where $C(R^d) = C([0,\infty), R^d), C(\Theta) = C([0,\infty), \Theta^{cl})$.

In the following let us show that for any given step function Y (2.1) has a unique solution.

2.5 Lemma . Let $(Y, Y_0) \in D \times \Theta^{cl}$, and Y is a step function, then (2.1) has a unique solution.
Proof. Let $T_0 = 0, \phi_0 = 0$,
$T_n = \inf \{t \geq T_{n-1} : Y_t + \phi_{T_{n-1}} \notin \Theta^{cl}\}, n \geq 1$,
$X_t = Y_t + \phi_{T_{n-1}}$,as $T_{n-1} \leq t < T_n$;
$X_t = [Y_{T_n} + \phi_{T_{n-1}}]_\partial$,as $t = T_n$;
$\phi_t = \phi_{T_{n-1}}$,as $T_{n-1} \leq t < T_n$;
$\phi_t = [Y_{T_n} + \phi_{T_{n-1}}]_\partial - Y_{T_n}$,as $t = T_n$;
where $[x]_\partial \in \Theta^{cl}$ is the unique vector satisfying the following equality
$|x - [x]_\partial| = Min \{|x - y| : y \in \Theta^{cl}\}$,
i.e. $[x]_\partial$ is the point, which arrives at the shortest distance from $x \in R^d$ to Θ^{cl}. It is easily verified that (X, ϕ) is a solution of (2.1) with (Y, Y_0). The uniqueness is derived by Corollary 2.4. ■

2.6 Lemma. Assume that $(Y^n, Y_0^n) \in D\times\ \Theta^{cl}$, and (X^n, ϕ^n) is the solution of (2.1) with (Y^n, Y_0^n). If
$1°$ there exists a $Y \in D$ and a $Y_0 \in \Theta^{cl}$ such that
$\lim_{n\to\infty} \sup_{t\leq T} |Y_t^n - Y_t| = 0$, for any $T > 0$;
$2°$ $\sup_n |\phi^n|_T \leq k_T$, for any $T > 0$,
where $k_T > 0$ is a constant depending on T and Y only;
then (2.1) has a unique solution (X, ϕ) with (Y, Y_0), and for any $T > 0$ it satisfies
$\lim_{n\to\infty} \sup_{t\leq T} |X_t^n - X_t| = 0$,
$\lim_{n\to\infty} \sup_{t\leq T} |\phi_t^n - \phi_t| = 0$.
Proof. By Lemma 2.3
$\sup_{t\leq T} |X_t^n - X_t^m|^2 \leq 2\sup_{t\leq T} |Y_t^n - Y_t^m|^2 + 4\sup_{t\leq T} |Y_t^n - Y_t^m| \cdot$
$\cdot(|\phi^n|_T + |\phi^m|_T)$.
Hence by $1°$ and $2°$
$\lim_{n,m\to\infty} \sup_{t\leq T} |X_t^n - X_t^m| + \lim_{n\to\infty} \sup_{t\leq T} |\phi_t^n - \phi_t^m| = 0$.
Hence there exists a $(X, \phi) \in D(\Theta) \times D$, such that for any $T > 0$
$\lim_{n\to\infty}(\sup_{t\leq T} |X_t^n - X_t| + \sup_{t\leq T} |\phi_t^n - \phi_t|) = 0$.

Now for any $t > 0$ let us make a subdivision in $[0, t]$:

$0 = t_0 < t_1 < ... < t_k = t.$

Then one has that

$\sum_{j=1}^{k} |\phi_{t_j} - \phi_{t_{j-1}}| \leq \sum_{j=1}^{k} \left|\phi_{t_j}^n - \phi_{t_{j-1}}^n\right| + 2\sum_{j=1}^{k} \left|\phi_{t_j} - \phi_{t_j}^n\right|.$

Hence

$|\phi|_t \leq \sup_n |\phi^n|_t \leq k_t$, for all $t \geq 0$.

Now let us show that 3') and 4') in Lemma 2.2 hold. In fact, by Lemma 1.3 we can choose step functions $Z^N \in D$ such that

$\sup_{s \leq t} \left|X_s - Z_s^N\right| \to 0$, as $N \to \infty$, for any $0 \leq t < \infty$.

For any $f : \Theta^{cl} \to R^d$, which is bounded and continuous such that $f\,|_{\partial\Theta} = 0$, one has that

$\left|\int_0^t f(X_s) \cdot d\phi_s\right| = \left|\int_0^t f(X_s) \cdot d\phi_s - \int_0^t f(X_s^n) \cdot d\phi_s^n\right|$

$\leq \left|\int_0^t (f(X_s) - f(X_s^n)) \cdot d\phi_s^n\right| + \left|\int_0^t f(Z_s^N) \cdot d(\phi_s - \phi_s^n)\right|$

$+ \left|\int_0^t (f(X_s) - f(Z_s^N)) \cdot d\phi_s\right| + \left|\int_0^t (f(X_s) - f(Z_s^N)) \cdot d\phi_s^n\right|$

$= \sum_{I=1}^{4} I_t^i.$

Since

$I_t^1 \leq \sup_{s \leq t} |f(X_s) - f(X_s^n)|\, k_t.$

Hence as $n \to \infty$

$I_t^1 \to 0.$

Similarly, as $N \to \infty$

$I_t^4 \to 0.$

By the Lebesgue dominated convergence theorem as $N \to \infty$

$I_t^3 \to 0.$

Finally, note that Z^N is a step function, for each N as $n \to \infty, I_t^2 \to 0$. Therefore 3') is derived.

Now for any given $\overline{X} \in D(\Theta)$ one has that for any $s \leq t$

$\int_s^t (\overline{X_r} - X_r^n) \cdot d\phi_s^n \geq 0$, as $n \geq 1$.

Take step functions $Z^N \in D$ such that

$\sup_{s \leq t} \left|\overline{X_s} - X_s - Z_s^N\right| \to 0$, as $N \to \infty$, for any $0 \leq t < \infty$.

Then

$\left|\int_s^t (\overline{X_r} - X_r) \cdot d\phi_r - \int_s^t (\overline{X_r} - X_r^n) \cdot d\phi_r^n\right| \leq \left|- \int_s^t (X_r - X_r^n) \cdot d\phi_r^n\right|$

$+ \left|\int_s^t Z_r^N \cdot d(\phi_r - \phi_r^n)\right| + \left|\int_s^t (\overline{X_r} - X_r - Z_r^N) \cdot d\phi_r\right|$

$+ \left|- \int_s^t (\overline{X_r} - X_r - Z_r^N) \cdot d\phi_r^n\right| = \sum_{I=1}^{4} I_t^i.$

Now in a similar way we find that as $n \to \infty$

$\sum_{I=1}^{4} I_t^i \to 0.$

Hence 4') is true for (X, ϕ). By

$X_t^n = Y_t^n + \phi_t^n$

letting $n \to \infty$ one finds that (X, ϕ) is a solution of (2.1) with (Y, Y_0). The uniqueness is derived from Corollary 2.4. ■

To show the existence of a solution to (2.1) the following assumption is needed:

2.7 Assumption. There exists a constant $c_0 > 0$ and a vector $e \in R^d, |e| = 1$ such that

$e \cdot n \geq c_0 > 0$, for all $n \in \cup_{x \in \partial\Theta} \aleph_x$.

From now on we always make Assumption 2.7. Let us establish one more estimate, which is better than Lemma 4 in Chapter 10 of Liptser and Shiryayev, 1986.

2.8 Lemma. Assume that (X, ϕ) is a solution of (2.1) with (Y, Y_0). Then as $s \leq t$

$$|X_t - X_s| + (|\phi|_t - |\phi|_s) \leq k_0 \sup_{s \leq r \leq t} |Y_r - Y_s|,$$

where $k_0 \geq 0$ is a constant depending on c_0 only. In particular,

$$|\phi|_t \leq 2k_0 \sup_{s \leq t} |Y_s|.$$

Proof. For $s < t$ let

$$\triangle_t^{Y_s} = \sup_{s \leq r \leq t} |Y_r - Y_s|.$$

Then

$$e \cdot (\phi_t - \phi_s) = e \cdot \int_s^t n_r d|\phi|_r \geq c_0(|\phi|_t - |\phi|_s).$$

Hence

$$|\phi|_t - |\phi|_s \leq c_0^{-1} e \cdot (\phi_t - \phi_s) \leq c_0^{-1}(|X_t - X_s| + |Y_t - Y_s|).$$

On the other hand, by Lemma 2.3

$$|X_t - X_s|^2 \leq 2(\left|\triangle_t^{Y_s}\right|^2 + 2\,\triangle_t^{Y_s}\,(|\phi|_t - |\phi|_s)) \leq 2\left|\triangle_t^{Y_s}\right|^2$$

$$+2^{-1}|X_t - X_s|^2 + 16c_0^{-2}\left|\triangle_t^{Y_s}\right|^2 + 2^{-1}|Y_t - Y_s|^2.$$

Hence

$$|X_t - X_s|^2 \leq (5 + 32c_0^{-2})\left|\triangle_t^{Y_s}\right|^2. \blacksquare$$

Now we are in a position to give an existence and uniqueness theorem for the solution of (2.1).

2.9 Theorem . Assume that $Y \in D$ with $Y_0 \in \Theta^{cl}$, then (2.1) has a unique solution (X, ϕ) with (Y, Y_0), and $\phi = \phi(Y)$ is a continuous map of $Y \in D$, i.e., if $Y_n \to Y$ (in D under Skorohod topology), then

$\phi(Y_n) \to \phi(Y) \in D$ (in D under Skorohod topology).

Moreover,

$$|\phi(Y)|_t \leq k_0' \sup_{s \leq t} |Y|_s,$$

where constant $k_0' \geq 0$ depends on c_0 only. Therefore $X = X(Y)$ is also a continuous map of $Y \in D$ such that

$$|X(Y)|_t \leq (k_0' + 1) \sup_{s \leq t} |Y|_s.$$

Proof. First, for any given $(Y, Y_0) \in D \times \Theta^{cl}$ by Lemma 1.3 one can take step functions $Y^n \in D$ such that

$\sup_{s \leq T} |Y_s - Y_s^n| \to 0$, as $n \to \infty$, for any $0 \leq T < \infty$.

Now by Lemma 2.5 (2.1) has a unique solution (X^n, ϕ^n) with (Y^n, Y_0^n) for each n. By Lemma 2.8

$|\phi^n|_T \leq 2k_0 \sup_{s\leq T} |Y_s^n|$.

Hence

$\sup_n |\phi^n|_T \leq 2k_0(\sup_{s\leq T} |Y_s| + \sup_n \sup_{s\leq T} |Y_s^n - Y_s|) \leq k_T$,

where $k_T \geq 0$ depends on c_0 (introduced in assumption 2.7), Y and T only. Therefore applying Lemma 2.6, one finds that (2.1) has a unique solution (X, ϕ) with (Y, Y_0) such that

$\lim_{n\to\infty} \sup_{s\leq T} |X_s^n - X_s| = 0$,

$\lim_{n\to\infty} \sup_{s\leq T} |\phi_s^n - \phi_s| = 0$.

Now assume that $(Y^n, Y_0^n) \in D\times \Theta^{cl}$, and (X^n, ϕ^n) is the solution of (2.1) with (Y^n, Y_0^n). Moreover, assume that

$Y_n \to Y$ (in D under Skorohod topology), as $n \to \infty$,

i.e. there exist continuous functions λ^n with $\lambda^n(0) = 0, n = 1, 2, ...$, and each component of λ^n is strictly increasing such that for each $T > 0$

$\lim_{n\to\infty} \sum_{j=1}^d \sup_{s\leq T} \left|\lambda_s^{nj} - s\right| = 0$,

$\lim_{n\to\infty} \sum_{j=1}^d \sup_{s\leq T} \left|Y_{\lambda_s^{nj}}^{nj} - Y_s^j\right| = 0$,

where $Y_s^T = (Y_s^1, ..., Y_s^d)$, etc. But

$X_{\lambda_s^{nj}}^{nj} = Y_{\lambda_s^{nj}}^{nj} + \phi_{\lambda_s^{nj}}^{nj}, j = 1, ..., d.$

If we denote

$\overline{Y}_t^{nj} = Y_{\lambda_s^{nj}}^{nj}, \overline{X}_t^{nj} = X_{\lambda_s^{nj}}^{nj}, \overline{\phi}_t^{nj} = \phi_{\lambda_s^{nj}}^{nj}$,

then

$\overline{X}_t^{nj} = \overline{Y}_t^{nj} + \overline{\phi}_t^{nj}$, $j = 1, ..., d$, for $t \geq 0$,

or,

$\overline{X}_t^n = \overline{Y}_t^n + \overline{\phi}_t^n$, for all $t \geq 0$,

and

$\left|\overline{\phi^n}\right|_T < \infty$, for any $T < \infty$.

Now by Lemma 2.8 one has that

$\left|\overline{\phi}^n\right|_t \leq 2k_0 \sup_{s\leq t} \left|\overline{Y}_s^n\right| \leq k_T$, for all n,

where $k_T \geq 0$ depending on c_0 (introduced in Assumption 2.7), Y and T only. Therefore, applying Lemma 2.6 again, one has that for any $T > 0$

$\lim_{n\to\infty} \sup_{t\leq T} \left|\overline{X}_t^n - X_t\right| = 0$,

$\lim_{n\to\infty} \sup_{t\leq T} \left|\overline{\phi}_t^n - \phi_t\right| = 0$,

i.e. for all $T > 0$

$\lim_{n\to\infty} \sum_{j=1}^d \sup_{s\leq T} \left|\phi_{\lambda_s^{nj}}^{nj} - \phi_s^j\right| = 0.$

By assumption

$\lim_{n\to\infty} \sum_{j=1}^d \sup_{s\leq T} \left|\lambda_s^{nj} - s\right| = 0.$

Hence as $n \to \infty$

$\phi(Y_n) \widehat{=} \phi^n \to \phi(Y) \widehat{=} \phi$ (in D under Skorohod topology). ■

By the above proof it is seen that the following corollary holds.

2.10 Corollary. Assume that (X^n, ϕ^n) is the solution of (2.1) with $(Y^n, Y_0^n) \in D\times \Theta^{cl}, n = 0, 1, 2, ...$ Denote $(X^0, \phi^0, Y^0) = (X, \phi, Y)$. If there exist continuous functions λ^n with $\lambda^n(0) = 0, n = 1, 2, ...,$ and each component of λ^n is strictly increasing such that for each $T > 0$

$\lim_{n\to\infty} \sum_{j=1}^d \sup_{s\le T} \left|\lambda_s^{nj} - s\right| = 0,$

$\lim_{n\to\infty} \sum_{j=1}^d \sup_{s\le T} \left|Y_{\lambda_s^{nj}}^{nj} - Y_s^j\right| = 0,$

where $Y_s^T = (Y_s^1, ..., Y_s^d)$, etc., then for each $T > 0$ for this $\{\lambda_n\}$

$\lim_{n\to\infty} \sum_{j=1}^d \sup_{s\le T} \left|X_{\lambda_s^{nj}}^{nj} - X_s^j\right| = 0,$

$\lim_{n\to\infty} \sum_{j=1}^d \sup_{s\le T} \left|\phi_{\lambda_s^{nj}}^{nj} - \phi_s^j\right| = 0.$

It must be remembered that from Lemma 2.8 up to now we have always made the assumption 2.7. Now let us relax the assumption to the following

2.11 Assumption. Θ is a convex region such that there exist constants $\varepsilon > 0$ and $\delta > 0$ such that for each point $x \in \partial\Theta$ there exists a x_0 satisfying that $|x - x_0| < \delta$ and

$B_\varepsilon(x_0) = \left\{y \in R^d : |y - x_0| < \varepsilon\right\} \subset \Theta.$

Obviously, a bounded convex region Θ satisfies Assumption 2.11 .

2.12 Theorem. Under assumption 2.11 the conclusion of Theorem 2.9 still holds but with

$|\phi(Y)|_T \le K_T(Y)$, for each $T > 0$,

where $K_T(Y)$ is a continuous function of Y in the uniform convergence topology given by (2.3) below, which is a fixed function of Y depending on the given T and constant $c_0 = \varepsilon/2\delta$ only

Proof. Denote by $\widetilde{H_y}(\Theta)$ the set of all hyperplanes of support of Θ at point $y \in \partial\Theta$, and by $H(\Theta)$ the closed half-space with the basic hyperplane H as a boundary, containing the region Θ. Define the region Θ_x as the interior of the set

$\bigcap_{y\in\partial\Theta\cap\overline{B}_{\varepsilon/2}(x)} \bigcap_{H\in\widetilde{H}_y(\Theta)} H(\Theta),$

where $\overline{B}_{\varepsilon/2}(x)$ is the closure of the ball $B_{\varepsilon/2}(x)$. Then the region Θ_x is convex and satisfies Assumption 2.7 with the vector

$e = (x_0 - x)/\left|x_0 - x\right|$

and the constant $c_0 = \varepsilon/(2\delta)$. Set now

$T_0 = \inf\left\{t \ge 0 : Y_t \notin \Theta^{cl}\right\}.$

For $0 < T_0 < \infty$, define

$$X_t^0 = \begin{cases} Y_t, 0 \le t < T, \\ [Y_{T_0}]_\partial, t = T_0. \end{cases}$$

For $m \ge 1$ set

$\widehat{Y}_t^m = X_{T_{m-1}}^{m-1} + Y_{t+T_{m-1}} - Y_{T_{m-1}},$

with

$X_{T_{m-1}}^{m-1} \in \partial\Theta.$

By Theorem 2.9 there exists a unique solution $(\widehat{X}_t^m, \widehat{\phi}_t^m)$ for (2.1) in the closed region $\Theta_{X_{T_{m-1}}^{m-1}}^{cl}$:

$\widehat{X}_t^m = \widehat{Y}_t^m + \widehat{\phi}_t^m,\ t \ge 0.$

Set

$$t_m = \inf\left\{t \geq T_{m-1} : \left|\widehat{X}^m_{t-T_{m-1}} - \widehat{X}^m_0\right| \geq \varepsilon/2\right\}.$$

For $T_{m-1} < t_m < \infty$ define on $[T_{m-1}, t_m]$ a function

$$X^m_t = \begin{cases} \widehat{X}^m_{t-T_{m-1}}, T_{m-1} \leq t < t_m, \\ [\widehat{X}^m_{t_m - T_{m-1}}]_\partial, t = t_m. \end{cases}$$

Then in the region $B_{\varepsilon/2}(x^{m-1}_{T_{m-1}}) \cap \Theta^{cl}$: as $T_{m-1} \leq t < t_m$

$$X^m_t = Y_t + X^{m-1}_{T_{m-1}} - Y_{T_{m-1}} + \widehat{\phi}^m_{t-T_{m-1}}.$$

Or, as $T_{m-1} \leq t < t_m$

$$X^m_t = Y_t + \phi^m_t,$$

where $\phi^m_t = X^{m-1}_{T_{m-1}} - Y_{T_{m-1}} + \widehat{\phi}^m_{t-T_{m-1}}$.

Next, set

$$T_m = \inf\left\{t \geq t_m : X^m_{t_m} + Y_t - Y_{t_m} \notin \Theta^{cl}\right\}$$

and for $t_m < T_m < \infty$ define on $[t_m, T_m]$ a function

$$X^m_t = \begin{cases} X^m_{t_m} + Y_t - Y_{t_m}, t_m \leq t < T_m, \\ [X^m_{t_m} + Y_t - Y_{t_m}]_\partial, t = T_m. \end{cases}$$

Then in the region Θ^{cl} :

$$X^m_t = Y_t + X^m_{t_m} - Y_{t_m}, t_m \leq t < T_m.$$

Or, as $t_m \leq t < T_m$

$$X^m_t = Y_t + \phi^m_t,$$

where $\phi^m_t = X^m_{t_m} - Y_{t_m}$.

Set

$$T_\infty = \lim_{m\to\infty} T_m.$$

Obviously, if $T_\infty < \infty$, then there exists a solution of (2.1) in $t \in [0, T_\infty)$ given by

$$X_t = Y_t I_{0 \leq t < T_0} + \sum_{m=1}^{\infty} I_{(T_{m-1} \leq t < T_m)} X^m_t.$$

Now we are going to show that $T_\infty = \infty$. By definition

$$\left|\widehat{X}^m_{t_m - T_{m-1}} - \widehat{X}^m_0\right| \geq \varepsilon/2.$$

By Lemma 2.8

$$\left|\widehat{X}^m_{t_m - T_{m-1}} - \widehat{X}^m_0\right| \leq k_0 \sup_{0 \leq t \leq t_m - T_{m-1}} \left|\widehat{Y}^m_t - \widehat{Y}^m_0\right| = k_0 \triangle^{Y_{T_{m-1}}}_{t_m}$$

$$\leq k_0 \triangle^{Y_{T_{m-1}}}_{T_m} \leq k_0 \triangle^Y_{T_{m-1}, T_m},$$

where

$$\triangle^{Y_s}_t = \sup_{s \leq s_1 \leq t} |Y_{s_1} - Y_s|, \; \triangle^Y_{s,t} = \sup_{s \leq s_1 \leq s_2 \leq t} |Y_{s_1} - Y_{s_2}|$$

Hence,

$$\triangle^Y_{T_{m-1}, T_m} \geq \varepsilon/(2k_0).$$

Let $T > 0$. By Lemma 1 in §14 of Ch.3 of Billingsley, 1968, there exists a partition of the interval $[0, T]$ at points $\{t'_i\}$ with

$$\inf_i (t'_i - t'_{i-1}) > 0,$$

and

$$\max_i \triangle^Y_{t'_{i-1}, t'_i} < \varepsilon/(4k_0).$$

This means that each interval $[t'_{i-1}, t'_i]$ contains at most one element of the sequence T_m, $m \geq 1$, i.e. $[0, T]$ contains at most finite elements of the sequence

T_m, $m \geq 1$. Since T is arbitrary, $T_\infty = \infty$. Therefore (2.1) has a solution. Indeed, as $t \in [T_{m-1}, T_m)$,

$$X_t^m = Y_t + \phi_t^m.$$

Moreover, $\forall t \geq 0$

$$\left|\widehat{\phi}^m\right|_{t-T_{m-1}} = \int_0^{t-T_{m-1}} I_{\partial\Theta}(\widehat{X}_s^m) d\left|\widehat{\phi}^m\right|_s.$$

Note that as $t \in [T_{m-1}, t_m)$

$$X_t^m = \widehat{X}_{t-T_{m-1}}^m, \ \phi_t^m = \widehat{\phi}_{t-T_{m-1}}^m,$$

and as $t \in [t_m, T_m)$, ϕ_t^m is a constant, i.e. in this case $|\phi^m|_t - |\phi^m|_{t_m} = 0$. Hence as $0 \leq t \leq T_m - T_{m-1}$,

$$|\phi^m|_{t+T_{m-1}} - |\phi^m|_{T_{m-1}} = \int_{T_{m-1}}^{t+T_{m-1}} I_{\partial\Theta}(X_s^m) d\left|\phi^m\right|_s.$$

Similarly, one shows that as $0 \leq t \leq T_m - T_{m-1}$,

$$\phi_{t+T_{m-1}}^m - \phi_{T_{m-1}}^m = \int_{T_{m-1}}^{t+T_{m-1}} n_s I_{\partial\Theta}(X_s^m) d\left|\phi^m\right|_s.$$

From this it is not difficult to derive that (X_t, ϕ_t) solves (2.1), where

$$X_t = Y_t I_{0 \leq t < T_0} + \sum_{m=1}^{\infty} I_{(T_{m-1} \leq t < T_m)} X_t^m, \text{ and } \phi_t = \sum_{m=1}^{\infty} I_{(T_{m-1} \leq t < T_m)}^m \phi_t^m.$$

The uniqueness of the solution is derived by Corollary 2.4. Finally, let us show that $\phi(Y)$ is continuous in Skorohod's topology. For this it suffices that

(2.2) $\text{Var}(\phi(Y))_T \leq K_T(Y)$,

for each $T > 0$, where $K_T(Y)$ is a continuous function of Y in the uniform convergence topology:

(2.3) $\lim_{k\to\infty} \sup_{t \leq T} \left|Y_t^k - Y_t\right| = 0 \Rightarrow \lim_{k\to\infty} K_T(Y^k) = K_T(Y)$.

If this can be shown, then for any $Y^k \to Y$ in D, one has that for each $T > 0$ there exists a constant $k_T \geq 0$ such that

$$\text{Var}(\phi(Y))_T \leq k_T.$$

Hence Lemma 2.6 applies, and one finds that $\phi(Y)$ is continuous in the Skorohod's topology, and so also is $X(Y)$.

Consider the case $0 \leq T \leq T_1$.

By construction

$$X_t^1 = \begin{cases} Y_t, \text{ as } 0 \leq t < T_0, \\ Y_t + X_{T_0}^0 - Y_{T_0} + \widehat{\phi}_{t-T_0}^1, \text{ as } T_0 \leq t < t_1, \\ \left[Y_{t_1} + X_{T_0}^0 - Y_{T_0} + \widehat{\phi}_{t_1-T_0}^1\right]_\partial, \text{ as } t = t_1, \\ Y_t + X_{t_1}^1 - Y_{t_1}, \text{ as } t_1 \leq t < T_1, \\ \left[Y_{T_1} + X_{t_1}^1 - Y_{t_1}\right]_\partial, \text{ as } t = T_1. \end{cases}$$

Hence,

$$\text{Var}(\phi(Y))_T \leq |Y_{T_0} - [Y_{T_0}]_\partial| + Var(\widehat{\phi}^1)_{(t_1-T_0)} -$$
$$+ \left|\left[Y_{t_1} + X_{T_0}^0 - Y_{T_0} + \widehat{\phi}_{t_1-T_0}^1\right]_\partial - (Y_{t_1} + X_{T_0}^0 - Y_{T_0} + \widehat{\phi}_{t_1-T_0}^1)\right|)$$
$$+ \left|\left[X_{t_1}^1 + Y_{T_1} - Y_{t_1}\right]_\partial - (X_{t_1}^1 + Y_{T_1} - Y_{t_1})\right|).$$

However, by Theorem 2.9 $\forall k$

$$\text{Var}(\widehat{\phi}^k)_{(t_k - T_{k-1})} \leq L \sup_{t \leq t_k - T_{k-1}} \left|\widehat{Y}_t^k\right|$$
$$\leq L(\left|X_{T_{k-1}}^{k-1} - Y_{T_{k-1}}\right| + \sup_{t \leq t_k} |Y_t|).$$

Hence, as $0 \leq T \leq T_1$

$\mathrm{Var}(\phi(Y))_T \le (L+1)(|Y_{T_0} - [Y_{T_0}]_\partial| + \sup_{t \le t_1} |Y_t|).$
$+\left|\left[Y_{t_1} + X^0_{T_0} - Y_{T_0} + \widehat{\phi}^1_{t_1 - T_0}\right]_\partial - (Y_{t_1} + X^0_{T_0} - Y_{T_0} + \widehat{\phi}^1_{t_1 - T_0})\right|)$
$+\left|\left[X^1_{t_1} + Y_{T_1} - Y_{t_1}\right]_\partial - \left(X^1_{t_1} + Y_{T_1} - Y_{t_1}\right)\right|) = K_T(Y),$
where $K_T(Y)$ is a fixed function of Y depending on the given T and constant $c_0 = \varepsilon/2\delta$ only, since constant L depends on c_0 only.
Note that $(\widehat{X}^m_t, \widehat{\phi}^m_t)$ is the unique solution for (2.1) in the closed region $\Theta^{cl}_{X^{m-1}_{T_{m-1}}}$:

$$\widehat{X}^m_t = \widehat{Y}^m_t + \widehat{\phi}^m_t,\ t \ge 0.$$

By Theorem 2.9 and Corollary 2.10 $\widehat{\phi}^m = \widehat{\phi}^m(\widehat{Y}^m)$ is continuous in $\widehat{Y}^m$ in the uniform convergence topology (2.3). However,

$$\widehat{Y}^1_t = |Y_{T_0} - [Y_{T_0}]_\partial| + Y_{t+T_0} - Y_{T_0}$$

is continuous in Y in the uniform convergence topology (2.3), and so are $\widehat{\phi}^1$ and $\widehat{X}^1_t$, and $K_T(Y)$, since $[x]_\partial$ is continuous in $x \in R^d$. Now by induction one can complete the proof of (2.2) and (2.3) for $\forall . T_{m-1} \le T < T_m, m \ge 2$. ■

2.13 Corollary. Assume that under Assumption 2.11 (X^n, ϕ^n) is the solution of (2.1) with $(Y^n, Y^n_0) \in D \times \Theta^{cl}, n = 0, 1, 2, ...$ Denote $(X^0, \phi^0, Y^0) = (X, \phi, Y)$. If for each $T > 0$

$$\lim_{n\to\infty} \sup_{s \le T} |Y^n_s - Y_s| = 0,$$

then for each $T > 0$

$$\lim_{n\to\infty} \sup_{s \le T} |X^n_s - X_s| = 0,\ \lim_{n\to\infty} \sup_{s \le T} |\phi^n_s - \phi_s| = 0.$$

Proof. Suppose that we are given a $T > 0$. By Theorem 2.12

$$\mathrm{Var}(\phi(Y^n))_T \le K_T(Y^n) \le |K_T(Y^n) - K_T(Y)| + |K_T(Y)|.$$

$K_T(Y)$ is a continuous function of Y in the uniform convergence topology, however, by assumption Y^n uniformly converges to Y on $[0, T]$. Hence it is easily shown that

$$\mathrm{Var}(\phi(Y^n))_T \le k_T < \infty, \forall n,$$

where k_T is a constant. Now applying Lemma 2.6, the conclusion is obtained. ■

2.3 Skorohod Problem with Jump Reflection in a Half Space

In this section we will discuss the Skorohod problem in a half space. For simplicity we discuss the case in R^1. All results obtained in this section are new, since the systems considered in (3.1) is new. When $\alpha_t = 2$, for all $t \ge 0$, then the system (3.1) reduces to the case considered in M. Chaleyat-Maurel et al. (1980).

3.1 Definition. For a given cadlag R^1-valued function y_t, if there exist two cadlag R^1-valued functions x_t and ϕ_t, such that

$$
(3.1)\begin{cases}
x_t = x_0 + y_t + \phi_t, t \geq 0; 0 \leq x_0 \text{ is given;} \\
x_t \geq 0, \text{ for all } t \geq 0, (y_0 = 0); \\
\phi_t - \text{real increasing function with } \phi_0 = 0 \text{ such that} \\
\int_0^t x_s \cdot d\phi_s^c = 0, \text{ for all } t \geq 0, \\
\text{where } \phi_t^c = \phi_t - \sum_{0<s\leq t} \Delta\phi_s, \text{ and} \\
\Delta\phi_t > 0 \Rightarrow \Delta\phi_t = \alpha_t x_t;
\end{cases}
$$

where $\alpha_t \geq 2$, for all $t \geq 0$, and it is continuous; then (x_t, ϕ_t) is called a solution of the Skorohod problem (3.1).

For the existence of solution to (3.1) we have the following

3.2 Theorem. There exists a unique solution (x_t, ϕ_t) satisfying (3.1) for arbitrary given cadlag function y_t and $x_0 \geq 0$ $(y_0 = 0)$ such that

(3.2) $a(t) \leq \phi_t \leq a(t) + \sum_{0<s\leq t}(\alpha_s - 1)^{-1} \Delta\, a(s)$,

where

$a(t) = \sup_{s\leq t}(-y(s), 0)$.

Furthermore, ϕ_t is continuous, if and only if

$x_{t-} + \Delta y_t \geq 0$.

If y_t is a $\Im_t$−adapted random process, then so is (x_t, ϕ_t) .

3.3 Remark. As $\alpha_t = 2$, for all $t \geq 0$, then the first conclusion is reduced to theorem 5 of M. Chaleyat-Maurel et al. (1980). Moreover, the second conclusion also includes Proposition 17 there.

Proof. Denote $y_0 = x_0$ for simplicity.

Case1. Assume that

$t_1 < t_2 < ... < t_n < ...$

is the totality of jump points for $a(t), t \geq 0$. Let $\phi(0) = 0$,

$\phi(t) = a(t)$, as $t \in [0, t_1)$.

Define $\phi(t)$ by induction as follows:

$$
(3.3)\begin{cases}
\phi(t) = \ a(t) \vee \phi(t_n), \text{as } t \in [t_n, t_{n+1}), \\
\phi(t_{n+1}) = [(\alpha_{t_{n+1}} - 1)^{-1} |y(t_{n+1}) + \phi(t_{n+1}-)| - y(t_{n+1})] \\
\vee\phi(t_{n+1}-)
\end{cases}
$$

Let

(3.4) $x(t) = y(t) + \phi(t)$.

We are going to show that (x_t, ϕ_t) satisfying (3.1) and (3.2). For this we need the following propositions. Denote

$S_\phi = \{t \geq 0 : \Delta\phi_t > 0\}$.

3.4 Proposition. If $t \in S_\phi$, then

$\Delta a(t) > 0$, $-y(t) > \phi(t-)$, $a(t) = -y(t)$,

$\phi(t) = [\alpha_t(\alpha_t - 1)^{-1}a(t) - (\alpha_t - 1)^{-1}\phi(t-)]$, $\Delta\phi_t = \alpha_t x_t$.

Proof. Assume that $t \in S_\phi$. If $t = t_n$, then $\Delta a(t_n) > 0$. If $t \in (t_n, t_{n+1})$, then $0 < \Delta\phi_t = \Delta a(t)$. However, this is impossible. Hence $t = t_{n+j}$, for some j. One still has $\Delta a(t_{n+j}) > 0$. Let us show

$-y(t) > \phi(t-)$.

If not, then $-y(t) \leq \phi(t-)$. By (3.3)

$$\phi(t) = (\alpha_t - 1)^{-1}(y(t) + \phi(t-)) - y(t) = (\alpha_t - 1)^{-1}(\phi(t-) - (\alpha_t - 2)y(t))$$
$$(\alpha_t - 1)^{-1}(\phi(t-) + (\alpha_t - 2)\phi(t-)) \leq \phi(t-).$$
It is a contradiction with $t \in S_\phi$. From this and by definition (3.3) and (3.4)
$$x(t) = (\alpha_t - 1)^{-1}(-y(t) - \phi(t-)) = (\alpha_t - 1)^{-1}(-x(t-) - \triangle y(t)).$$
Hence,
(3.5) $\triangle\phi_t = \alpha_t x_t$.
Now all conclusions of Proposition 3.4 are easily derived. ■

By Proposition 3.4 we can rewrite (3.3) as

$$(3.3)' \begin{cases} \phi(t) = a(t) \vee \phi(t_n), \text{as } t \in [t_n, t_{n+1}), \\ \phi(t_{n+1}) = (\alpha_{t_{n+1}} - 1)^{-1}(a(t_{n+1}) - \phi(t_{n+1}-)) + a(t_{n+1}), \\ \qquad \text{as } t_{n+1} \in S_\phi, \\ \phi(t_{n+1}) = [(\alpha_{t_{n+1}} - 1)^{-1}\,|-y(t_{n+1}) - \phi(t_{n+1}-)| - y(t_{n+1})] \\ \qquad \vee\phi(t_{n+1}-), \text{as } t_{n+1} \notin S_\phi. \end{cases}$$

3.5 Proposition. If t is a strictly increasing point of $\phi^c(.)$, i.e.
$$\phi^c(t-\varepsilon) < \phi^c(t), \text{ as } \varepsilon > 0 \text{ is small enough;}$$
then it is also such a point of $a(.)$. In this case
$$a(t-) = \phi(t-),$$
and $x(t-) = 0$.

Proof. Assume that t is a strictly increasing point of $\phi^c(.)$. By (3.3)' and assumption as $\varepsilon > 0$ is small enough;
$$a(t-\varepsilon) \leq \phi(t-\varepsilon) < \phi(t-) = a(t-).$$
This implies that
$$a(t-) = -y(t-).$$
Therefore
$$x(t-) = y(t-) + \phi(t-) = 0. \blacksquare$$
Now let us return to the proof of Theorem 3.2 in case 1. By Proposition 3.5
(3.6) $\int_0^t x_s d\phi_s^c = \int_0^t x_{s-} d\phi_s^c = 0$.

(3.4)-(3.6) shows that (x_t, ϕ_t) satisfying (3.1). Furthermore, as $t \in [0, t_1]$
(3.7) $a(t) \leq \phi(t) \leq a(t) + \tilde{a}_t^d$,
where
$$\tilde{a}^d(t) = \textstyle\sum_{0<s\leq t}(\alpha_s - 1)^{-1} \triangle a_s.$$
Now assume that (3.7) holds for $t \in [t_{n-1}, t_n]$. Then as $t \in [t_n, t_{n+1})$
$$\phi(t) = a(t) \vee \phi(t_n) \leq a(t) \vee (a(t_n) + \tilde{a}^d(t_n)) \leq a(t) + \tilde{a}^d(t).$$
In case $t = t_{n+1}$ and $\phi(t_{n+1}-) = \phi(t_{n+1})$ by the above we have
$$\phi(t_{n+1}) = \phi(t_{n+1}-) \leq a(t_{n+1}-) + \tilde{a}^d(t_{n+1}-) \leq a(t_{n+1}) + \tilde{a}^d(t_{n+1}).$$
In case $t = t_{n+1}$ and $\phi(t_{n+1}-) < \phi(t_{n+1})$ then
$$\phi(t_{n+1}) = (\alpha_{t_{n+1}} - 1)^{-1}(a(t_{n+1}) - \phi(t_{n+1}-)) + a(t_{n+1})$$
$$\leq a(t_{n+1}) + (\alpha_{t_{n+1}} - 1)^{-1} \triangle a(t_{n+1}) \leq a(t_{n+1}) + \tilde{a}^d(t_{n+1}).$$
Therefore (3.2) holds.

Case 2. (General case). For any $\varepsilon > 0$ define
$$\{t_1^\varepsilon < t_2^\varepsilon < ... < t_n^\varepsilon < ...\} = \{t \geq 0 : \triangle a_t \geq \varepsilon\}.$$
As case 1 define (denote $t_0^\varepsilon = 0$)

$$\begin{cases} \phi^\varepsilon(t) = a(t) \vee \phi^\varepsilon(t_n^\varepsilon), \text{as } t \in [t_n^\varepsilon, t_{n+1}^\varepsilon), \\ \phi^\varepsilon(t_{n+1}^\varepsilon) = [(\alpha_{t_{n+1}^\varepsilon} - 1)^{-1} \left| y(t_{n+1}^\varepsilon) + \phi^\varepsilon(t_{n+1}^\varepsilon -)\right| - y(t_{n+1}^\varepsilon)] \\ \qquad \vee \phi^\varepsilon(t_{n+1}^\varepsilon -), \end{cases}$$

and

$$x_t^\varepsilon = y_t + \phi_t^\varepsilon.$$

Then similarly to case 1 one can derive the following propositions. Denote

$$S_\varepsilon = S_{\phi^\varepsilon} \cap \{t \geq 0 : \Delta a_t \geq \varepsilon\}.$$

3.6 Proposition. (i) $t \in S_{\phi^\varepsilon} \Rightarrow \Delta a_t > 0$;
(ii) $t \in S_\varepsilon \Rightarrow a_t = -y(t)$, $a(t) > \phi^\varepsilon(t-)$, $\Delta \phi_t^\varepsilon = \alpha_t x_t^\varepsilon$,

$$\phi^\varepsilon(t) = \alpha_t(\alpha_t - 1)^{-1} a(t) - (\alpha_t - 1)^{-1} \phi^\varepsilon(t-).$$

3.7 Proposition. If t is a strictly increasing point of $(\phi^\varepsilon)^c(.)$, then it is also such a point of $a(.)$. Moreover,

$$x^\varepsilon(t-) = 0.$$

Furthermore, one also has the following propositions.

3.8 Proposition. (i) $t \in S_\varepsilon \Rightarrow$
1) $\phi^\varepsilon(t) + \phi^\varepsilon(t-) = (\alpha_t - 1)^{-1}(\alpha_t a(t) + (\alpha_t - 2)\phi^\varepsilon(t-))$,
2) $\Delta \phi_t^\varepsilon = \alpha_t(\alpha_t - 1)^{-1}(a(t) - \phi^\varepsilon(t-))$,
3) $\Delta \phi_t^\varepsilon \leq \alpha_t(\alpha_t - 1)^{-1} \Delta a(t)$,
4) $\phi^\varepsilon(t-) = -(\alpha_t - 1)\alpha_t^{-1} \Delta \phi_t^\varepsilon + a(t)$;
(ii) $t \in \bar{S}_\varepsilon \Rightarrow \Delta \phi_t^\varepsilon \leq \Delta a(t)$;
(iii) $t \in S_{\phi^\varepsilon} \Rightarrow \phi^\varepsilon(t) + \phi^\varepsilon(t-) \leq 2a(t) \leq 2\phi^\delta(t)$, for any $\delta > 0$;
where ϕ^δ is similarly defined, and

$$\bar{S}_\varepsilon = S_{\phi^\varepsilon} \cap \{t \geq 0 : 0 < \Delta a(t) < \varepsilon\}.$$

Proof. 1) in (i) is derived by (ii) in Proposition 3.6. Obviously, 1) $\Rightarrow$ 2) $\Rightarrow$ 3) and 2) $\Rightarrow$ 4). Let us show (ii). If $t \in \bar{S}_\varepsilon$, then there exists a n such that $t_n^\varepsilon < t < t_{n+1}^\varepsilon$. However, by the definition of $\phi^\varepsilon(t)$ and $t \in \bar{S}_\varepsilon$ one has that $\phi^\varepsilon(t) = a(t)$. Note that $\phi^\varepsilon(t-) \geq a(t-)$. Hence,

$$\Delta \phi_t^\varepsilon \leq \Delta a(t).$$

Now let us show (iii). If $t \in S_\varepsilon$, then by 1) and 4) in (i) (the latter shows that $\phi^\varepsilon(t-) \leq a(t)$) the conclusion is derived. If $t \in \bar{S}_\varepsilon$, then there exists an n such that

$$t_n^\varepsilon < t < t_{n+1}^\varepsilon,$$

then, since $\Delta \phi^\varepsilon(t) > 0$,

$$\phi^\varepsilon(t) = a(t).$$

However,

$$\phi^\varepsilon(t-) \leq \phi^\varepsilon(t) = a(t).$$

Hence, the conclusion is also true. ■

Denote

$$S_{\phi^\delta}^c = \text{the complement of } S_{\phi^\varepsilon}.$$

3.9 Proposition. There exist $x(t)$ and $\phi(t)$, $t \geq 0$, such that for any $T \geq 0$ as $\varepsilon \to 0$,

$$x_t^\varepsilon \to x_t, \text{ and } \phi_t^\varepsilon \to \phi_t, \text{ uniformly for } 0 \leq t \leq T.$$

Proof. Let $0 < \delta < \varepsilon$. By Ito's formula

$$(x_t^\varepsilon - x_t^\delta)^2 = (\phi_t^\varepsilon - \phi_t^\delta)^2 = 2\int_0^t(\phi_{s-}^\varepsilon - \phi_{s-}^\delta)d(\phi_s^\varepsilon - \phi_s^\delta)$$
$$+\sum_{0<s\leq t}\left|\Delta(\phi_s^\varepsilon - \phi_s^\delta)\right|^2 = -2\int_0^t x_s^\varepsilon d(\phi^\delta)_s^c - 2\int_0^t x_s^\delta d(\phi^\varepsilon)_s^c$$
$$+\sum_{0<s\leq t}(\phi_s^\varepsilon - \phi_s^\delta + \phi_{s-}^\varepsilon - \phi_{s-}^\delta)\triangle(\phi_s^\varepsilon - \phi_s^\delta)$$
$$= -2\int_0^t x_s^\varepsilon d(\phi^\delta)_s^c - 2\int_0^t x_s^\delta d(\phi^\varepsilon)_s^c$$
$$+\sum_{0<s\leq t} I_{S_{\phi^\varepsilon}\cap S_{\phi^\delta}}(\phi_s^\varepsilon - \phi_s^\delta + \phi_{s-}^\varepsilon - \phi_{s-}^\delta)\triangle(\phi_s^\varepsilon - \phi_s^\delta)$$
$$-\sum_{0<s\leq t} I_{S_{\phi^\varepsilon}\cap S_{\phi^\delta}^c}(2\phi_s^\delta - \phi_s^\varepsilon - \phi_{s-}^\varepsilon)\triangle\phi_s^\varepsilon$$
$$-\sum_{0<s\leq t} I_{S_{\phi^\delta}\cap S_{\phi^\varepsilon}^c}(2\phi_s^\varepsilon - \phi_s^\delta - \phi_{s-}^\delta)\triangle\phi_s^\delta = \sum_{i=1}^5 I_t^i.$$

Obviously,

$$I_t^1,\ I_t^2 \leq 0.$$

By (iii) of Proposition 3.8 we also have

$$I_t^4,\ I_t^5 \leq 0.$$

Denote

(3.8) $I_t^3 = \sum_{0<s\leq t} I_{S_{\phi^\varepsilon}\cap S_{\phi^\delta}}\Phi_s,$

where

$$\Phi_s = (\phi_s^\varepsilon - \phi_s^\delta + \phi_{s-}^\varepsilon - \phi_{s-}^\delta)\triangle(\phi_s^\varepsilon - \phi_s^\delta).$$

Then by 1) and 4) of (i) in Proposition 3.8 as $s \in S_\varepsilon$

$$0 \leq \phi_s^\varepsilon + \phi_{s-}^\varepsilon = 2a(s) - (\alpha_s - 2)\alpha_s^{-1}\triangle\phi_s^\varepsilon.$$

A similar formula also holds for ϕ^δ. Hence

$$I_{S_\varepsilon\cap S_\delta}\Phi_s = -(\alpha_s - 2)\alpha_s^{-1}(\Delta(\phi_s^\varepsilon - \phi_s^\delta))^2 \leq 0.$$

Note that

$$S_{\phi^\varepsilon} = S_\varepsilon \cup \bar{S}_\varepsilon,\ S_{\phi^\delta} = S_\delta \cup \bar{S}_\delta.$$

However, by (iii) of Proposition 3.8 and by $a(s) \leq \phi_s^\varepsilon$, as $s \in S_{\phi^\varepsilon}\cap S_{\phi^\delta}$

(3.9) $|\Phi_s| \leq (2a(s) - a(s) - a(s-))(\Delta\phi_s^\varepsilon + \Delta\phi_s^\delta) = (\Delta\phi_s^\varepsilon + \Delta\phi_s^\delta)\triangle a(s).$

Hence by 3) of (i) and (ii) in Proposition 3.8 as $s \in \bar{S}_\delta \cap S_{\phi^\varepsilon}$

$$|\Phi_s| \leq (1 + (\alpha_s - 1)^{-1}\alpha_s)(\Delta a(s))^2 \leq 3\delta\triangle a(s) \leq 3\varepsilon\triangle a(s).$$

A similar inequality holds for $s \in \bar{S}_\varepsilon \cap S_{\phi^\delta}$. Now note that as $s \in \bar{S}_\varepsilon \cap \overline{S}_\delta$ by (3.9) and (ii) in Proposition 3.8

$$|\Phi_s| \leq 2(\Delta a(s))^2 \leq 2\varepsilon\triangle a(s).$$

Hence

$$(x_t^\varepsilon - x_t^\delta)^2 = (\phi_t^\varepsilon - \phi_t^\delta)^2 \leq 8\varepsilon\sum_{0<s\leq T}\Delta a(s) \leq 8\varepsilon a(T).$$

Therefore the conclusion of Proposition 3.9 is derived. ■

Applying Proposition 3.6-3.9 it is seen that (x_t, ϕ_t) found in Proposition 3.9 is a solution of (3.1).

In fact, by the uniform convergence if $t \in S_\phi$, i.e. $\Delta\phi_t > 0$, then there exists a $\varepsilon' > 0$ such that as $0 < \varepsilon < \varepsilon'$

$$\Delta\phi_t^\varepsilon > 0.$$

Hence

$$\Delta\phi_t^\varepsilon = \alpha_t x_t^\varepsilon.$$

Let $\varepsilon \to 0$. Then we get

$$\Delta\phi_t = \alpha_t x_t.$$

Now assume that t is such a point $x_t > 0$. By the right continuity of x_t there exists a $h > 0$ and $b > 0$ such that as $s \in [t, t+h]$

$$x_s > b > 0.$$

Again, by the uniform convergence there exists an $\varepsilon' > 0$ such that as $0 < \varepsilon < \varepsilon'$ for all $s \in [t, t+h]$

$$x_s^\varepsilon > b/2 > 0.$$

However, from this we have that for all $s \in [t, t+h]$ as $0 < \varepsilon < \varepsilon'$

$$\phi_t^{\varepsilon c} = \phi_s^{\varepsilon c} = \phi_{t+h}^{\varepsilon c}.$$

Let us show that as $\varepsilon \to 0$ $\phi_t^{\varepsilon c}$ converges to ϕ_t^c uniformly in $t \in [0, T]$. For this one only needs to show that as $\varepsilon \to 0$

(3.10) $I_t^\varepsilon = \sum_{0<s<t} |\Delta\phi_s - \Delta\phi_s^\varepsilon| \to 0.$

In fact, by 3) of (i) and (ii) in proposition 3.8 $\forall t \in [0, T]$

$$\Delta\phi_t^\varepsilon \le 2 \triangle a_t.$$

However, denote $S_a \cap [0, T] = \{s_1, ..., s_n, ...\}$, then

$$\sum_{0<s\le T} \Delta a_s = \sum_{n=1}^\infty \Delta a_{s_n}.$$

Hence for any given $\delta > 0$ one can take an N large enough such that

$$\sum_{n=N}^\infty \Delta a_{s_n} < \delta/4.$$

Now take an $\varepsilon' > 0$ small enough such that as $0 < \varepsilon < \varepsilon'$

$$\sum_{n=1}^N \left|\Delta\phi_{s_n} - \Delta\phi_{s_n}^\varepsilon\right| < \delta/2.$$

Then (3.10) is proved. Hence let $\varepsilon \to 0$ for all $s \in [t, t+h]$ $\phi_s^{\varepsilon c} \to \phi_s^c$. Therefore

$$\phi_t^c = \phi_s^c = \phi_{t+h}^c.$$

Hence for any $0 < T < \infty$

$$\int_0^T x_s d\phi_s^c = 0$$

is derived. Therefore it is proved that (x_t, ϕ_t) is a solution of (3.1).

Now let us show the uniqueness. That is true by the following

3.10 Lemma. If (x_t, ϕ_t) and $(\hat{x}_t, \hat{\phi}_t)$ are two solutions of (3.1) for the same y_t and x_0, then $\left|\phi_t - \hat{\phi}_t\right|^2 = |x_t - \hat{x}_t|^2 = 0$.

Proof. By Ito's formula

$$|x_t - \hat{x}_t|^2 = 2\int_0^t (x_{s-} - \hat{x}_{s-}) d(\phi_s - \hat{\phi}_s) + \sum_{0<s\le t} \left|\Delta(\phi_s - \hat{\phi}_s)\right|^2$$

$$= -2\int_0^t x_s d\hat{\phi}_s^c - 2\int_0^t \hat{x}_s d\phi_s^c - \sum_{0<s\le t} \left|\Delta(\phi_s - \hat{\phi}_s)\right|^2$$

$$+2\sum_{0<s\le t}(x_s - \hat{x}_s) \triangle (\phi_s - \hat{\phi}_s) \le -4\sum_{0<s\le t}[(x_s)^2 I_{S_\phi} + (\hat{x}_s)^2 I_{S_{\hat{\phi}}}]$$

$$+8\sum_{0<s\le t} x_s\hat{x}_s I_{S_\phi \cap S_{\hat{\phi}}} + 4\sum_{0<s\le t}[(x_s)^2 I_{S_\phi} + (\hat{x}_s)^2 I_{S_{\hat{\phi}}}]$$

$$-4\sum_{0<s\le t} x_s\hat{x}_s(I_{S_\phi} + I_{S_{\hat{\phi}}}) = -4\sum_{0<s\le t} x_s\hat{x}_s(I_{S_\phi} + I_{S_{\hat{\phi}}})^2 \le 0.$$

Hence

$$\left|\phi_t - \hat{\phi}_t\right|^2 = |x_t - \hat{x}_t|^2 = 0. \blacksquare$$

Now by construction it is seen that if y_t is adapted or continuous, then so is (x_t, ϕ_t). The proof of Theorem 3.2 is complete.

Chapter 3

Reflecting Stochastic Differential Equations with Jumps

In this chapter we will discuss the existence of solutions to reflecting stochastic differential equations (RSDE) with jumps in the general convex domain with continuous reflection and in the half space with jump reflections. We have obtained the results for RSDE with jumps and with non-Lipschitzian coefficients. The Yamada-Watanabe Theorem and Girsanov Theorem and so on for RSDE with jumps are also examined.

3.1 Yamada-Watanabe Theorem, Tanaka Formula and Krylov Estimate

Consider the d-dimensional reflecting stochastic differential equation (RSDE) with Poisson jumps as follows:

(1.1) $dx_t = b(t, x_t, \omega)dt + \sigma(t, x_t, \omega)dw_t + \int_Z c(t, x_{t-}, \omega)q(dt, dz) + d\phi_t,$

$x_0 = x \in \overline{\Theta},$

where (x_t, ϕ_t) satisfies one of the following conditions (1.2), (1.2)':

(1.2): 1) $x_t \in \overline{\Theta}, \forall t \geq 0$, ($x_t$ is a R^d-valued $\Im_t-$adapted cadlag process),

2) ϕ_t is a R^d-valued $\Im_t-$adapted cadlag process with bounded variation $|\phi|_t$ on each finite interval $[0, t]$ such that $\phi_0 = 0$,

3) $|\phi|_t = \int_0^t I_{\partial\Theta}(x_s)d\,|\phi|_s\,,$

4) $\phi_t = \int_0^t n(s)d\,|\phi|_s\,,$

$n(t) \in \aleph_{x(t)}$, as $x(t) \in \partial\Theta,$

where

$$\aleph_x = \cup_{r=0}\aleph_{x,r}, \aleph_{x,r} = \{n \in R^d : |n| = 1, B(x - nr, r) \cap \Theta = \emptyset\},$$

w_t is a d-dimensional standard Brownian motion process (BM), $q(dt, dz)$ is a Poisson martingale measure with a compensator $\pi(dz)dt$, $\pi(.)$ is a $\sigma-$finite measure on a measurable space $(Z, \Re(Z))$ such that

$$q(dt, dz) = p(dt, dz) - \pi(dz)dt,$$

and Θ is a given convex domain in R^d, $\partial\Theta$ is its boundary, $\overline{\Theta}$ is its closure, where in (1.2) we assume that Θ satisfies Assumption 2.11 in Chapter 2, which is also called

the uniform inner sphere condition: there exist constants $\varepsilon > 0$ and $\delta > 0$ that define at each point $x \in \partial\Theta$ an open ball

$$B_\varepsilon(x_0) = \{y \in R^d : |y - x_0| < \varepsilon\}$$

with the center at x_0, which may depend on x, such that $|x - x_0| < \delta$ and $B_\varepsilon(x_0) \subset \Theta$. (Obviously, the uniform inner sphere condition is the same as Assumption 2.11 in Chapter 2).

(1.2)' 1) is the same as that in (1.2),

2) ϕ_t is a R^1-valued $\Im_t$ - adapted cadlag process with bounded variation $|\phi|_t$ on each finite interval $[0, t]$ such that $\phi_0 = 0$, and $\phi_t^i = 0$, as $1 \leq i \leq d-1$; $\phi_t^d = \phi_t$ in (1.1),

3) $\int_0^t x_s^d \cdot d\phi_s^c = 0$, as $t \geq 0$; $\phi_t^c = \phi_t - \sum_{0<s\leq t} \triangle\phi_s$,

$$\triangle\phi_t > 0 \Rightarrow \triangle\phi_t = \alpha_t x_t^d,$$

where $\Theta = \{x \in R^d : x = (x^1, \cdots, x^d), x^d > 0\}$, $\alpha_t \geq 2$ is a non-random continuous function.

1.1 Remark. 1) in case $\Theta = R^d$ one can take $\phi_t \equiv 0$, then (1.1) reduces to the usual SDE (without reflection).

2) By Chapter 2 it is known that 3) and 4) in (1.2) are equivalent to

3'): for all f: $\Theta^{cl} \to R^d$, bounded and continuous such that $f|_{\partial\Theta} = 0$, and for all $t > 0$

$$\int_0^t f(X_s) \cdot d\phi_s = 0;$$

4') for all $Y(.) \in D^d([0, \infty), \overline{\Theta})$

$$\int_0^t (Y_s - X_s) \cdot d\phi_s \text{ is increasing, as } t \text{ does.}$$

For the solution of RSDE (1.1) (with (1.2) or (1.2)') we have the following

1.2 Definition. 1) We say that (1.1) has a weak solution, if there exist a probability space $(\Omega, \Im, (\Im_t), P)$ and two $\Im_t-$adapted cadlag processes (x_t, ϕ_t) with a BM w_t and a Poisson martingale measure $q(dt, dz)$ which has the same compensator $\pi(dz)dt$ as the given one in (1.1) defined on it such that $(x_t, \phi_t, w_t, q(dt, dz))$ satisfy (1.1) (with (1.2) or (1.2)') on $(\Omega, \Im, (\Im_t), P)$. In case (x_t, ϕ_t) are $\Im_t^{w,q}-$adapted, where $\Im_t^{w,q} = \sigma(w_s, q((0, s\}, U); s \leq t, U \in \Re(Z))$, then (x_t, ϕ_t) is said to be a strong solution of (1.1).

2) We say that the pathwise uniqueness for solutions of (1.1) holds, if for any two solutions $(x_t^i, \phi_t^i), i = 1, 2$, of (1.1) (with (1.2) or (1.2)'), which are defined on the same probability space $(\Omega, \Im, (\Im_t), P)$ with the same BM w_t and Poisson martingale measure $q(dt, dz)$,

$P(\sup_{t\geq 0}\left|x_t^1 - x_t^2\right| = 0) = 1,\ P(\sup_{t\geq 0}\left|\phi_t^1 - \phi_t^2\right| = 0) = 1.$

3) We say that the uniqueness in law for solutions of (1.1) holds, if for any two solutions $(x_t^i, \phi_t^i), i = 1, 2,$of (1.1) (with (1.2) or (1.2)'), which can be defined on different probability spaces $(\Omega^i, \Im^i, (\Im_t^i), P^i)$, with different BM w_t^i and Poisson martingale measures $q^i(dt, dz)$, $i = 1, 2$, respectively, but with Poisson martingale measures $q^i(dt, dz)$, $i = 1, 2$, having the same compensator $\pi(dz)dt$ as the given one, the probability law of $\{(x_t^1, \phi_t^1), t \geq 0\}$ and $\{(x_t^1, \phi_t^1), t \geq 0\}$ are the same.

1.3 Theorem. (Yamada-Watanabe type theorem). Assume that the coefficients b, σ, and c in (1.1) do not depend on ω, and for simplicity assume that $Z = R^d - \{0\}, \pi(dz) = dz/\left|z\right|^{d+1}$. Then we have the following two assertions.

1) If (1.1) has a weak solution and the pathwise uniqueness holds for (1.1), then (1.1) has a pathwise unique strong solution, $t \geq 0$.

2) If (1.1) has two weak solutions $(x_t^i, \phi_t^i, w_t^i, q^i(dt, dz)), i = 1, 2$, defined on two different probability spaces $(\Omega^i, \Im^i, (\Im_t^i), P^i), i = 1, 2$, respectively, where w_t^i is a P^i-BM, $q^i(dt, dz)$ is a P^i-Poisson martingale measure with the same compensator $\pi(dz)dt, i = 1, 2$, then there exist a probability space $(\Omega, \Im, (\Im_t), P)$ and four $\Im_t-$adapted cadlag processes $(\widetilde{x}_t^i, \widetilde{\phi}_t^i), i = 1, 2$, with a BM $\widetilde{w}_t$ and a Poisson martingale measure $\widetilde{q}(dt, dz)$ having the same compensator $\pi(dz)dt$ as the given one defined on it such that $(\widetilde{x}_t^i, \widetilde{\phi}_t^i), i = 1, 2$, are adapted to $\Im_t$, and the probability law of $(\widetilde{x}_t^i, \widetilde{\phi}_t^i, \widetilde{w}_t, \widetilde{q}(dt, dz))$ coincides with that of $(x_t^i, \phi_t^i, w_t^i, q^i(dt, dz)), i = 1, 2$. Moreover, $(\widetilde{x}_t^i, \widetilde{\phi}_t^i, \widetilde{w}_t, \widetilde{q}(dt, dz))$ satisfy (1.1) (with (1.2) or (1.2)') on the same probability space $(\Omega, \Im, (\Im_t), P), i = 1, 2$.

To show Theorem 1.3 let us prepare a Lemma first.

1.4 Lemma. Assume that (x_t, ϕ_t) is a solution of (1.1) with (1.2) (or (1.2)') with BM w_t and Poisson martingale measure $q(dt, dz)$, which has the compensator $\pi(dz)dt$ such that

$$q(dt, dz) = p(ds, dz) - \pi(dz)dt,$$

denote

$$\zeta_t = \int_0^t \int_{|z|\leq 1} zq(ds, dz) + \int_0^t \int_{|z|>1} zp(ds, dz).$$

Suppose that four $\Im_t-$adapted cadlag processes $(\widetilde{x}_t, \widetilde{\phi}_t, \widetilde{w}_t, \widetilde{\zeta}_t)$ have the same finite-dimensional probability distributions as that of $(x_t, \phi_t, w_t, \zeta_t)$. Denote

$$\widetilde{p}((0, t], U) = \sum_{0<s\leq t} I_{0=\Delta\widetilde{\zeta}_s \in U}, \text{ for } t \geq 0, U \in \Re(Z),$$

$$\widetilde{q}(dt, dz) = \widetilde{p}(ds, dz) - \pi(dz)dt.$$

Then $(\widetilde{x}_t, \widetilde{\phi}_t)$ is also a solution of (1.1) with (1.2) (or (1.2)') with BM $\widetilde{w}_t$ and Poisson martingale measure $\widetilde{q}(dt, dz)$, which has the compensator $\pi(dz)dt$.

Proof. ζ_t is the so-called Cauchy process. p is the counting measure induced by counting the jumps of ζ_t with compensator $\pi(dz)dt$. Hence, it is a Poisson counting measure. Since $\widetilde{\zeta}_t, t \geq 0$, and $\zeta_t, t \geq 0$, have the same finite-dimensional probability distributions, $\widetilde{\zeta}_t$ is also a Cauchy process. $\widetilde{p}$ is also a Poisson counting measure with the same compensator $\pi(dz)dt$. Hence,

$\widetilde{q}(dt,dz) = \widetilde{p}(ds,dz) - \pi(dz)dt$ is also a Poisson martingale measure. By the coincidence of finite-dimensional probability distributions of $(\widetilde{x}_t, \widetilde{\phi}_t, \widetilde{w}_t, \widetilde{\zeta}_t), t \geq 0$, and $(x_t, \phi_t, w_t, \zeta_t), t \geq 0$, applying the Dynkin system technique (Lemma 5.1 in Chapter 1, Ikeda & Watanabe, 1989) or the monotone class method in measure theory, it is easily seen that the probability laws of $(\widetilde{x}_t, \widetilde{\phi}_t, \widetilde{w}_t, \widetilde{\zeta}_t), t \geq 0$, and $(x_t, \phi_t, w_t, \zeta_t), t \geq 0$, also coincide. There exists a Baire function F such that

$$F(x_s, \phi_s, w_s, \zeta_s; s \leq t) = x_t - x - \int_0^t b(s, x_s)ds - \int_0^t \sigma(s, x_s)dw_s$$
$$- \int_0^t \int_Z c(s, x_{s-}, z)q(ds, dz).$$

Hence, one has that

$$P(F(\widetilde{x}_s, \widetilde{\phi}_s, \widetilde{w}_s, \widetilde{\zeta}_s; s \leq t) = 0) = P(F(x_s, \phi_s, w_s, \zeta_s; s \leq t) = 0) = 1.$$

A similar idea shows that $\left|\widetilde{\phi}\right|_t$ is finite for all $t \geq 0$.

Now, suppose that $(x_t, \phi_t, w_t, q(dt, dz))$ satisfies (1.1) with (1.2). To show $(\widetilde{x}_t, \widetilde{\phi}_t)$ satisfying 3) and 4) in (1.2) by Remark 1.1 one only needs to show that it satisfies 3)' and 4)' in Remark 1.1. However, this can be done just as above. Hence $(\widetilde{x}_t, \widetilde{\phi}_t, \widetilde{w}_t, \widetilde{q}(dt, dz))$ also satisfies (1.1) and (1.2). If $(x_t, \phi_t, w_t, q(dt, dz))$ satisfies (1.1) with (1.2)', one only needs to show that

$$\Delta\widetilde{\phi}_t > 0 \Rightarrow \Delta\widetilde{\phi}_t = \alpha_t \widetilde{x}_t^d,$$

since the others can be proved similarly. However, if $P(\Delta\widetilde{\phi}_t > 0) > 0$, then $P(\Delta\phi_t > 0) = P(\Delta\widetilde{\phi}_t > 0) > 0$. In this case

$$P(\Delta\widetilde{\phi}_t = \alpha_t \widetilde{x}_t^d \,\Big|\, \Delta\widetilde{\phi}_t > 0)$$
$$= P(\left\{\Delta\widetilde{\phi}_t = \alpha_t \widetilde{x}_t^d\right\} \cap \left\{\Delta\widetilde{\phi}_t > 0\right\})/P(\Delta\widetilde{\phi}_t > 0)$$
$$= P(\{\Delta\phi_t = \alpha_t x_t^d\} \cap \{\Delta\phi_t > 0\})/P(\Delta\phi_t > 0)$$
$$= P(\Delta\phi_t = \alpha_t x_t^d \,|\Delta\phi_t > 0) = 1.$$

Hence (1.2)' is satisfied. The proof of Lemma 1.4 is complete. ■

Now let us show Theorem 1.3.

Proof. The proof is similar to that of Ikeda & Watanabe (1989), which is for the case of SDE without reflection and without jumps. Let us show 2) first. Suppose that the weak solutions $(x_t^i, \phi_t^i, w_t^i, q^i(dt, dz)), i = 1, 2$, are defined on probability spaces $(\Omega^i, \mathfrak{F}^i, (\mathfrak{F}_t^i), P^i)$, $i = 1, 2$, respectively. By the assumption of 2) in Theorem 1.3 we can let

$$\zeta_t^i = \int_0^t \int_{|z| \leq 1} zq^i(ds, dz) + \int_0^t \int_{|z|>1} zp^i(ds, dz), i = 1, 2.$$

Construct space $\Omega = D^d \times D^d \times D^d \times D^d \times W_0^d \times D^d$, where $D^d = D([0, \infty); R^d)$ (see chapter 2), and W_0^d is the set of continuous R^d-valued functions $w(t)$ defined on $t \geq 0$ with $w(0) = 0$.

Map $(x^1(.,\omega), \phi^1(.,\omega), w^1(.,\omega), \varsigma^1(.,\omega))$ and $(x^2(.,\omega), \phi^2(.,\omega), w^2(.,\omega), \varsigma^2(.,\omega))$ into the $\Omega^1 = 1st \times 2nd \times 5th \times 6th$ component space of Ω and $\Omega^2 = 3rd \times 4th \times 5th \times 6th$ component space of Ω, respectively. From these two maps we get the probability laws P_x^1 and P_x^2 on Ω^1 and Ω^2, respectively. Then both marginal distributions $\pi(P_x^1)$ and $\pi(P_x^2)$ coincide with $P^w \times P^\varsigma$, where P^w is the Wiener measure on W_0^d, P^ζ is the measure on D^d induced by ζ^1 or ζ^2 (both of them have the same probability distribution), and $\pi : D^d \times D^d \times W_0^d \times D^d \to W_0^d \times D^d$ is the projection. Let $Q_i^{(w_5,w_6)}(dw_1dw_2)$ be the regular conditional distribution of (w_1, w_2) given (w_5, w_6). Define a probability measure Q on the space Ω by

$$Q(dw_1dw_2dw_3dw_4dw_5dw_6)$$
$$= Q_i^{(w_5,w_6)}(dw_1dw_2)Q_i^{(w_5,w_6)}(dw_3dw_4)P^w(dw_5)P^\zeta(dw_6).$$

Let $\Im$ be the completion of the $\sigma-$field $\Re(\Omega)$ by Q, and $\Im_t = \cap_{\varepsilon>0}(\Re_{t+\varepsilon} \vee \aleph)$, where $\Re_t = \Re_t(D^d) \times \Re_t(D^d) \times \Re_t(W_0^d) \times \Re_t(D^d)$ and $\aleph$ is the set of all $Q-$null sets. Then clearly (w_1, w_2, w_5, w_6) and $(x^1, \phi^1, w^1, \zeta^1)$ have the same distribution, and so do (w_3, w_4, w_5, w_6) and $(x^2, \phi^2, w^2, \zeta^2)$. Now let us show the following facts.

Fact A. For $A \in \Re_t(D^d \times D^d)$, $(w_5, w_6) \in W_0^d \times D^d \to Q_1^{(w_5,w_6)}(A)$ and $Q_2^{(w_5,w_6)}(A)$ are $\Re_t(W_0^d \times D^d)^{P^{w,\zeta}}-$ measurable, where $P^{w,\zeta} = P^w \times P^\zeta$, and $\Re_t(W_0^d \times D^d)^{P^{w,\zeta}}$ is the completion of $\Re_t(W_0^d \times D^d)$ by $P^{w,\zeta}$.

Indeed, for fixed $t > 0$ and $A \in \Re_t(D^d \times D^d)$, there exists a conditional probability $Q_{1t}^{(w_5,w_6)}(A)$ such that $(w_5, w_6) \in W_0^d \times D^d \to Q_{1t}^{(w_5,w_6)}(A)$ is $\Re_t(W_0^d \times D^d)^{P^{w,\zeta}}-$ measurable, and

$$P_x^1(A \times C) = \int_C Q_{1T}^{(w_5,w_6)}(A)P^w(dw_5)P^\zeta(dw_6), \text{ for any } C \in \Re_t(W_0^d \times D^d).$$

Now let

$$C = \{(w_5, w_6) \in W_0^d \times D^d : \rho_t(w_5, w_6) \in A_1, \theta_t(w_5, w_6) \in A_2\},$$
$$A_1, A_2 \in \Re_t(W_0^d \times D^d),$$

where θ_t is defined by $\theta_t(w_5, w_6)(s) = (w_5(t + s) - w_5(t), w_6(t + s) - w_6(t))$, $s \geq 0$,and ρ_t is defined by $\rho_t(w_5, w_6)(s) = (w_5(t \wedge s), w_6(t \wedge s))$. Since $\theta_t(w_5, w_6)$ is independent of $\Re_t(W_0^d \times D^d)$ with respect to $P^{w,\zeta}$ we have

$$\int_C Q_{1t}^{(w_5,w_6)}(A)P^w(dw_5)P^\zeta(dw_6)$$
$$= \int_{\rho_t(w_5,w_6)\in A_1} Q_{1t}^{(w_5,w_6)}(A)P^w(dw_5)P^\zeta(dw_6)P^{w,\zeta}(\theta_t(w_5, w_6) \in A_2)$$
$$= P_x^1(A \times \{\rho_t(w_5, w_6) \in A_1\})P^{w,\zeta}(\theta_t(w_5, w_6) \in A_2)$$
$$= P^1((x^1(.), \phi^1(.)) \in A, \rho_t(w, \zeta) \in A_1)P^1(\theta_t(w, \zeta) \in A_2)$$
$$= P^1((x^1(.), \phi^1(.)) \in A, \rho_t(w, \zeta) \in A_1, \theta_t(w, \zeta) \in A_2)$$
$$= P^1((x^1(.), \phi^1(.)) \in A, (w, \zeta) \in C) = P_x^1(A \times C),$$

where we have used the result that $\{(x^1(.), \phi^1(.)) \in A, \rho_t(w, \zeta) \in A_1\} \in \Im_t$, moreover, , $\theta_t(w, \zeta)$ and $\Im_t$ are independent. Hence it is easily shown that

$$Q_{1t}^{(w_5,w_6)}(A) = Q_1^{(w_5,w_6)}(A), a.a.(w_5, w_6) \ (P^{w,\zeta}).$$

Fact A is proved.

Fact B. $w_5(t)$ is an d-dimensional $\Im_t-$BM on $(\Omega, \Im, Q)$, and $p(dt, dz)$ is a Poisson counting measure with the same compensator $\pi(dz)dt$ on $(\Omega, \Im, Q)$, where

$$p((0, t], U) = \sum_{0<s\leq t} I_{0=\triangle\zeta_s \in U}, \text{ for } t \geq 0, U \in \Re(Z).$$

Indeed, by using Fact A we have that

$$E^Q[e^{i(\lambda,w_5(t)-w_5(s))}I_{A_1\times A_2\times A_3\times A_4\times A_5\times A_6}]$$
$$= \int_{A_5\times A_6} e^{i(\lambda,w_5(t)-w_5(s))}Q_1^{(w_5,w_6)}(A_1\times A_2)Q_2^{(w_5,w_6)}(A_3\times A_4)P^w(dw_5)P^\zeta(dw_6)$$
$$= e^{-(|\lambda|^2/2)(t-s)} \int_{A_5\times A_6} Q_1^{(w_5,w_6)}(A_1\times A_2)Q_2^{(w_5,w_6)}(A_3\times A_4)P^w(dw_5)P^\zeta(dw_6)$$

$= e^{-(|\lambda|^2/2)(t-s)} Q(A_1 \times A_2 \times A_3 \times A_4 \times A_5 \times A_6),$

for $\lambda \in R^d, A_i \in \Re_s(D^d), i = 1,2,3,4,6; A_5 \in \Re_s(W_0^d), 0 \le s \le t.$

Therefore the first conclusion of Fact B is true. Now for $t > s \ge 0$, disjoint $U_1, \cdots, U_m \in \Re(Z)$ such that $p((0,t], U_i) < \infty, \forall i = 1, \cdots, m$ and $\lambda_i > 0, i = 1, \cdots, m$; $A_i \in \Re_s(D^d), i = 1,2,3,4,6; A_5 \in \Re_s(W_0^d), 0 \le s \le t$

$E^Q[\exp[-\sum_{i=1}^m \lambda_i p((s,t], U_i)] I_{A_1 \times A_2 \times A_3 \times A_4 \times A_5 \times A_6}]$

$= \int_{A_5 \times A_6} \exp[-\sum_{i=1}^m \lambda_i p((s,t], U_i)] Q_1^{(w_5,w_6)}(A_1 \times A_2)$

$\cdot Q_2^{(w_5,w_6)}(A_3 \times A_4) P^w(dw_5) P^\zeta(dw_6)$

$= \exp[(t-s) \sum_{i=1}^m (e^{-\lambda_i} - 1)\pi(U_i)]$

$\cdot \int_{A_5 \times A_6} Q_1^{(w_5,w_6)}(A_1 \times A_2) Q_2^{(w_5,w_6)}(A_3 \times A_4) P^w(dw_5) P^\zeta(dw_6)$

$= \exp[(t-s) \sum_{i=1}^m (e^{-\lambda_i} - 1)\pi(U_i)] Q(A_1 \times A_2 \times A_3 \times A_4 \times A_5 \times A_6).$

Therefore the second conclusion is also true.

Now let us return to the proof of Theorem 1.3. From Fact B it is not difficult to show that $(w_1, w_2, w_5, q(\cdot,\cdot))$ and $(w_3, w_4, w_5, q(\cdot,\cdot))$ are solutions of (1.1) (with (1.2) or (1.2)') on the same space $(\Omega, \Im, (\Im_t), Q)$, where

$q(dt,dz) = p(dt,dz) - \pi(dz)dt$

is a Poisson martingale measure with the compensator $\pi(dz)dt$. Hence 2) of Theorem 1.3 is proved. Now assume that the condition in 1) holds. Then the pathwise uniqueness implies that $w_1 = w_2, w_3 = w_4, Q - a.s.$ This implies that

$Q_1^{(w_5,w_6)} \times Q_2^{(w_5,w_6)}((w_1,w_2) = (w_3,w_4)) = 1,\ P^{w,\zeta} - a.s.$

Now it is easy to see that there exists a function $(w_5, w_6) \in W_0^d \times D^d \to F_x(w_5,w_6) \in D^d \times D^d$ such that $Q_1^{(w_5,w_6)} = Q_2^{(w_5,w_6)} = \delta_{\{F_x(w_5,w_6)\}}, P^{w,\zeta} - a.s.$ By Fact A this function is $\Re_t(W_0^d \times D^d)^{P^{w,\varsigma}} / \Re_t(D^d \times D^d)$–measurable. Clearly, $F_x(w_5,w_6)$ is uniquely determined up to $P^{w,\zeta}$–measure 0. Now by the assumption of 1) (1.1) with (1.2) or (1.2)' has a weak solution. Denote it by $(x_t, \phi_t, w_t, q(dt,dz))$, where $q(dt,dz)$ is a Poisson martingale measure with the compensator $\pi(dz)dt$. By above $F_x(w_s, \zeta_s, s \le t)$, where $\zeta_t = \int_0^t \int_{|z| \le 1} z q(ds,dz) + \int_0^t \int_{|z|>1} z p(ds,dz)$ is also a weak solution on the same probability space, since it has the same probability distributions as that of $F_x(w_5(s), w_6(s), s \le t)$. Then by pathwise uniqueness $(x_t, \phi_t) = F_x(w_s, \varsigma_s, s \le t)$. Hence (x_t, ϕ_t) is a strong solution of (1.1) (with (1.2) or (1.2)', respectively). ■

Now let us discuss the Tanaka type formula for RSDE (1.1) with (1.2) only.

1.5 Theorem. (Tanaka type formula). Assume that

1° $\|\sigma_{ik}(t,x,\omega) - \sigma_{ik}(t,y,\omega)\|^2 \le k_N(t)\rho_N(|x_i - y_i|^2)$, as $|x|, |y| \le N, \forall i,k$; where $0 \le k_N(t)$, it is non-random such that $\int_0^t k_N(s)ds < \infty$, for any $t > 0$, and $\rho_N(u)$ is a concave and strictly increasing function of u in $u \ge 0$ with $\rho_N(0) = 0$ satisfying $\int_{0+} du/\rho_N(u) = +\infty$,for each $N = 1, 2, ...$;

2° b^1, b^2, σ and $\int_Z |c(t,x,z,\omega)|^2 \pi(dz)$ are locally bounded, i.e., for all $t \ge 0, \omega \in \Omega$

$\int_Z |c(t,x,z,\omega)|^2 \pi(dz) + |h(t,x,\omega)| \le k_r$, as $|x| \le r, h = b^1, b^2, \sigma$,

where k_r is a constant depending on r only.

If $(x^i(t), \phi^i(t))$ satisfies (1.1) with (1.2) and with b^i, σ, c, $i = 1, 2$, respectively, and $\phi^i(t), i = 1, 2$, are continuous, then

$$(x_i^1(t) - x_i^2(t))^+ = (x_i^1(0) - x_i^2(0))^+ + \int_0^t I_{(x_i^1(s-) > x_i^2(s-))}(s) d(x_i^1(s) - x_i^2(s)) + J(t),$$

where

$$J(t) = \int_0^t \int_Z (x_i^1(s-) - x_i^2(s-) + c_i(s, x^1(s-), z, \omega) - c_i(s, x^2(s-), z, \omega))^+ - (x_i^1(s-) - x_i^2(s-))^+ - I_{(x_i^1(s-) > x_i^2(s-))}(s)(c_i(s, x^1(s-), z, \omega) - c_i(s, x^2(s-), z, \omega)))p(ds, dz), \ 1 \leq i \leq d.$$

Furthermore, if $\forall (t, x, y, z, \omega) \in [0, \infty) \times \overline{\Theta} \times \overline{\Theta} \times Z \times \Omega$

$$x_i \geq y_i \Longrightarrow x_i + c_i(t, x, z, \omega) \geq y_i + c_i(t, y, z, \omega), \ 1 \leq i \leq d,$$

then for all $t \geq 0$

$$J(t) = 0.$$

Proof. Since

$$x_i^1(t) = x_i^1(0) + \int_0^t b_i^1(s, x^1(s), \omega) ds + \sum_{k=1}^d \int_0^t \sigma_{ik}(s, x^1(s), \omega) dw_k(s) + \int_0^t \int_Z c_i(s, x^1(s-), z, \omega) q(ds, dz) + \phi_i^1(t), \ 1 \leq i \leq d;$$

and $x_i^2(t)$ satisfies a similar SDE. Hence the conclusion follows from Theorem 10.8 of Chapter 1. ■

Now let us quote Theorem 1 of Anulova (1978) as the following lemma.

1.6 Lemma. Suppose we are given a bounded Borel measurable function $0 \leq f = f(t, x) : \overline{R}_+^1 \times R^d \to \overline{R}_+^1 (= [0, \infty))$ and $\lambda > 0$, then for any $\varepsilon > 0$ there exists a smooth function $u^\varepsilon(t, x) : R \times R^d \to [0, \infty)$ such that

1) $\sum_{i,j=1}^d h_i h_j (\partial^2 / \partial x_i \partial x_j) u^\varepsilon \leq \lambda u^\varepsilon, \ \forall h \in R^d, |h| \leq 1,$

2) $\forall p \geq d + 1, (t, x) \in \overline{R}_+^1 \times R^d$

$$u^\varepsilon(t, x) \leq k(p, \lambda, d, \varepsilon) \left\| e^{-\lambda s (d+1)/p} f(t + s, y) \right\|_{p, (s, y) \in \overline{R}_+^1 \times R^d},$$

where

$$k(p, \lambda, d, \varepsilon) = e^{\varepsilon \lambda (d+1)/p} p^{d/p} (V_d d!)^{1/p} \lambda^{d/2p} (\lambda(d + 1))^{(1/p) - 1},$$

V_d is the volume of $d-$dimensional unit sphere, and

$$\|g(s, y)\|_{p, (s, y) \in \overline{R}_+^1 \times R^d} = (\int_{\overline{R}_+^1 \times R^d} |g(s, y)|^p \, ds dx)^{1/p} (= \|g(s, y)\|_{p, \overline{R}_+^1 \times R^d}),$$

3) $|grad_x u^\varepsilon| \leq (\lambda)^{1/2} u^\varepsilon,$

4) for all non-negative definite symmetric matrices $A = (a_{i,j})_{d \times d}$

$$\sum_{i,j=1}^d a_{i,j} (\partial^2 u^\varepsilon / \partial x_i \partial x_j) - \lambda (tr.A) u^\varepsilon \leq 0,$$

$$\sum_{i,j=1}^d a_{i,j} (\partial^2 u^\varepsilon / \partial x_i \partial x_j) - \lambda (tr.A + 1) u^\varepsilon + (\partial / \partial t) u^\varepsilon \leq -(\det A)^{1/(d+1)} f_\varepsilon,$$

where f_ε is the smoothness function of f, i.e.

$$f_\varepsilon = f * \omega^\varepsilon = \int_{-\infty}^\infty ds \int_{R^d} \widetilde{f}(t - \varepsilon s, x - \varepsilon y) \omega(s, y) dy,$$

$$\widetilde{f}(t, x) = \begin{cases} f(t, x), & \text{as } t \geq 0, \\ 0, & \text{otherwise,} \end{cases}$$

$\omega(t, x)$ is a smooth function such that $\omega(t, x) : R^1 \times R^d \to [0, \infty)$,

$$\omega(t, x) = 0, \text{ as } (t, x) \notin [-1, 1] \times [-1, 1]^d,$$

$$\int_{-\infty}^\infty dt \int_{R^d} \omega(t, x) dx dt = 1.$$

By using Lemma 1.6 we can prove the following

1.7 Theorem. (Krylov type estimate). Suppose that $b = b(t,x) : [0,T] \times R^d \to R^d$, $\sigma = \sigma(t,x) : [0,T] \times R^d \to R^{d\otimes d}$, $c = c(t,x,z) : [0,T] \times R^d \times Z \to R^d$ are jointly measurable and satisfy

$|b|^2 + \|\sigma\|^2 + \int_Z |c|^2 \pi(dz) \leq k_0.$

Assume that $(x(t), \phi(t))$ satisfies (1.1) with (1.2) or (1.2)', moreover, assume that $\forall \lambda > 0$

(1.3) $E \int_0^T e^{-\lambda \eta_s - \lambda(1-(d+1)/p)s} d|\phi|_s < \infty$,

where $\eta_t = \int_0^t tr.A_s ds$, $A = \frac{1}{2}\sigma\sigma^*$, and

(1.4) $E \sum_{0<s\leq t} (\Delta x_i(s))^2 < \infty, 1 \leq i \leq d.$

Then for all $0 \leq f$, which is a bounded Borel measurable function defined on $[0,T] \times R^d$, and $p > d+1$

$E \int_0^T (\det A(s,x_s))^{1/(d+1)} f(s,x_s) ds \leq k(p,k_0,d,T) \|f\|_{p,[0,\infty)\times R^d}$,

where k is a constant depending on p, k_0, d and T only.

Proof. Let $f(t,x) = 0$, as $(t,x) \in (T,\infty) \times R^d$. According to Lemma 1.6 for $f \geq 0$ there exists a u^ε satisfying 1)-4) there. Applying Ito's formula to $u^\varepsilon(t,x_t)e^{-\lambda(\eta_t+t)}$ on $[0,T]$ we get that

(1.5) $E\, u^\varepsilon(T,x_T)e^{-\lambda(\eta_T+T)} - Eu^\varepsilon(0,x_0) = E\int_0^T e^{-\lambda(\eta_s+s)} grad_x u^\varepsilon(s,x_s)\cdot$
$\cdot b(s,x_s)ds + E\int_0^T e^{-\lambda(\eta_s+s)}((\partial/\partial s)u^\varepsilon + \sum_{i,j=1}^d a_{ij}(\partial^2 u^\varepsilon/\partial x_i \partial x_j)$
$-\lambda(tr.A_s + 1)u^\varepsilon)ds + E\sum_{0<s\leq T} e^{-\lambda(\eta_s+s)}(u^\varepsilon(s,x_s) - u^\varepsilon(s,x_{s-})$
$-grad_x u^\varepsilon(s,x_{s-}) \cdot \Delta x_s) + R_T = \sum_{i=1}^4 I_T^i$,

where

$R_T = E\int_0^T e^{-\lambda(\eta_s+s)} grad_x u^\varepsilon(s,x_s) \cdot d\phi_s.$

Now we set

$\beta = k(p,\lambda,d,\varepsilon) \left\|e^{-\lambda s(d+1)/p} f(s,y)\right\|_{p,[0,\infty)\times R^d}.$

By 2)

$E\, u^\varepsilon(0,x_0) \leq \beta.$

Note that

$(\int_{R^d}\int_0^\infty \left|e^{-\lambda r(d+1)/p} f(r+s,y)\right|^p drdy)^{1/p}$
$\leq e^{\lambda s(d+1)/p} \left\|e^{-\lambda r(d+1)/p} f(r,y)\right\|_{p,(r,y)\in \overline{R}_+^1 \times R^d}.$

Hence by 3) and 2)

$I_T^1 \leq k_0 \beta \lambda^{\frac{1}{2}} E\int_0^T e^{-\lambda\eta_s - \lambda(1-(d+1)/p)s} ds \leq k_0\beta\lambda^{\frac{1}{2}} T.$

Now by 1) and (1.4) we have

$I_T^3 \leq \frac{1}{2} E\sum_{0<s\leq T}\sum_{i,j=1}^d e^{-\lambda(\eta_s+s)}$
$\cdot(\partial^2 u^\varepsilon(s,x_{s-} + \theta\Delta x_s)/\partial x_i \partial x_j)\Delta x_i(s)\Delta x_j(s)$
$\leq \frac{1}{2}\lambda\beta e^{\lambda T(d+1)/p} E\sum_{0<s\leq T}(\Delta x(s))^2 \leq k'_{d,\lambda,T}\beta.$

Applying (1.3) we have

$R_T \leq \lambda^{\frac{1}{2}} k_{d,\lambda,T}\beta.$

Therefore the conclusion is valid for $f_\varepsilon(s,x_s)$. Let $\varepsilon \to 0$. By Fatou's lemma the conclusion is derived for f. ■

In case that σ is uniformly non-degenerate then we can get a better Krylov estimate as follows:

1.8 Corollary. (Krylov type estimate). Under assumption of Theorem 1.7 if, in addition, there exists a $\delta_0 > 0$ such that for all $\mu \in R^d$

$$\langle A(t,x)\mu,\mu\rangle \geq |\mu|^2 \delta_0,$$

where $\langle \cdot,\cdot \rangle$ is the inner product in R^d, then the conclusion of Theorem 1.7 can be strengthened to

$$E\int_0^T f(s,x_s)ds \leq k(p,k_0,d,\delta,T)\,\|f\|_{p,[0,\infty)\times R^d},$$

where k is a constant depending on p, k_0, d, δ and T only.

In case that σ is locally uniformly non-degenerate then we can only get the following

1.9 Corollary. (Krylov type estimate). Under assumption of Theorem 1.7 suppose that, in addition, for each $N = 1,2,\cdots$there exists a $\delta_N > 0$ such that for all $\mu \in R^d$

$$\langle A(t,x)\mu,\mu\rangle \geq |\mu|^2 \delta_N, \text{ as } |x| \leq N.$$

Let

$$\tau_N = \inf\{t \geq 0 : |x_t| > .N\}.$$

Then the conclusion of Theorem 1.7 can be strengthened to be

$$E\int_0^{\tau_N} f(s,x_s)ds \leq k(p,k_0,d,\delta_N,T)\,\|f\|_{p,[0,\infty)\times[-N,N]},$$

where k is a constant depending on p, k_0, d, δ_N and T only.

The derivation of Corollary 1.8 from Theorem 1.7 is obvious. Let us show Corollary 1.9.

Proof. By Theorem 1.7

$$E\int_0^{\tau_N} (\det A(s,x_s))^{1/(d+1)} f(s,x_s)ds$$
$$\leq E\int_0^T (\det A(s,x_s))^{1/(d+1)} f(s,x_s)I_{(|x_s|\leq N)}ds$$
$$\leq k(p,k_0,d,T)\,\|f\|_{p,[0,\infty)\times[-N,N]}.$$

From this the conclusion is easily derived. ■

Let us quote a result due to Skorohod, A.V., 1965 as follows, which is useful in the discussion of existence of solutions to RSDE:

1.10 Lemma. Suppose that random processes x_t^n $(t \geq 0, n = 1,\cdots)$ satisfy for each $0 \leq T < \infty, \varepsilon > 0$

$$\lim_{N\to\infty} \sup_n \sup_{t\leq T} P(|x_t^n| > N) = 0,$$
$$\lim_{h\downarrow\infty} \sup_n \sup_{\substack{t_1,t_2\leq T\\ |t_1-t_2|\leq h}} P(|x_{t_1}^n - x_{t_2}^n| > \varepsilon) = 0$$

Then there exist a subsequence $\{n_k\}$, denote it by $\{n\}$ again, a probability space, and random processes $\widetilde{x}_t^n$ $(t \geq 0, n = 0,1,\cdots)$ defined on this new probability space such that all finite-dimensional probability distributions of $\widetilde{x}_t^n$ $(t \geq 0, n = 1,\cdots)$ coincide with the corresponding finite-dimensional probability distributions of x_t^n $(t \geq 0, n = 1,\cdots)$, and as $n \to \infty, \forall t \geq 0$,

$$\widetilde{x}_t^n \to \widetilde{x}_t^0, \text{ in probability.}$$

3.2 Moment Estimates and Existence of Solutions for Random Coefficients

3.2.1 Moment Estimates and Uniqueness of Solutions

For the solutions of (1.1) and (1.2) we have the following estimates for their moments.

2.1 Theorem. If (x_t, ϕ_t) is a solution of (1.1) and (1.2) with $E|x_0|^2 < \infty$, and suppose that
(2.1) $|b(t,x,\omega)|^2 + \|\sigma(t,x,\omega)\|^2 + \int_Z |c(t,x,z,\omega)|^2 \pi(dz) \leq k_0(1+|x|^2)$,
(2.2) one of the following conditions is satisfied:
(i) Assumption 2.7 in Chapter 2 holds, and
$E\sum_{s<r\leq t} |\Delta x_r|^2 \leq k_0 E\int_s^t (1+|x_r|^2)dr$;
(ii) ϕ_t is continuous;
(iii) Assumption 2.7 in Chapter 2 holds, and
$|b(t,x,\omega)|^2 + \|\sigma(t,x,\omega)\|^2 + \int_Z |c(t,x,z,\omega)|^2 \pi(dz) \leq k_0$;
then for any $0 \leq T < \infty$
$E\sup_{0\leq t\leq T} |x_t|^2 + E\sup_{0\leq t\leq T} |\phi_t|^2 \leq k_T$,
$E\sup_{s\leq r\leq t} |x_r - x_s|^2 + E\sup_{s\leq r\leq t} |\phi_r - \phi_s|^2 \leq k_T |t-s|$, as $0 \leq s \leq t \leq T$,
where $0 \leq k_T$ is a constant depending on k_0, c_0 (appeared in Assumption 2.7 in Chapter 2), and T only.
Moreover, in case that (i) of (2.2) is satisfied then we also have
$E|\phi|_T^2 \leq k_T < \infty$, $E(|\phi|_t - |\phi|_s)^2 \leq k_T |t-s|$.
Furthermore, in the case that (iii) of (2.2) is satisfied, we will have
$E|\phi|_T^2 \leq k_T < \infty$, $E(|\phi|_t - |\phi|_s)^2 \leq k_0' |t-s|$,
where $0 \leq k_0'$ is a constant depending on k_0 and c_0 only.

2.2 Remark. By Lemma 2.2 in Chapter 2 for all $0 \leq s \leq t$
$\int_s^t (x_r - x_s) \cdot d\phi_r \leq 0.$
Let us show Theorem 2.1.
Proof. By Ito's formula
$|x_t - x_s|^2 = 2\int_s^t (x_r - x_s) \cdot b(r,x_r,\omega)dr + 2\int_s^t (x_r - x_s) \cdot \sigma(r,x_r,\omega)dw_r$
$+ \int_s^t \|\sigma(r,x_r,\omega)\|^2 dr + 2\int_s^t (x_r - x_s) \cdot \int_Z c(r,x_{r-},z,\omega)q(dr,dz)$
$+2\int_s^t (x_{r-} - x_s) \cdot d\phi_r + \sum_{s<r\leq t} |\Delta x_r|^2$. (A)
For the condition (i) of (2.2) by Remark 2.2
$2\int_s^t (x_{r-} - x_s) \cdot d\phi_r \leq -2\sum_{s<r\leq t} \Delta x_r \cdot \Delta\phi_r.$
By Lemma 2.8 in Chapter 2
$(|\phi|_t - |\phi|_s) \leq k_0' \sup_{s<r\leq t} |y_r - y_s|$, (B)
where
$y_t = x_0 + \int_0^t b(r,x_r,\omega)dr + \int_0^t \sigma(r,x_r,\omega)dw_r + 2\int_0^t \int_Z c(r,x_{r-},z,\omega)q(dr,dz).$
Hence
$2\int_s^t (x_{r-} - x_s) \cdot d\phi_r \leq \sum_{s<r\leq t} |\Delta x_r|^2 + k_0' \sup_{s<r\leq t} |y_r - y_s|$. (C)

Let $s=0$. By the martingale inequality, Gronwall's inequality, and Fatou's lemma (introduce stopping times $\tau_N = \inf\{t \in [0,T] : |x_t| > N\}, N = 1,2,...,$ if necessary), it is easily seen that

$E \sup_{0\leq t\leq T} |x_t - x_0|^2 \leq k_T.$

Hence

$E \sup_{0\leq t\leq T} |x_t|^2 \leq k_T.$

By (B) one also finds that

$E |\phi|_T^2 \leq k'_T.$

Again by (A)-(C) one finds that as $0 \leq s \leq t \leq T$

$E \sup_{s\leq r\leq t} |x_r - x_s|^2 \leq k''_T |t-s|.$

Applying (B) again one also has that

$E \left||\phi|_t - |\phi|_s\right|^2 \leq k'''_T |t-s|$, as $0 \leq s \leq t \leq T.$

For the condition (ii) of (2.2) the proof is even easier. In fact, in this case the 5th term in the right side of (A) ≤ 0.

Moreover, one has that

$$|\phi_t - \phi_s|^2 \leq 4(|x_t - x_s|^2 + \left|\int_s^t b(r,x_r,\omega)dr\right|^2 + \left|\int_s^t \sigma(r,x_r,\omega)dw_r\right|^2$$
$$+ \left|\int_s^t \int_Z c(r,x_{r-},z,\omega)q(dr,dz)\right|^2.$$

Therefore all conclusions can be derived similarly or easily.

For the condition (iii) of (2.2) one still has (B). Hence the estimation for ϕ can be immediately obtained from (B) as above. Note that

$|x_t|^2 \leq 2(|y_t|^2 + |\phi_t|^2).$

Hence it also follows that

$E \sup_{s\leq t} |x_s|^2 \leq \widetilde{k}_T$, as $t \leq T$.

The second conclusion of Theorem 2.1 is also derived from (B) and SDE (1.1). Moreover, in this case

$E \sup_{s\leq r\leq t} |x_r - x_s|^2 + E \left||\phi|_t - |\phi|_s\right|^2 \leq k'_0 |t-s|$, as $0 \leq s \leq t$,

where $0 \leq k'_0$ is a constant depending on k_0 and c_0 (introduced in Assumption 2.7 in Chapter 2) only. ■

For the uniqueness of solutions to (1.1) and (1.2)we have the following

2.3 Theorem. Assume that $(x_t^i, \phi_t^i), i = 1,2$, satisfy (1.1) and (1.2) with the same BM and the same Poisson martingale measure on the same probability space, and assume that (2.1) - (2.2) hold. Moreover, assume that

$$(2.3)\ E[2\int_0^{t\wedge\tau_N}(x_s^1 - x_s^2)\cdot(b(s,x_s^1,\omega) - b(s,x_s^2,\omega))ds$$
$$+ \int_0^{t\wedge\tau_N} \|\sigma(s,x_s^1,\omega) - \sigma(s,x_s^1,\omega)\|^2 ds + \sum_{0<s\leq t\wedge\tau_N} |\Delta x_s^1 - \Delta x_s^2|^2]$$
$$\leq k_0 E \int_0^{t\wedge\tau_N} k_N(s)\rho_N(|x_s^1 - x_s^2|^2)ds,$$

where

$\tau_N = \inf\{t \geq 0 : |x_t^1| + |x_t^2| > N\}, N = 1,2,...,$

$0 \leq k_N(t), \int_0^T k_N(s)ds < \infty$, for any $0 \leq T < \infty$, and $\rho_N(u)$ is concave and strictly increasing in $u \geq 0$ such that $\rho_N(0) = 0$, and $\int_{0+} du/\rho_N(u) = \infty$, for $N = 1,2,...,$

(2.4) $E\int_0^t (x_{s-}^1 - x_{s-}^2)\cdot d(\phi_s^1 - \phi_s^2) \leq 0.$
Then

$$P(\sup_{t\geq 0}\left|x_t^1 - x_t^2\right| + \sup_{t\geq 0}\left|\phi_t^1 - \phi_t^2\right| = 0) = 1.$$

2.4 Remark. In case ϕ_t^i, $i = 1, 2$, are continuous, then by Remark 2.2 (2.4) is satisfied, and (2.3) can be reduced to

(2.3)' $E[2\int_0^{t\wedge\tau_N}(x_s^1 - x_s^2)\cdot(b(s, x_s^1, \omega) - b(s, x_s^2, \omega))ds$

$$+\int_0^{t\wedge\tau_N}\left\|\sigma(s, x_s^1, \omega) - \sigma(s, x_s^1, \omega)\right\|^2 ds$$

$$+\int_0^{t\wedge\tau_N}\int_Z\left|c(s, x_s^1, z, \omega) - c(r, x_s^2, z, \omega)\right|^2\pi(dz)ds]$$

$$\leq k_0 E\int_0^{t\wedge\tau_N} k_N(s)\rho_N(\left|x_s^1 - x_s^2\right|^2)ds.$$

Let us show Theorem 2.3.

Proof. By Ito's formula and by assumption we get

$$E\left|x_{t\wedge\tau_N}^1 - x_{t\wedge\tau_N}^2\right|^2 \leq K_0 E\int_0^{t\wedge\tau_N} k_N(s)\rho_N(\left|x_s^1 - x_s^2\right|^2)ds.$$

Hence

$$P(\sup_{t\geq 0}\left|x_t^1 - x_t^2\right| = 0) = 1.$$

From this and (1.1) it is also seen that $P-a.s.$

$$\phi_t^1 = \phi_t^2, \text{ for all } t \geq 0.$$

Hence the conclusion follows. ■

The condition of Theorem 2.3 can be given more precisely, if we note the following

2.5 Remark. 1) Assume that (x_t, ϕ_t) is a solution of (1.1) and (1.2), then for each $t > 0$

$$\triangle\phi_t = 0 \Rightarrow x_{t-} + \int_Z c(t, x_{t-}, z, \omega)p(\{t\}, dz) \in \overline{\Theta}\ .$$

Indeed,

$$\triangle x_t = \int_Z c(t, x_{t-}, z, \omega)p(\{t\}, dz) + \triangle\phi_t.$$

Hence by $x_t \in \overline{\Theta}$ the conclusion is derived.

2) For any solution (x_t, ϕ_t) of (1.1) and (1.2) if

$$x_{t-} + c(t, x_{t-}, z, \omega) \in \overline{\Theta}, \text{ for all } z, \omega,$$

then we have that

$$\triangle\phi_t = 0\ .$$

In fact, denote by y_t the expression under the above (B), then by assumption

$$x_{t-} + \triangle y_t \in \overline{\Theta}.$$

However,

$$x_t - (x_{t-} + \triangle y_t) = \triangle\phi_t.$$

Hence by 4)' of Lemma 2.2 in Chapter 2 one has

$$\left|\triangle\phi_t\right|^2 = \triangle\phi_t\cdot[x_t - (x_{t-} + \triangle y_t)] \leq 0.$$

The proof is complete.

Because of Remark 2.5 it is natural to give the following

2.6 Theorem. For RSDE (1.1) with (1.2) and with
(2.5) ϕ_t is a continuous process,
if (2.1) holds and
(2.6) $x + c(t, x, z, \omega) \in \overline{\Theta}$, for all $t \geq 0, z \in Z, \omega \in \Omega$ and $x \in \overline{\Theta}$;

(2.7) $2(x-y)\cdot(b(s,x,\omega)-b(s,y,\omega))+\|\sigma(s,x,\omega)-\sigma(s,y,\omega)\|^2$
$+\int_Z |c(s,x,z,\omega)-c(s,y,z,\omega)|^2\,\pi(dz) \le k_N(s)\rho_N(|x-y|^2),$
as $|x|,|y| \le N$, where $0 \le k_N(t), \int_0^T k_N(s)ds < \infty$, for any $0 \le T < \infty$, and $\rho_N(u)$ is concave and strictly increasing in $u \ge 0$ such that $\rho_N(0)=0$, and $\int_{0+} du/\rho_N(u) = \infty$, for $N = 1,2,\cdots,$
then the pathwise uniqueness for solutions of (1.1) with (1.2) and (2.5) holds.

Obviously, assumption of Theorem 2.6 implies conditions (2.1)-(2.4). Hence the conclusion follows directly from Theorem 2.3.

Furthermore, for RSDE (1.1) with (1.2) and (2.5) Theorem 2.1 can also be stated as follows:

2.7 Corollary. 1) For RSDE (1.1) with (1.2) and (2.5) if (2.1) is satisfied, then for any $0 \le T < \infty$
$E\sup_{0\le t\le T}|x_t|^2 + E\sup_{0\le t\le T}|\phi_t|^2 \le k_T,$
$E\sup_{s\le r\le t}|x_r - x_s|^2 + E\sup_{s\le r\le t}|\phi_r-\phi_s|^2 \le k_T|t-s|$, as $0 \le s \le t \le T$,
where $0 \le k_T$ is a constant depending on k_0, c_0 (introduced in Assumption 2.7 in Chapter 2), and T only.
2) For RSDE (1.1) with (1.2), if (2.1),(2.6) and (2.7) are satisfied, then the conclusion in 1) is true, and ϕ_t is a continuous process, i.e. it satisfies (2.5).

For the proof of 2) one only needs to use the fact that (2.6)$\Longrightarrow$ (2.5) by 2) of Remark 2.5.

3.2.2 Existence of Solutions for Continuous RSDE with Random Lipschitzian Coefficients

In this subsection we will discuss the existence of solution for continuous RSDE with random coefficients.

Consider RSDE (1.1) with (1.2) with the following assumption:
(2.8) (x_t, ϕ_t) are continuous processes.

Then we have the following

2.8 Theorem. Assume that for all $t \ge 0$ $\omega \in \Omega, z \in Z, x \in \overline{\Theta}$
1° $|b(t,x,\omega)|^2 + \|\sigma(t,x,\omega)\|^2 \le k_0(1+|x|^2), c(t,x,z,\omega)=0,$
where $k_0 \ge 0$ is a constant, moreover, $b(t,x,\omega)$ and $\sigma(t,x,\omega)$ are $\Im_t^w$−adapted, as $x \in \overline{\Theta}$ is fixed;
2° $|b(t,x,\omega)-b(t,y,\omega)|^2 + \|\sigma(t,x,\omega)-\sigma(t,y,\omega)\|^2 \le k_N|x-y|^2,$
as $|x|,|y| \le N, N = 1,2,...$
where N is a constant depending on N only.
Then (1.1) with (1.2) and (2.8) has a pathwise unique strong solution.
Proof. Case 1. Assume that $k_N = k_0$ in 2° is independent of N, and $|b|^2 + \|\sigma\|^2 \le k_0$. By Theorem 2.12 of Chapter 2 there exists a unique (x_t^n, ϕ_t^n) satisfying for each $n = 1,2,...$
$x_t^n = x_0 + \int_0^t b(s,x_s^{n-1},\omega)ds + \int_0^t \sigma(s,x_s^{n-1},\omega)dw_s + \phi_t^n = y_t^n + \phi_t^n,$
$x_0^n = x_0,$

and the statement in (1.2) for (x_t^n, ϕ_t^n) holds. Moreover, (x_t^n, ϕ_t^n) are continuous, since y_t^n is also continuous. Hence, by Ito's formula

$$\left|x_t^n - x_t^{n-1}\right|^2 \leq 2\int_0^t (x_s^n - x_s^{n-1}) \cdot (b(s, x_s^{n-1}, \omega) - b(s, x_s^{n-2}, \omega))ds$$
$$+2\int_0^t (x_s^n - x_s^{n-1}) \cdot (\sigma(s, x_s^{n-1}, \omega) - \sigma(s, x_s^{n-2}, \omega))dw_s$$
$$+\int_0^t \left\|\sigma(s, x_s^{n-1}, \omega) - \sigma(s, x_s^{n-2}, \omega)\right\| ds.$$

It follows by the martingale inequality that as $0 \leq t \leq T$

$$I_t^n = E\sup_{s\leq t}\left|x_t^n - x_t^{n-1}\right|^2 \leq k_T\int_0^t I_s^{n-1}ds \leq k_T(k_Tt)^{n-1}/(n-1)!,$$
$$I_t'^n = E\sup_{s\leq t}\left|\phi_t^n - \phi_t^{n-1}\right|^2 \leq k_T'(k_T't)^{n-1}/(n-1)!.$$

Therefore,

$$\sum_{n=1}^\infty P(\sup_{s\leq t}\left|x_s^n - x_s^{n-1}\right|^2 > 1/n^2) + P(\sup_{s\leq t}\left|\phi_s^n - \phi_s^{n-1}\right|^2 > 1/n^2)$$
$$< \infty.$$

By the Borel-Cantelli lemma there exist (x_t, ϕ_t) such that $P-a.s.$

$x_t = \lim_{n\to\infty} x_t^n$, $\phi_t = \lim_{n\to\infty}\phi_t^n$, uniformly in $t \in [0,T]$, for any given $T < \infty$.

Since by Theorem 2.1

$$E\sup_{s\leq t}\left|x_t^n\right|^2 + E\sup_{s\leq t}\left|\phi_t^n\right|^2 \leq k_T, \forall n. \ (@)_1$$

Hence by Fatou's lemma

$$E\sup_{s\leq t}\left|x_t\right|^2 + E\sup_{s\leq t}\left|\phi_t\right|^2 \leq k_T.$$

On the other hand, by Theorem 2.12 of Chapter 2 there exists a unique $(\widetilde{x}_t, \widetilde{\phi}_t)$ satisfying

$$\widetilde{x}_t = x_0 + \int_0^t b(s, x_s, \omega)ds + \int_0^t \sigma(s, x_s, \omega)dw_s + \widetilde{\phi}_t = y_t + \widetilde{\phi}_t,$$

and the statement in (1.2) for $(\widetilde{x}_t, \widetilde{\phi}_t)$ holds. Moreover, $(\widetilde{x}_t, \widetilde{\phi}_t)$ are continuous. Now let us show that there exists at least a subsequence $\{n_k\}$ such that as $k \to \infty$, then $n_k \to \infty$ and

$$y_t^{n_k} \to y_t, \text{ uniformly in } t \in [0,T], \ P-a.s. \ (@)_2$$

In fact, as $n \to \infty$

$$\sup_{r\leq T}\int_0^r \left|b(s, x_s^{n-1}, \omega) - b(s, x_s, \omega)\right| ds \leq k_0T\sup_{s\leq t}\left|x_s^n - x_s\right| \to 0,$$
$$E\sup_{r\leq T}\left|\int_0^r (\sigma(s, x_s^{n-1}, \omega) - \sigma(s, x_s, \omega))dw_s\right| \leq k_TE\sup_{s\leq t}\left|x_s^n - x_s\right| \to 0,$$

where we have applied the uniform integrability of $\sup_{s\leq t}\left|x_s^n - x_s\right|$, which is derived by $(@)_1$. Hence $(@)_2$ is true. Applying Corollary 2.10 in Chapter 2 there exists at least a subsequence $\{n_{k_l}\}$selected from $(@)_2$ such that as $l \to \infty$

$$\sup_{s\leq t}\left|x_s^{n_{k_l}} - \widetilde{x}_s\right| + \sup_{s\leq t}\left|\phi_s^{n_{k_l}} - \widetilde{\phi}_s\right| \to 0$$

By the uniqueness of a limit for $\forall t \in [0,T]$

$$\widetilde{x}_t = x_t, \ \widetilde{\phi}_t = \phi_t.$$

Therefore (x_t, ϕ_t) satisfies (1.1) with (1.2) and (2.8). By Theorem 2.6 the pathwise uniqueness of solutions to (1.1) with (1.2) and (2.8) holds.

Case 2. General case. Let

$$b^N(s, x, \omega) = \begin{cases} b(s, x, \omega), & \text{as } |x| \leq N, \\ b(s, Nx/|x|, \omega), & \text{as } |x| > N, \end{cases}$$

and σ^N is similarly defined.

Then by case 1 there exists a pathwise unique strong solution (x_t^N, ϕ_t^N) satisfying for each $N = 1, 2, ...$

$x_t^N = x_0 + \int_0^t b^N(s, x_s^N, \omega)ds + \int_0^t \sigma^N(s, x_s^N, \omega)dw_s + \phi_t^N$,

and the statement in (1.2) for (x_t^N, ϕ_t^N) holds. Moreover, (x_t^N, ϕ_t^N) are continuous. Set now

$\tau_N = \inf\{t \geq 0 : |x_t^N| > N\}$,

$x_t = x_t^N$, as $0 \leq t \leq \tau_N$,

$\phi_t = \phi_t^N$, as $0 \leq t \leq \tau_N$.

By pathwise uniqueness it is easily seen that $\tau_N \leq \tau_{N+1}, P-a.s.$ Otherwise, if $P(\tau_N > \tau_{N+1}) > 0$, then by pathwise uniqueness as $\omega \in \{\tau_N > \tau_{N+1}\}$ one has that $N \geq x^N_{\tau_{N+1}(\omega)} = x^{N+1}_{\tau_{N+1}(\omega)} = N+1$. This is a contradiction. Hence, the above definition is well posed, and

$x_{t\wedge\tau_N} = x_0 + \int_0^{t\wedge\tau_N} b(s, x_s, \omega)ds + \int_0^{t\wedge\tau_N} \sigma(s, x_s, \omega)dw_s + \phi_{t\wedge\tau_N}, \forall t \geq 0.$

Now we show that $P-a.s.$

$\lim_{N\to\infty} \tau_N = \infty.$

In fact, for any $T < \infty$

$N^2 P(\tau_N \leq T) \leq E\left|x^N_{T\wedge\tau_N}\right|^2 \leq k_T, \forall T \geq 0$ and $\forall N = 1, 2, ...$

It means that as $N \to \infty$

$P(\tau_N > T) \to 1, \forall T < \infty.$

Take now $T_k \uparrow \infty$. For T_k there exists Λ_k such that $P(\Lambda_k) = 0$, and as $\omega \notin \Lambda_k$

$\lim_{N\to\infty} \tau_N > T_k.$

Denote

$\Lambda = \cup_{k=1}^{\infty} \Lambda_k.$

Then as $\omega \notin \Lambda$

$\lim_{N\to\infty} \tau_N = \infty.$

Now the proof is easily completed. ∎

3.2.3 Existence of Solutions for Continuous RSDE with Random Continuous Coefficients

In this subsection we will discuss the existence of solutions to continuous RSDE (1.1) (with (1.2) and (2.8)),with continuous random non-Lipschitzian coefficients. Moreover, the diffusion coefficients can be degenerate.

2.9 Theorem. Consider RSDE (1.1) (with (1.2) and (2.8)). Assume that for all $t \geq 0, \omega \in \Omega$

1° $|b(t, x, \omega)|^2 + \|\sigma(t, x, \omega)\|^2 \leq k_0(1 + |x|^2)$, for all $x \in \overline{\Theta}$,

where k_0 is a constant, moreover, $b(t, x, \omega)$, $\sigma(t, x, \omega)$ are $\Im_t^w$-adapted, as x is fixed;

2° $b(t, x, \omega)$ and $\sigma(t, x, \omega)$ are continuous in x, and $c(t, x, z, \omega) = 0$;

3° (2.7) holds.

Then (1.1) (with (1.2) and (2.8)) has a pathwise unique strong solution.

Proof. Case 1. Assume that $k_N(t)$ and $\rho_N(u)$ in 3° are independent of N, i.e. $k_N(t) = k(t)$, $\rho_N(u) = \rho(u)$, for all $N = 1, 2, ...$, and assume that

$|b|^2 + \|\sigma\|^2 \leq k_0.$

Let for $z \in R^d$

$$f(z) = c\exp[(1-|z|^2)^{-1}], \text{ as } |z|<1;\ f(z)=0, \text{ otherwise};$$

where c is a constant such that

$$\int_{R^d} f(z)dz = 1.$$

For $n = 1, 2, ..., (t, \omega) \in [0, \infty) \times \Omega$ set

$$b^n(t,x,\omega) = \int_{R^d} b(t, x - n^{-1}y, \omega) f(y) dy.$$

$\sigma^n(t,x,\omega)$ is defined similarly. Then by Theorem 2.8 there exists a pathwise unique strong solution (x_t^n, ϕ_t^n) satisfying

$$\begin{cases} x_t^n = x_0 + \int_0^t b^n(s, x_s^n, \omega)ds + \int_0^t \sigma^n(s, x_s^n, \omega)dw_s + \phi_t^n, \\ \quad \text{and the other statements in (1.2) hold for } (x_t^n, \phi_t^n). \end{cases}$$

By Ito's formula one verifies that

$$E|x_t^n - x_t^m|^2 \leq \int_0^t \int_{R^d} k(s)\rho(E|x_s^n - x_s^m - n^{-1}y + m^{-1}y|^2) f(y)dyds$$
$$+\bar{k}_T(n^{-1} + m^{-1}).$$

Since $\rho(u)$ is increasing and by Theorem 2.1 as $t \leq T$

$$E(\sup_{s\leq t} |x_t^n|^2 + |\phi_t^n|^2) \leq k_T < \infty, \text{ for all } n = 1, 2, ...(@)$$

Hence by Fatou's' lemma

$$\overline{\lim}_{n,m\to\infty} E|x_t^n - x_t^m|^2 \leq \int_0^t \int_{R^d} k(s)\rho(\overline{\lim}_{n,m\to\infty} E|x_s^n - x_s^m|^2) f(y)dyds$$
$$= \int_0^t k(s)\rho(\overline{\lim}_{n,m\to\infty} E|x_s^n - x_s^m|^2)ds.$$

Therefore

$$\overline{\lim}_{n,m\to\infty} E|x_t^n - x_t^m|^2 = 0.$$

Again by (@) and by Lebesgue's dominated convergence theorem one sees that for any given $0 < T < \infty$ as $m, n \to \infty$

$$E\int_0^T |x_t^n - x_t^m|^2 dt = \int_0^T E|x_t^n - x_t^m|^2 dt \to 0.$$

There exists a $x_t(\omega) \in L^2([0,T] \times \Omega, \Re([0,T]) \times \Im, dt \times dP)$, such that

$$\lim_{n\to\infty} E\int_0^T |x_t^n - x_t|^2 dt = 0.$$

The measurability and $\Im_t$-adaptness of x_t follow from the limit. Now take a subsequence $\{n_k\}$ of $\{n\}$, denote it by $\{n\}$ again such that as $n \to \infty$

$$x_t^n(\omega) \to x_t(\omega), \text{ in } R^d,\ dt \times dP - a.e.\ (t,\omega) \in [0,T] \times \Omega.$$

Then by the continuity assumption 2° applying Lebesgue's dominated convergence theorem one obtains

$$\overline{\lim}_{n\to\infty} \int_0^t |b^n(s, x_s^n, \omega) - b(s, x_s, \omega)|^2 ds$$
$$\leq \overline{\lim}_{n\to\infty} 2\int_0^t |b^n(s, x_s^n, \omega) - b^n(s, x_s, \omega)|^2 ds$$
$$\leq \overline{\lim}_{n\to\infty} 2\int_0^t \int_{R^d} \left|b(s, x_s^n - n^{-1}y, \omega) - b(s, x_s - n^{-1}y, \omega)\right|^2 f(y)dyds = 0.$$

Similarly,

$$\lim_{n\to\infty} E\left|\int_0^t (\sigma^n(s, x_s^n, \omega) - \sigma(s, x_s, \omega))dw_s\right|^2$$
$$= \lim_{n\to\infty} E\int_0^t |\sigma^n(s, x_s^n, \omega) - \sigma(s, x_s, \omega)|^2 ds = 0.$$

Denote

$$\phi_t = x_t - (x_0 + \int_0^t b(s, x_s, \omega)ds + \int_0^t \sigma(s, x_s, \omega)dw_s).$$

Then as $n \to \infty$

$$E|\phi_t^n - \phi_t| \to 0.$$

Now denote

$$y_t = x_0 + \int_0^t b(s, x_s, \omega)ds + \int_0^t \sigma(s, x_s, \omega)dw_s.$$

Similarly, discussing as the proof of Theorem 2.8 one easily derives that (x_t, ϕ_t) satisfies (1.1) with (1.2) and (2.8) on $[0, T]$. By Theorem 2.6 the pathwise uniqueness of solution to (1.1) with (1.2) and (2.8) holds. Since $0 < T < \infty$ is arbitrarily given, hence we get a pathwise unique strong solution (x_t, ϕ_t) satisfying (1.1) with (1.2) and (2.8) on $t \geq 0$.

General case. The general case can be reduced to the special case 1 by the following Lemma 2.10 (cf. the proof of general case for Theorem 2.8). ■

2.10 Lemma. If b and σ satisfy conditions 1° and 3° in Theorem 2.9, then there exist $b^N, \sigma^N, k^N(t)$, and $\rho^N(u)$ such that

$$f^N(t,x,\omega) = \begin{cases} f(t,x,\omega), & \text{as } |x| \leq N, \\ 0, & \text{as } |x| > N+3, \end{cases}$$

where $f = b, \sigma$; and

$$|b^N|^2 \leq (|b|^2 \wedge c_{N+3}),\ \|\sigma^N\|^2 \leq (dc_N \wedge \|\sigma\|^2),$$

where $c_N = k_0(1 + N^2)$, moreover,

$$(2.9)\ 2(x-y)\cdot(b^N(t,x,\omega) - b^N(t,y,\omega)) + \|\sigma^N(t,x,\omega) - \sigma^N(t,x,\omega)\|^2$$
$$\leq k^N(t)\rho^N(|x-y|^2), \text{ for all } x, y \in R^d;$$

where $0 \leq k^N(t)$, $\int_0^t k^N(s)ds < \infty$, and $\rho^N(u)$ is concave and strictly increasing in $u \geq 0$ such that $\rho^N(0) = 0$, and $\int_{0+} du/\rho^N(u) = \infty$, for each $N = 1, 2, ...$ Furthermore, if b and σ are continuous in x, then one can also choose b^N and σ^N to be continuous in x.

To show Lemma 2.10 we need some preparation. Denote

$$k_{ij}^N(t,\omega) = \sup_{|x| \leq N} |\sigma_{ij}(t,x,\omega)|, i, j = 1, 2, ..., d.$$

Let

$$h_{ij}^N(t,x,\omega) = \begin{cases} k_{ij}^N(t,\omega), & \text{as } |x| \leq N, \\ k_{ij}^N(t,\omega)[(N+1) - |x|], & \text{as } N < |x| \leq N+1, \\ 0, & \text{as } N+1 < |x|. \end{cases}$$

Set

$$\sigma_{ij}^N = Min(h_{ij}^N, (\sigma_{ij})_+) - Min(h_{ij}^N, (\sigma_{ij})_-),$$

where

$$(\sigma_{ij})_+ = Max(\sigma_{ij}, 0),\ (\sigma_{ij})_- = Max(-\sigma_{ij}, 0).$$

We have the following

2.11 Lemma. $\sigma_{ij}^N = \begin{cases} \sigma_{ij}, & \text{as } |x| \leq N, \\ 0, & \text{as } |x| \geq N+1, \end{cases}$

$$|\sigma^N|^2 \leq \text{Min}(|\sigma|^2, k_0(1+N^2)),$$

$$(2.10)\ |\sigma_{ij}^N(t,x,\omega) - \sigma_{ij}^N(t,y,\omega)|^2 \leq |\sigma_{ij}(t,x,\omega) - \sigma_{ij}(t,y,\omega)|^2$$
$$+k_0(1+N^2)|x-y|^2.$$

Moreover, if σ_{ij} is continuous in x and $\Im_t$-adapted, then so is σ_{ij}^N.

Proof. We only need to show (2.10). By inequality

$$|Min(a,b) - Min(c,d)| \leq Max(|a-c|, |b-d|), \text{ for all } a, b, c, d \in R^d,$$

one has

$$\begin{aligned}
&\left| \sigma_{ij}^N(t,x,\omega) - \sigma_{ij}^N(t,y,\omega)\right| \\
&\leq Max(\left| h_{ij}^N(t,x,\omega) - h_{ij}^N(t,y,\omega)\right|, \left| \sigma_{ij}(t,x,\omega) - \sigma_{ij}(t,y,\omega)\right|) \\
&\leq Max(c_N^{1/2} \left| x-y\right|, \left| \sigma_{ij}(t,x,\omega) - \sigma_{ij}(t,y,\omega)\right|),
\end{aligned}$$

in case that

$$\sigma_{ij}(t,x,\omega) \geq 0, \text{ and } \sigma_{ij}(t,y,\omega) \geq 0.$$

Hence (2.10) follows in this case. For

$$\sigma_{ij}(t,x,\omega) \leq 0, \text{ and } \sigma_{ij}(t,y,\omega) \leq 0$$

the proof is similar. Now assume that

$$\sigma_{ij}(t,x,\omega) \geq 0, \text{ and } \sigma_{ij}(t,y,\omega) \leq 0.$$

Then

$$\begin{aligned}
&\left| \sigma_{ij}^N(t,x,\omega) - \sigma_{ij}^N(t,y,\omega)\right| \leq (\sigma_{ij})_+(t,x,\omega) + (\sigma_{ij})_-(t,y,\omega) \\
&\leq \left| \sigma_{ij}(t,x,\omega) - \sigma_{ij}(t,y,\omega)\right|.
\end{aligned}$$

Hence (2.10) is proved. ■

Now we are in a position to show Lemma 2.10.

Proof. Let $w^N(x) \in c_0^\infty(R^d)$ such that

$$w^N(x) = \begin{cases} 1, \text{as } |x| \leq N+2, \\ 0, \text{ as } |x| \geq N+3, \end{cases}$$

and $0 \leq w^N(x) \leq 1$. Set

$$b^N(t,x,\omega) = b(t,x,\omega)w^N(x).$$

Then

$$b^N(t,x,\omega) = \begin{cases} b(t,x,\omega), \text{as } |x| \leq N+2, \\ 0, \text{ as } |x| \geq N+3. \end{cases}$$

Let us take σ_{ij}^N from Lemma 2.11.

Case 1. Assume that $|x|, |y| \leq N+2$. Then

$$w^N(x) = w^N(y) = 1.$$

By (2.10) and condition 3° of Theorem 2.9 one has that

$$\begin{aligned}
&\text{The L.H.S. of (2.9)} \leq 2(x-y)\cdot(b(t,x,\omega) - b(t,y,\omega)) \\
&+ \left\|\sigma(t,x,\omega) - \sigma(t,x,\omega)\right\|^2 + dk_0(1+|N|^2)\left|x-y\right|^2 \\
&\leq k_{N+2}(t)\rho_{N+2}(|x-y|^2) + dk_0(1+|N|^2)\left|x-y\right|^2.
\end{aligned}$$

Case 2. Assume that $|x|, |y| \geq N+1$. Then

$$\sigma^N(t,x,\omega) = \sigma^N(t,y,\omega) = 0.$$

Hence

$$\text{The L.H.S. of (2.9)} = 2(x-y)\cdot(b^N(t,x,\omega) - b^N(t,y,\omega)).$$

(i) If $|x| > N+3, N+1 \leq |y| \leq N+3$, then $w^N(x) = 0$. Hence

$$\begin{aligned}
&\text{The L.H.S. of (2.9)} = -2(x-y)\cdot b(t,y,\omega)w^N(y) \\
&\leq 2c_{N+3}^{1/2}\left|x-y\right|\left|w^N(x) - w^N(y)\right| \leq c_{N+3}^{1/2}\tilde{k}_N\left|x-y\right|^2.
\end{aligned}$$

(ii) If $|x|, |y| > N+3$, then

$$\text{The L.H.S. of (2.9)} = 0.$$

(iii) If $N+1 \leq |x|, |y| \leq N+3$, then

$$\begin{aligned}
&\text{The L.H.S. of (2.9)} \leq 2(x-y)\cdot(b(t,x,\omega) - b(t,y,\omega))w^N(x) \\
&+ |b(t,y,\omega)|\left|w^N(x) - w^N(y)\right| 2\left|x-y\right| \\
&\leq w^N(x)k_{N+3}(t)\rho_{N+3}(|x-y|^2) + 2c_{N+3}^{1/2}\widetilde{k}_N\left|x-y\right|^2
\end{aligned}$$

Case 3. Assume that $|x| < N+1, |y| > N+2$. Then

$$|x-y| > 1, \sigma^N(t,y,\omega) = 0,$$

and

$$\left\|\sigma^N(t,x,\omega)\right\|^2 \leq c_{N+1} \leq c_{N+1}|x-y|^2.$$

(i) If $|y| > N+3, |x| < N+1$, then

$$w^N(y) = 0.$$

Hence

$$2(x-y)\cdot b(t,x,\omega)w^N(x) \leq 2c_{N+3}^{1/2}\widetilde{k}_N |x-y|^2.$$

(ii) If $|y| \leq N+3, |x| < N+1$, then

$$2(x-y)\cdot(b(t,x,\omega)-b(t,y,\omega))w^N(x)$$
$$+|b(t,y,\omega)|\left|w^N(x)-w^N(y)\right| 2|x-y|$$
$$\leq w^N(x)k_{N+3}(t)\rho_{N+3}(|x-y|^2) + 2c_{N+3}^{1/2}\widetilde{k}_N|x-y|^2.$$

Therefore, in any case one has

The L.H.S. of (2.9)$\leq \tilde{k}_N(t)\tilde{\rho}_N(|x-y|^2)$, for all $x,y\in R^d$,

where

$$\tilde{k}_N(t) = \widetilde{k}_N k_{N+3}(t) + k_{N+2}(t) + 2c_{N+3}^{1/2}\widetilde{k}_N + c_{N+1} + d\cdot c_N,$$
$$\tilde{\rho}_N(u) = \tilde{\rho}_{N+2}(u) + \tilde{\rho}_{N+3}(u) + u;$$

and $\widetilde{k}_N \geq 0$ is the Lipschitzian constant for the Lipschitzian continuous function w^N, i.e.,

$$\left|w^N(x)-w^N(y)\right| \leq \widetilde{k}_N|x-y|, \forall x,y\in R^d.$$

The proof of Lemma 2.10 is now complete. ■

3.3 Existence of Solutions for RSDE with Jumps

3.3.1 Existence of Weak and Strong Solutions for RSDE with Continuous Coefficients

For RSDE (1.1) with (1.2) and with jumps we discuss the case for non-random coefficients in this subsection 3.3.1 only. However, for simplicity in the whole section 3.3 we make the following assumption: Assume now that Θ satisfies Assumption 2.7 in Chapter 2, i.e., there exists a constant $c_0 > 0$ and a vector $e\in R^d, |e| = 1$ such that

$$e\cdot n \geq c_0 > 0, \text{ for all } n \in \cup_{x\in\partial\Theta}\aleph_x.$$

3.1 Theorem. Assume that for $t\geq 0, x\in\overline{\Theta}, z\in Z$

1° $b(t,x),\sigma(t,x),c(t,x,z)$ are jointly measurable, and there exists a constant $k_0>0$ such that

$$|b|^2 + \|\sigma\|^2 + \int_Z |c|^2\pi(dz) \leq k_0,$$

2° $b(t,x),\sigma(t,x)$ are jointly continuous, and as $|x-y|\to 0, |t-s|\to 0$

$$\int_Z |c(t,x,z)-c(s,y,z)|^2\pi(dz)\to 0,$$

3° $\pi(dz) = dz/|z|^{d+1}$, $(Z = R^d - \{0\})$.

Then (1.1) with (1.2) has a weak solution.

Proof. Let

$h_n(0) = 0,$

$h_n(t) = (k-1)2^{-n}$, as $(k-1)2^{-n} < t \leq k2^{-n}$.

Then by Theorem 2.12 in Chapter 2 there exists a unique solution (x_t^n, ϕ_t^n) satisfying

$$(3.1)\begin{cases} x_t^n = x_0 + \int_0^t b^n(s, x_s^n)ds + \int_0^t \sigma^n(s, x_s^n)dw_s \\ \quad + \int_0^t \int_Z c^n(s, x_{s-}^n, z)q(ds, dz) + \phi_t^n, \\ \text{and the other statements in (1.2) for } (x_t^n, \phi_t^n), \end{cases}$$

where

$b^n(s, x_s) = b(h_n(s), x_{h_n(s)})$, etc.

In fact, once x_t^n is obtained for $0 \leq t \leq k2^{-n}$, then (x_t^n, ϕ_t^n) is uniquely determined as the solution of the Skorohod problem:

$$(3.2)\begin{cases} x_t^n = x_{k2^{-n}}^n + b(k2^{-n}, x_{k2^{-n}}^n)(t - k2^{-n}) + \sigma(k2^{-n}, x_{k2^{-n}}^n) \\ \cdot(w(t) - w(k2^{-n})) + \int_Z c(k2^{-n}, x_{k2^{-n}}^n, z)q((k2^{-n}, t], dz) + \phi_t^n, \\ \text{and the other statements in (1.2) hold for } (x_t^n, \phi_t^n) \\ \text{on } k2^{-n} \leq t \leq (k+1)2^{-n}. \end{cases}$$

Now we show that for arbitrary $\varepsilon > 0$ and $T > 0$

$\lim_{c\to\infty} \sup_n \sup_{t\leq T} P(|\eta_t^n| > c) = 0,\ (@)_1$

$\lim_{h\to\infty} \sup_n \sup_{t,s\leq T, |t-s|\leq h} P(|\eta_t^n - \eta_s^n| > \varepsilon) = 0,\ (@)_2$

hold for $\eta_t^n = x_t^n, \phi_t^n, w_t, \zeta_t$, where

$\zeta_t = \int_0^t \int_{|z|\leq 1} zq(ds, dz) + \int_0^t \int_{|z|>1} zp(ds, dz) = I_t^1 + I_t^2.$

Obviously,

$E \sup_{t\leq T} \left|I_t^1\right|^2 \leq \int_0^t \int_{|z|\leq 1} dz/\left|z\right|^{d-1} < \infty.$

Denote

$\bar{I}_t^2 = \int_0^t \int_{|z|>1} |z|\, p(ds, dz).$

Since $\bar{I}_t^2$ is cadlag, $\{0 < s \leq T : \triangle \bar{I}_s^2 > 1\}$ is a finite set, and

$\sum_{0<s\leq T} \triangle \bar{I}_s^2 I_{(\triangle \bar{I}_s^2 > 1)} = \sum_{k=1}^{n(\omega)} |z_k(\omega)| < \infty.$

Hence

$\mathrm{P}(\sup_{t\leq T} \left|I_t^2\right| < \infty) = 1.$

Therefore this shows that ζ_t satisfies $(@)_1$.On the other hand, for arbitrary $\varepsilon > 0$

$P(|\zeta_t - \zeta_s| > \varepsilon) \leq P(\left|\int_s^t \int_{|z|\leq 1} zq(ds, dz)\right| > \varepsilon/2)$

$+ P(\left|\int_s^t \int_{|z|>1} zq(ds, dz)\right| > \varepsilon/2) = J_1 + J_2.$

It is evident that as $|t - s| \leq h \to 0$

$J_1 \leq (4/\varepsilon)^2 E\left|\int_s^t \int_{|z|\leq 1} zq(ds, dz)\right|^2 \leq (4/\varepsilon)^2 \int_{|z|\leq 1} dz/\left|z\right|^{d-1} |t - s| \to 0\ .$

Since $p(dt, dz)$ is a Poisson random measure, as $|t - s| \leq h \to 0$

$J_2 \leq P(p((s, t], |z| > 1) > 0) = 1 - \exp(-\int_s^t \int_{|z|>1} dz/\left|z\right|^{d+1} dr)$

$\leq 1 - \exp(-\pi(|z| > 1)h) \to 0\ .$

Hence ζ_t satisfies $(@)_2$. By Assumption 2.7 in Chapter 2 Lemma 2.8 of Chapter 2 holds. Hence one can apply Theorem 2.1 and

$E\left|w_t - w_s\right|^2 = |t - s|\,,$

to show that $(@)_1$ and $(@)_2$ also hold for $\eta_t^n = x_t^n, \phi_t^n, w_t$. Now by Lemma 1.9 it is routine (cf. Krylov, 1980) to show that there exist a probability space $(\widetilde{\Omega}, \widetilde{\Im}, \widetilde{P})$ and cadlag processes $(\widetilde{x}_t^n, \widetilde{\phi}_t^n, \widetilde{w}_t^n, \widetilde{\zeta}_t^n)$ defined on it with the same finite probability distributions as that of $(x_t^n, \phi_t^n, w_t, \zeta_t)$ and there exist $(\widetilde{x}_t^0, \widetilde{\phi}_t^0, \widetilde{w}_t^0, \widetilde{\zeta}_t^0)$, where $\widetilde{w}_t^0$ is a BM on $(\widetilde{\Omega}, \widetilde{\Im}, \widetilde{P})$, and there exists a subsequence $\{n_k\}$ of $\{n\}$, denote it by $\{n\}$ again, such that as $n \to \infty$

$\widetilde{\eta}_t^n \to \widetilde{\eta}_t^0$, in probability, as $\widetilde{\eta}_t^n = \widetilde{x}_t^n, \widetilde{\phi}_t^n, \widetilde{w}_t^n, \widetilde{\zeta}_t^n, n = 0, 1, 2, ...$

Now set (Skorohod, 1965)

$\widetilde{p}^n(dt, dz) = \sum_{s \in dt} I_{(0 = \Delta \widetilde{\zeta}_s^n \in dz)}(s)$, $\widetilde{q}^n(dt, dz) = \widetilde{p}^n(dt, dz) - \pi(dz)dt$, $n = 0, 1, 2, ...$

Then $\widetilde{p}^n(dt, dz)$ is a Poisson random point measure with compensator $\pi(dz)dt$ for each $n = 0, 1, 2, ...$, and it verifies

$\widetilde{\zeta}_t^n = \int_0^t \int_{|z| \leq 1} z\widetilde{q}^n(ds, dz) + \int_0^t \int_{|z| > 1} z\widetilde{p}^n(ds, dz), n = 0, 1, 2, ...$

By the coincidence of finite probability distributions one easily sees that $(\widetilde{x}_t^n, \widetilde{\phi}_t^n)$ satisfies (1.1) and (1.2) with $\widetilde{w}_t^n$ and $\widetilde{q}^n(dt, dz)$ on $(\widetilde{\Omega}, \widetilde{\Im}, \widetilde{P})$. Now let us show that as $n \to \infty$

$(3.2)_1$ $\left| \int_0^t (b^n(s, \widetilde{x}_s^n) - b(s, \widetilde{x}_s^0))ds \right| \to 0$, in probability.

$(3.2)_2$ $\int_0^t \sigma^n(s, \widetilde{x}_s^n) d\widetilde{w}_s^n \to \int_0^t \sigma(s, \widetilde{x}_s^0) d\widetilde{w}_s^0$, in probability,

$(3.2)_3$ $\int_0^t \int_Z c^n(s, \widetilde{x}_{s-}^n, z)\widetilde{q}^n(ds, dz) \to \int_0^t \int_Z c(s, \widetilde{x}_{s-}^0, z)\widetilde{q}^0(ds, dz)$, in probability,

$(3.2)_4$ $\int_0^t \int_Z \left| c^n(s, \widetilde{x}_s^n, z) - c(s, \widetilde{x}_s^0, z) \right|^2 \pi(dz)ds \to 0$, in probability.

For this let us first establish a lemma.

Lemma 3.2. Under assumption 1° and 2° one has (recall that $Z = R^d - \{0\}$)

$\lim_{\varepsilon \to 0} J_\varepsilon^{N,T} = \lim_{\varepsilon \to 0} \sup_{|x| \leq N, t \in [0,T]} \int_{|z| \leq \varepsilon} |c(t, x, z)|^2 \pi(dz) = 0.$

Proof. By assumption 2° for any given $\delta > 0$ for each $(t, x) \in [0, T] \times [-N, N]$ there exists a $\eta > 0$ such that

$\int_Z |c(t, x, z) - c(s, y, z)|^2 \pi(dz) < \delta,$

as $(s, y) \in U_{(t,x)}(\eta) = \{(s, y)[0, T] \times [-N, N] : |y - x| < \eta, |s - t| < \eta\}$.

Hence by Hiene-Borel finite cover theorem there exist finite $U_{(t_i, x_i)}(\eta), i = 1, \cdots, n_0$, which cover $[0, T] \times [-N, N]$, i.e. $[0, T] \times [-N, N] \subset \cup_{i=1}^{n_0} U_{(t_i, x_i)}(\eta)$.

Therefore, as $(s, y) \in U_{(t_i, x_i)}(\eta)$

$\int_{|z| \leq \varepsilon} |c(t_i, x_i, z) - c(s, y, z)|^2 \pi(dz) \leq \int_Z |c(t_i, x_i, z) - c(s, y, z)|^2 \pi(dz) < \delta,$

$\int_{|z| \leq \varepsilon} |c(s, y, z)|^2 \pi(dz) \leq 2\delta + 2\int_{|z| \leq \varepsilon} |c(t_i, x_i, z)|^2 \pi(dz),$

$i = 1, 2, \cdots, n_0.$

Thus

$\sup_{|y| \leq N, s \in [0,T]} \int_{|z| \leq \varepsilon} |c(s, y, z)|^2 \pi(dz)$

$\leq 2\delta + 2\sup_{1 \leq i \leq n_0} \int_{|z| \leq \varepsilon} |c(t_i, x_i, z)|^2 \pi(dz).$

Letting $\varepsilon \to 0$, one obtains

$\lim_{\varepsilon \to 0} \sup_{|y| \leq N, s \in [0,T]} \int_{|z| \leq \varepsilon} |c(s, y, z)|^2 \pi(dz) \leq 2\delta.$

The proof is complete. ■

Note now that for each $n^0 = 1, 2, ...$

$\int_0^t \int_Z \left|c^n(s,\widetilde{x}_s^n,z) - c(s,\widetilde{x}_s^0,z)\right|^2 \pi(dz)ds$
$\leq 3\int_0^t \int_Z \left|c^n(s,\widetilde{x}_s^n,z) - c^{n^0}(s,\widetilde{x}_s^n,z)\right|^2 \pi(dz)ds$
$+3\int_0^t \int_Z \left|c^{n^0}(s,\widetilde{x}_s^n,z) - c^{n^0}(s,\widetilde{x}_s^0,z)\right|^2 \pi(dz)ds$
$+3\int_0^t \int_Z \left|c^{n^0}(s,\widetilde{x}_s^0,z) - c(s,\widetilde{x}_s^0,z)\right|^2 \pi(dz)ds = 3(I_1^{n,n^0} + I_2^{n,n^0} + I_3^{n^0}).$
However, for any $\varepsilon > 0$
$\tilde{P}(I_1^{n,n^0} > \varepsilon) \leq \tilde{P}(\sup_{s\leq T} |\widetilde{x}_s^n| > N) + 2TJ_{\varepsilon'}^{N,T}$
$+\tilde{P}(\int_0^T \int_{|z|>\varepsilon'} \left|c^n(s,\widetilde{x}_s^n,z) - c^{n^0}(s,\widetilde{x}_s^n,z)\right|^2 I_{|\widetilde{x}_s^n|\leq N}\pi(dz)ds > \varepsilon)$
$= I_{1,1}^{n,n^0} + I_{1,2}^{n,n^0} + I_{1,3}^{n,n^0}.$
To show that as $n, n^0 \to \infty$
(3.3) $I_{1,3}^{n,n^0} \to 0.$
We need the following

Lemma 3.3. For any given $\eta > 0$
$\lim_{\delta\to 0} \sup_n \sup_{|t_1-t_2|\leq\delta} \widetilde{P}(\left|\widetilde{x}_{t_1}^n - \widetilde{x}_{t_2}^n\right| > \eta) = 0,$
$\lim_{\delta\to 0} \sup_n \sup_{|t_1-t_2|\leq\delta} \widetilde{P}(\left|\widetilde{\phi}_{t_1}^n - \widetilde{\phi}_{t_2}^n\right| > \eta) = 0,$

Proof. Since $(\tilde{x}_t^n, \tilde{\phi}_t^n)$ satisfies (3.1). Hence, by the assumption on coefficients applying Lemma 2.8 in Chapter 2 and Theorem 2.1 one has that
(3.4) $\tilde{E}\sup_{t\leq T} |\tilde{x}_t^n|^2 \leq k_T$, and $\tilde{E}\sup_{s\leq r\leq t\leq T} |\tilde{x}_r^n - \tilde{x}_s^n|^2 \leq k_T |t-s|$, for all n.
Therefore, the first conclusion is derived. The proof of the second conclusion is similar. ■

Now note that for any given $\eta > 0$
$I_{1,3}^{n,n^0} \leq \sup_n \sup_{s\in[0,T]} \widetilde{P}(\left|\widetilde{x}_{h_n(s)}^n - \widetilde{x}_{h_{n^0}(s)}^n\right| > \eta)$
$+\varepsilon^{-1}\widetilde{E}(\int_0^T \int_{|z|>\varepsilon'} \left|c^n(s,\widetilde{x}_s^n,z) - c^{n^0}(s,\widetilde{x}_s^n,z)\right|^2 I_{|\widetilde{x}_s^n|\leq N}$
$\cdot I_{\left|\widetilde{x}_{h_n(s)}^n - \widetilde{x}_{h_{n^0}(s)}^n\right|\leq\eta} \pi(dz)ds).$
Hence applying condition 2° one easily sees that (3.3) is true. Now by (3.4), Lemma 3.2, and (3.3) it follows that as $n, n^0 \to \infty$
$\tilde{P}(I_1^{n,n^0} > \varepsilon) \to 0.$
Similarly, one can show that as $n^0 \to \infty$
$\tilde{P}(I_3^{n^0} > \varepsilon) \to 0.$
Now let us show that for each given n^0 as $n \to \infty$
$\tilde{P}(I_2^{n,n^0} > \varepsilon) \to 0.$
In fact, for any $\varepsilon > 0$
$\tilde{P}(I_2^{n,n^0} > \varepsilon) \leq \tilde{P}(\sup_{s\leq T} |\widetilde{x}_s^n| + \sup_{s\leq T} |\widetilde{x}_s^0| > N) + 2TJ_{\varepsilon'}^{N,T}$
$+\tilde{P}(\int_0^T \int_{|z|>\varepsilon'} \left|c^{n^0}(s,\widetilde{x}_s^n,z) - c^{n^0}(s,\widetilde{x}_s^0,z)\right|^2 I_{\sup_{s\leq T}(|\widetilde{x}_s^n|+|\widetilde{x}_s^0|)\leq N}\pi(dz)ds > \varepsilon)$

$= I_{2,1}^{n,n^0} + I_{2,2}^{n,n^0} + I_{2,3}^{n,n^0}.$

Denote

$$J_s = \int_{|z|>\varepsilon'} \left| c^{n^0}(s, \widetilde{x}_s^n, z) - c^{n^0}(s, \widetilde{x}_s^0, z) \right|^2 I_{\sup_{s \le T}(|\widetilde{x}_s^n| + |\widetilde{x}_s^0|) \le N} \pi(dz).$$

Then

$$\widetilde{P}(J_s > \varepsilon) \le \widetilde{P}(\left| \widetilde{x}_{h_{n^0}(s)}^n - \widetilde{x}_{h_{n^0}(s)}^0 \right| > \eta)$$
$$+ \widetilde{P}(\int_{|z|>\varepsilon'} \left| c^{n^0}(s, \widetilde{x}_s^n, z) - c^{n^0}(s, \widetilde{x}_s^0, z) \right|^2 I_{\left| \widetilde{x}_{h_{n^0}(s)}^n - \widetilde{x}_{h_{n^0}(s)}^0 \right| \le \eta} \pi(dz) > \varepsilon).$$

From this it follows that for each given n^0 as $n \to \infty$

$$\tilde{P}(J_s > \varepsilon) \to 0.$$

Applying condition 1° and 2° one finds that for each given n^0 as $n \to \infty$

$$\tilde{P}(I_2^{n,n^0} > \varepsilon) \to 0.$$

Therefore $(3.2)_4$ is proved. To show $(3.2)_3$ let us prove that for each fixed n^0 as $n \to \infty$

$$I_t^n = \tilde{P}(\left| \int_0^t \int_Z c^{n^0}(s, \widetilde{x}_s^n, z) \tilde{q}^n(ds, dz) - \int_0^t \int_Z c^{n^0}(s, \widetilde{x}_s^0, z) \tilde{q}^0(ds, dz) \right| > \varepsilon) \to 0. \ (@)$$

Indeed,

$$I_t^n \le \tilde{P}(\sup_{s \le T} |\widetilde{x}_s^n| > N) + \tilde{P}(\sup_{s \le T} |\widetilde{x}_s^0| > N) + 2(2/\varepsilon)^2 T J_{\varepsilon'}^{N,T}$$
$$+ (2/\varepsilon)^2 \tilde{E} \int_0^t \int_{|z|>\varepsilon'} \left| c^{n^0}(s, \widetilde{x}_s^n, z) - c^{n^0}(s, \widetilde{x}_s^0, z) \right|^2 \pi(dz) ds$$
$$+ \widetilde{P}(\left| \int_0^t \int_Z c^{n^0}(s, \widetilde{x}_{s-}^0, z) \tilde{q}^n(ds, dz) - \int_0^t \int_Z c^{n^0}(s, \widetilde{x}_{s-}^0, z) \tilde{q}^0(ds, dz) \right|$$
$$\cdot I_{\sup_{s \le T} |\widetilde{x}_s^0| \le N} > \varepsilon/2) = \sum_{i=1}^5 I_t^{in}.$$

It can be shown as above that for any given $\tilde{\delta} > 0$ there exist large enough n^0 and $\widetilde{N}$ such that as $n \ge \tilde{N}$

$$I_t^{in} < \tilde{\delta}, \ i = 1, 2, 3, 4.$$

To show that for each fixed n^0 as $n \to \infty$

$$I_t^{5n} \to 0, \ (@)_1$$

we need the following

3.4 Lemma. Assume that $p^n(dt, dz)$ $n = 0, 1, ...,$are the Poisson counting measures with the same compensator $\pi(dz)dt$ such that

$$q^n(dt, dz) = p^n(dt, dz) - \pi(dz)dt, \ n = 0, 1, ...,$$

set

$$\zeta_t^n = \int_0^t \int_{|z| \le 1} z q^n(ds, dz) + \int_0^t \int_{|z|>1} z p^n(ds, dz), n = 0, 1, 2, ...$$

If as $n \to \infty$

$$\zeta_t^n \to \zeta_t^0, \text{ in probability, for all } t \ge 0,$$

then for any finite Borel measurable function $f(z)$ and for any $\varepsilon > 0$ as $n \to \infty$

$$\int_0^t \int_{|z|>\varepsilon} f(z) p^n(ds, dz) \to \int_0^t \int_{|z|>\varepsilon} f(z) p^0(ds, dz), \text{ in probability,}$$

for all $t \ge 0$. Furthermore, if $g(t, x, z)$ is bounded, Borel measurable, and continuous in (t, x); and x_t is a $\Im_t-$adapted cadlag process, then for any $\varepsilon > 0$ any $h > 0$ as $n \to \infty$

$\int_t^{t+h}\int_{|z|>\varepsilon} g(t,x_t,z)p^n(ds,dz) \to \int_t^{t+h}\int_{|z|>\varepsilon} g(t,x_t,z)p^0(ds,dz)$, in probability, for all $t \geq 0$.

Lemma 3.4 can be found in Skorohod 1965 (see Lemma 4 and 5 in §3 of chapter 3 there).

Note that as $t \leq T$

$\tilde{E}\int_0^t \int_Z \left|c^{n^0}(s,\tilde{x}_s^0,z)\right|^2 \pi(dz)ds \leq \tilde{k}_T < \infty.$

Hence, for any given $\delta > 0$ one can take a $\varepsilon' > 0$ such that

$(4/\varepsilon)^2 \tilde{E}\int_0^t \int_{|z|\leq\varepsilon'} \left|c^{n^0}(s,\tilde{x}_s^0,z)\right|^2 \pi(dz)ds \leq \delta/8.$

Obviously,

$I_t^{5n} \leq \delta/2 + \check{P}(\Big|\int_0^t \int_{|z|>\varepsilon'} c^{n^0}(s,\tilde{x}_s^0,z)\tilde{q}^n(ds,dz)$

$- \int_0^t \int_{|z|>\varepsilon'} c^{n^0}(s,\tilde{x}_s^0,z)\tilde{q}^0(ds,dz)\Big| \cdot I_{\sup_{s\leq T}|\tilde{x}_s^0|\leq N} > \varepsilon/4)$

However, by Lemma 3.4 for each fixed n^0 as $n \to \infty$

(3.5) $\int_0^t \int_{|z|>\varepsilon'} c^{n^0}(s,\tilde{x}_s^0,z)I_{|\tilde{x}_s^0|\leq N}\tilde{q}^n(ds,dz)$

$= \sum_{k=0}^{\infty}\int_{(2^{-n^0}k\wedge t,2^{-n^0}(k+1)\wedge t]} c(2^{-n^0}k,\tilde{x}^0_{2^{-n^0}k},z)I_{\left|\tilde{x}^0_{2^{-n^0}k}\right|\leq N}\tilde{q}^n(ds,dz)$

$\to \sum_{k=0}^{\infty}\int_{(2^{-n^0}k\wedge t,2^{-n^0}(k+1)\wedge t]} c(2^{-n^0}k,\tilde{x}^0_{2^{-n^0}k},z)I_{\left|\tilde{x}^0_{2^{-n^0}k}\right|\leq N}\tilde{q}^0(ds,dz)$

$= \int_0^t \int_{|z|>\varepsilon'} c^{n^0}(s,\tilde{x}_s^0,z)I_{|\tilde{x}_s^0|\leq N}\tilde{q}^0(ds,dz)$, in probability.

Hence, $(@)_1$ is proved. Therefore, (@) is true. Now by the proof of (3.3) and (@) one obtains that $(3.2)_3$ is true. $(3.2)_2$ is true because of the following lemma.

3.5 Lemma Assume that $f_n(t)$ satisfies the following conditions: for any $0 \leq T < \infty$

1° $\lim_{N\to\infty} P(\sup_{t\leq T}|f_n(t)| > N) = 0$

2° for any $\varepsilon > 0$

$\lim_{\delta\to 0}\sup_n \sup_{|t-s|<\delta} P(|f_n(t) - f_n(s)| > \varepsilon) = 0.$

Moreover, if as $n \to \infty$

$f_n(t) \to f(t)$, in probability,

where the stochastic integrals $\int_0^T f_n(t)dw_n(t)$ and $\int_0^T f(t)dw(t)$ exist, $w_n(t)$ and $w(t)$ are BM's, $0 \leq T < \infty$ is arbitrarily given, and as $n \to \infty$, $w_n(t) \to w(t)$, in probability, then as $n \to \infty$

$\int_0^T f_n(t)dw_n(t) \to \int_0^T f(t)dw(t)$, in probability.

Lemma 3.5 can be found in Skorohod, 1965.

$(3.2)_1$ is also true because of the following fact:

$\tilde{E}\left|\int_0^t (b^n(s,\tilde{x}_s^n) - b(s,\tilde{x}_s^0))ds\right| \to 0$, as $n \to \infty$,

where $\left\{\int_0^t (b^n(s,\tilde{x}_s^n) - b(s,\tilde{x}_s^0))ds\right\}_{n=1}^{\infty}$ is uniformly integrable, since

$\tilde{E}\sup_{t\leq T}\left|\int_0^t (b^n(s,\tilde{x}_s^n) - b(s,\tilde{x}_s^0))ds\right|^2 \leq \hat{k}_T < \infty.$

Now in the rest by Corollary 2.10 in Chapter 2 as the proof in Theorem 2.9 one only needs to show that there exists a subsequence $\{n_k\}$ of $\{n\}$,denote it by $\{n\}$ again, such that for each $0 \leq T < \infty$

(3.6) $\lim_{n\to\infty} E\sup_{t\leq T} \left|\widetilde{y}_t^n - \widetilde{y}_t^0\right|^2 = 0,$

where

$\widetilde{y}_t^n = \int_0^t b^n(s,\widetilde{x}_s^n)ds + \int_0^t \sigma^n(s,\widetilde{x}_s^n)d\widetilde{w}_s^n + \int_0^t \int_Z c^n(s,\widetilde{x}_{s-}^n,z)\widetilde{q}^n(ds,dz).$

For this we need the following

3.6 Lemma. Assume that $f_n, n = 0,1,2,...$ are random variables. If as $n \to \infty$

$f_n \to f_0,\ a.s.$

$E\left|f_n\right|^2 \to E\left|f_0\right|^2 < \infty,$

then as $n \to \infty$

$E\left|f_n - f_0\right|^2 \to 0.$

Proof. Since

$0 \leq \left|f_n\right|^2 \to \left|f_0\right|^2, a.s.$

and

$E\left|f_n\right|^2 \to E\left|f_0\right|^2 < \infty.$

By Theorem 10.16 of J. Yeh, 1976 $\left\{\left|f_n\right|^2\right\}_{n=1}^{\infty}$ is uniformly integrable, so is $\left\{\left|f_n\right|^2 + \left|f_0\right|^2\right\}_{n=1}^{\infty}$. Hence as $n \to \infty$

$E\left|f_n - f_0\right|^2 \to 0.$ ■

Now by $(3.2)_3$ for each given $0 \leq t$ there exists a subsequence $\{n_k\}$ denote it by $\{n\}$ again, such that as $n \to \infty$

$\int_0^t \int_Z c^n(s,\widetilde{x}_s^n,z)\widetilde{q}^n(ds,dz) \to \int_0^t \int_Z c(s,\widetilde{x}_s^0,z)\widetilde{q}^0(ds,dz),\ P-a.s.$

However, by condition 1° and $(3.2)_4$ as $n \to \infty$

$E\int_0^t \int_Z \left|c^n(s,\widetilde{x}_s^n,z) - c(s,\widetilde{x}_s^0,z)\right|^2 \pi(dz)ds \to 0.$

Hence as $n \to \infty$

$E\left|\int_0^t \int_Z c^n(s,\widetilde{x}_s^n,z)\widetilde{q}^n(ds,dz)\right|^2 = E\int_0^t \int_Z \left|c^n(s,\widetilde{x}_s^n,z)\right|^2 \pi(dz)ds$

$\to E\int_0^t \int_Z \left|c(s,\widetilde{x}_s^0,z)\right|^2 \pi(dz)ds = E\left|\int_0^t \int_Z c(s,\widetilde{x}_s^0,z)\widetilde{q}^0(ds,dz)\right|^2.$

Therefore, applying Lemma 3.6, one has that as $n \to \infty$ (if necessary, take a subsequence)

$\tilde{E}\left|M_t^d\right|^2 \widehat{=} \tilde{E}\left|\int_0^t \int_Z c^n(s,\widetilde{x}_{s-}^n,z)\widetilde{q}^n(ds,dz) - \int_0^t \int_Z c(s,\widetilde{x}_{s-}^0,z)\widetilde{q}^0(ds,dz)\right|^2 \to 0.$

Since M_t^d is a martingale. One has that as $n \to \infty$

$\tilde{E}\left\langle M^d\right\rangle_t = \tilde{E}\left|M_t^d\right|^2 \to 0.$

Similarly, one can show that as $n \to \infty$ (if necessary, take a subsequence)

$\tilde{E}\left\langle M^c\right\rangle_t = \tilde{E}\left|M_t^c\right|^2 \widehat{=} \tilde{E}\left|\int_0^t \sigma^n(s,\widetilde{x}_s^n)d\widetilde{w}_s^n - \int_0^t \sigma(s,\widetilde{x}_s^0)d\widetilde{w}_s^0\right|^2 \to 0.$

More easily, one shows that as $n \to \infty$

$\tilde{E}\left|\int_0^t (b^n(s,\widetilde{x}_s^n) - b(s,\widetilde{x}_s^0))ds\right|^2 \to 0.$

Therefore it follows that for each $0 \leq T < \infty$ as $n \to \infty$ (if necessary, take a subsequence)

$\hat{E}\sup_{t\leq T}\left|\widetilde{y}_t^n - \widetilde{y}_t^0\right|^2 \leq \widetilde{k}_T(\widetilde{E}\left|\int_0^T (b^n(s,\widetilde{x}_s^n) - b(s,\widetilde{x}_s^0))ds\right|^2$

$+\tilde{E}\left\langle M^c\right\rangle_T + \tilde{E}\left\langle M^d\right\rangle_T) \to 0.$

Therefore (3.6) is established. The proof of Theorem 3.1 is now complete. ■

Applying Theorem 3.1 we easily derive the following

3.7 Theorem. Under the assumption of Theorem 3.1 if for $t \geq 0, x \in \overline{\Theta}, z \in Z$

1° $2(x-y) \cdot (b(t,x) - b(t,y)) + \|\sigma(t,x) - \sigma(t,y)\|^2$

$+\int_Z |c(t,x,z) - c(t,y,z)|^2 \pi(dz) \leq k_N(t)\rho_N(|x-y|^2)$, as $|x|, |y| \leq N$,

where $0 \leq k_N(t), \int_0^T k_N(t)dt < \infty$, for any $T < \infty$, and $\rho_N(u)$ is concave and strictly increasing in $u \geq 0$ such that $\rho_N(0) = 0$, and

$\int_{0+} du/\ \rho_N(u) = \infty$, for each $N = 1, 2, ...,$

2° $x + c(t,x,z) \in \overline{\Theta}$,

then (1.1) with (1.2) has a pathwise unique strong solution. Moreover, in this case $\phi(t)$ is continuous.

Proof. By Theorem 3.1 (1.1) with (1.2) has a weak solution. Applying Remark 2.5, by condition 2° one finds that $\phi(t)$ is continuous. Now the proof can be completed by using Theorem 2.3 and Theorem 1.3. ■

Now let us relax the bounded condition for coefficients b and σ to the less than linear growth condition.

3.8 Theorem. Assume that for $t \geq 0, x \in \overline{\Theta}, z \in Z$

1° $b(t,x), \sigma(t,x), c(t,x,z)$ are jointly measurable, and there exists a constant $k_0 > 0$ such that

$|b|^2 + \|\sigma\|^2 \leq k_0(1 + |x|^2), \int_Z |c|^2 \pi(dz) \leq k_0,$

2° $b(t,x), \sigma(t,x)$ are jointly continuous, and as $|x-y| \to 0, |t-s| \to 0$

$\int_Z |c(t,x,z) - c(s,y,z)|^2 \pi(dz) \to 0,$

3° $\pi(dz) = dz/|z|^{d+1}$, $(Z = R^d - \{0\})$.

4° $2(x-y) \cdot (b(t,x) - b(t,y)) + \|\sigma(t,x) - \sigma(t,y)\|^2$

$+\int_Z |c(t,x,z) - c(t,y,z)|^2 \pi(dz) \leq k_N(t)\rho_N(|x-y|^2)$, as $|x|, |y| \leq N$,

where $0 \leq k_N(t), \int_0^t k_N(t)ds < \infty$, and $\rho_N(u)$ is concave and strictly increasing in $u \geq 0$ such that $\rho_N(0) = 0$, and

$\int_{0+} du/\ \rho_N(u) = \infty$, for each $N = 1, 2, ...,$

5° $x + c(t,x,z) \in \overline{\Theta}$,

then (1.1) with (1.2) has a pathwise unique strong solution. Moreover, in this case $\phi(t)$ is continuous.

Proof. Let

$b^N(t,x) = b(t,x)$, as $|x| \leq N$; and $b^N(t,x) = b(t, Nx/|x|)$, as $|x| > N$;

σ^N is similarly defined. Then by Theorem 3.7 there exists a pathwise unique strong solution (x_t^N, ϕ_t^N), where ϕ_t^N is continuous, satisfying (1.1) and (1.2) with the coefficients b^N, σ^N and c. Set now

$\tau_N = \inf\{t \geq 0 : |x_t^N| > N\},$

$x_t = x_t^N,\ \phi_t = \phi_t^N$, as $0 \leq t \leq \tau_N$.

By the pathwise uniqueness the above definition is well posed (see the proof for Theorem 2.8 in general case), and

$$x_{t\wedge\tau_N} = x_0 + \int_0^{t\wedge\tau_N} b(s,x_s)ds + \int_0^{t\wedge\tau_N} \sigma(s,x_s)dw_s$$
$$+ \int_0^{t\wedge\tau_N} \int_Z c(s,x_{s-},z)q(ds,dz) + \phi_{t\wedge\tau_N}, \text{ for all } t \geq 0.$$

Note that for any $T < \infty$

$$N^2 P(\tau_N \leq T) \leq E\left|x_{T\wedge\tau_N}\right|^2 \leq k_T, \text{ for all } N = 1,2,...$$

Hence, it follows that

$$P(\lim_{N\to\infty} \tau_N = \infty) = 1.$$

The proof is complete. ■

3.3.2 Girsanov Theorem and Existence of Weak Solutions for RSDE with Discontinuous Coefficients

By Ito's formula it is easy to verify that

$$z_t^w = \exp[\int_0^t \theta_s^w \cdot dw_s - \tfrac{1}{2}\int_0^t |\theta_s^w|^2 ds]$$

solves the following SDE

(3.7) $dz_t^w = \theta_t^w \cdot z_t^w dw_t,\ z_0^w = 1.$

and

$$z_t^N = \exp[\int_0^t \theta_s^N \cdot d\widetilde{N}_s] \cdot \sqcap_{0<s\leq t}(1+\theta_s^N \triangle N_s)e^{-\theta_s^N \triangle N_s}$$
$$= \sqcap_{0<s\leq t}(1+\theta_s^N \triangle N_s)e^{-\int_0^t \theta_s^N \lambda_s ds}$$

solves the following SDE

(3.8) $dz_t^N = \theta_t^N \cdot z_{t-}^N d\widetilde{N}_t,\ z_0^N = 1,$

where w_t is a d-dimensional BM, N_t is a 1-dimensional Poisson process with integrable non-random, non-negative density function λ_t such that its compensator is $EN_t = \int_0^t \lambda_s ds$, i.e.,

$$\widetilde{N}_t = N_t - \int_0^t \lambda_s ds$$

is a martingale, θ_t^w is assumed to be a d-dimensional $\Im_t-$ adapted process, and θ_t^N a 1-dimensional $\Im_t-$ predictable process. We always make the following assumption:

$$\theta_t^N > -1.$$

Hence

$$0 \leq z_t, \forall t \geq 0.$$

We have

3.9 Lemma. Let $\forall 0 \leq r \leq t$

$$z_t^0 = z_t = z_t^w z_t^N,$$
$$z_t^r = \exp[\int_r^t \theta_s^w \cdot dw_s - \tfrac{1}{2}\int_r^t |\theta_s^w|^2 ds] \cdot \sqcap_{r<s\leq t}(1+\theta_s^N \triangle N_s)e^{-\int_r^t \theta_s^N \lambda_s ds}.$$

1) z_t^r satisfies the following properties:

(i) $E(z_t^r|\Im_r) \leq 1, \forall 0 \leq r \leq t,$

(ii) z_t is a non-negative $P-$supermartingale.

(iii) If $E(z_t) = 1$, then

$$E(z_t^r|\Im_r) = 1, \forall 0 \leq r \leq t,$$

hence z_t is a $P-$martingale.

2) Denote

(3.9) $d\widetilde{P}_t = z_t dP = z_t^w z_t^N dP.$

If

$\left|\theta_t^w\right| + \left|\theta_t^N\right| \leq k_0$,for all $0 \leq t \leq T$

then $\forall t \geq 0$

$E\left|z_t\right|^2 \leq e^{k_0(t+\int_0^t \lambda_s ds)}$,

$\hat{P}_T$ is a probability measure, $\forall 0 \leq T < \infty$; and z_t is a $P-$martingale.

Proof. 1): By Ito's formula

$$dz_t = z_{t-}^w dz_t^N + z_t^N dz_t^w + 0 = z_t^N \theta_t^w \cdot z_t^w dw_t + z_{t-}^w \theta_t^N z_{t-}^N d\widetilde{N}_t$$
$$= \theta_t^w \cdot z_t dw_t + \theta_t^N z_{t-} d\widetilde{N}_t.$$

Set

$$\widetilde{\tau}_N^r = \inf\left\{s \geq r : \int_r^s (\left|\theta_t^w z_t\right|^2 + \left|\theta_t^N z_{t-}\right|^2 \lambda_t) dt > N\right\}.$$

Then $\forall 0 \leq r \leq t$

$$E(z_{t\wedge\widetilde{\tau}_N^r}^r \left|\Im_r\right.) = 1.$$

Hence, by Fatou's lemma letting $n \to \infty$ one obtains (i). Note that

$$z_r z_t^r = z_t.$$

Hence, by (i)

$$E(z_t \left|\Im_r\right.) \leq z_r.$$

(ii) is obtained. Now note that $\forall 0 \leq r \leq t$

$$E(z_t) = E(z_r E(z_t^r \left|\Im_r\right.)).$$

Hence, $E(z_t^r \left|\Im_r\right.) < 1$ implies that $E(z_t) < 1$. Therefore, (iii) is obtained.

2): In this case, applying Gronwall's inequality, one gets that by (3.7) and (3.8) for any given $0 \leq T < \infty$

$$E\left|z_s^w\right|^2 \leq k_T,\ E\left|z_s^N\right|^2 \leq k_T, \text{ as } 0 \leq t \leq T.$$

In fact, let

$$\tau_N = \inf\left\{t \in [0,T] : \left|z_t^w\right| > N\right\};\ \tau_N = T, \text{ for } \inf\{\phi\}.$$

Then by Ito's formula and (3.7)

$$\left|z_{t\wedge\tau_N}^w\right|^2 = 1 + 2\int_0^{t\wedge\tau_N} \theta_s^w \cdot (z_s^w)^2 dw_s + \int_0^{t\wedge\tau_N} \left|\theta_s^w z_s^w\right|^2 ds.$$

Hence, by assumption

$$E\left|z_{t\wedge\tau_N}^w\right|^2 = 1 + E\int_0^{t\wedge\tau_N} \left|\theta_s^w z_s^w\right|^2 ds \leq 1 + k_0 \int_0^t E\left|z_{s\wedge\tau_N}^w\right|^2 ds.$$

Therefore by Gronwall's inequality and Fatou's lemma as $0 \leq t$

$$E\left|z_t^w\right|^2 \leq e^{k_0 t}.$$

Again by Ito's formula and (3.8)

$$\left|z_{t\wedge\tau_n}^N\right|^2 = 1 + 2\int_0^{t\wedge\tau_n} \theta_s^N (z_{s-}^N)^2 d\widetilde{N}_s + \int_0^{t\wedge\tau_n} \left|\theta_s^N z_{s-}^N\right|^2 dN_s.$$

Hence, as $0 \leq t$

$$E\left|z_t^N\right|^2 \leq e^{k_0 \int_0^t \lambda_s ds}.$$

Remember that λ_t is non-random. Therefore z_t^w and z_t^N are $P-$martingale.

Now by Ito's formula

$$\left|z_{t\wedge\tau_n}\right|^2 = 1 + 2\int_0^{t\wedge\tau_n} z_{s-} dz_s + [z]_s = 1 + 2\int_0^{t\wedge\tau_n} \theta_s^N (z_{s-})^2 d\widetilde{N}_s$$
$$+2\int_0^{t\wedge\tau_N} \theta_s^w \cdot (z_s)^2 dw_s + \int_0^{t\wedge\tau_N} \left|\theta_s^w z_s\right|^2 ds + \int_0^{t\wedge\tau_n} \left|\theta_s^N z_{s-}\right|^2 dN_s,$$

where

$$\tau_n = \inf\left\{t \in [0,T] : \left|z_t\right| > n\right\};\ \tau_n = T, \text{ for } \inf\{\phi\}.$$

Thus

$$E\left|z_{t\wedge\tau_n}\right|^2 = 1 + E\int_0^{t\wedge\tau_N} \left|\theta_s^w z_s\right|^2 ds + E\int_0^{t\wedge\tau_n} \left|\theta_s^N z_{s-}\right|^2 dN_s$$

$\leq 1 + k_0 \int_0^t E\left|z_{s\wedge\tau_N}\right|^2 ds + k_0 \int_0^t E\left|z_{s\wedge\tau_N}\right|^2 \lambda_s ds.$

By Gronwall's inequality and Fatou's lemma $\forall t \geq 0$

$E\left|z_t\right|^2 \leq e^{k_0(t+\int_0^t \lambda_s ds)}.$

Recall that

$dz_t = \theta_t^w \cdot z_t dw_t + \theta_t^N z_{t-} d\hat{N}_t.$

Hence, z_t is a $P-$martingale. This shows that

$Ez_t = Ez_t^w z_t^N = 1.$

Therefore $\hat{P}_T$ is a probability measure. ■

3.10 Theorem.(Girsanov type Theorem). Assume now that $\hat{P}_T$, defined by (3.9), for each $0 \leq T < \infty$ is a probability measure, and assume that $(\Omega, \Im)$ is a standard measurable space. Then there exists a probability measure $\widetilde{P}$ defined on $(\Omega, \Im)$ such that $\widetilde{P}\mid_{\Im_T} = \hat{P}_T$, for each $0 \leq T < \infty$. Furthermore, the following facts are true:

1) If

$\left|\theta_t^w\right| + \left|\theta_t^N\right| \leq k_0$, for all $0 \leq t \leq T$;

then

$w_t' = w_t - \int_0^t \theta_s^w ds, 0 \leq t,$

is a BM under probability $\hat{P}$;

$\widetilde{N_t'} = \tilde{N}_t - \int_0^t \lambda_s \theta_s^N ds$

$= N_t - \int_0^t \lambda_s(1+\theta_s^N)ds,\ 0 \leq t,$

is a $\hat{P}-$martingale.

2) If

$\int_0^T \left|\theta_t^w\right|^2 dt < \infty,\ \int_0^T \left|\theta_t^N\right|^2 dt < \infty,\ \widetilde{P}_T - a.s.$

then $w_t', t \in [0,T]$, is still a $\widetilde{P}_T-$BM; however, $\widetilde{N_t'}, t \in [0,T]$, is only a local $\widetilde{P}_T-$martingale.

Proof. Since $\left\{\widetilde{P}_T, 0 \leq T < \infty\right\}$ is a consistent probability measure system, and $(\Omega, \Im)$ is a standard measurable space. Hence, such extension probability measure $\widetilde{P}$ defined on $(\Omega, \Im)$ exists (see 4.1 in Chapter 4 of Ikeda & Watanabe, 1989). Now by Ito's formula for any $A \in \Im_s, \eta \in R^d$

$d(e^{i\eta\cdot(w_t'-w_s')} z_t I_A) = I_A e^{i\eta\cdot(w_t'-w_s')}[z_t^N \theta_t^w \cdot z_t^w dw_t + z_{t-}^w \theta_t^N \cdot z_{t-}^N d\tilde{N}_t$

$+ i\eta \cdot (dw_t - \theta_t^w dt) z_t^N z_t^w - \frac{1}{2}\left|\eta\right|^2 z_t^N z_t^w dt + i\eta \cdot \theta_t^w z_t^N z_t^w dt]$

$= dM_t - \frac{1}{2}\left|\eta\right|^2 z_t I_A e^{i\eta(w_t'-w_s')} dt,$

where M_t is a local $P-$martingale, and we have used the result that

$\sum_{r\in dt}[e^{i\eta\cdot(w_{r-}'-w_s')} z_{r-}^w (z_{r-}^N + \triangle z_r^N) - e^{i\eta\cdot(w_{r-}'-w_s')} z_{r-}^w z_{r-}^N$

$- e^{i\eta\cdot(w_{r-}'-w_s')} z_{r-}^w \triangle z_{r-}^N] = 0.$

1): If θ_t^w and θ_t^N are bounded, then by Lemma 3.9 M_t is a $P-$martingale. $EM_t = 0$. Hence, after solving an ordinary differential equation one finds that

$E[e^{i\eta\cdot(w_t'-w_s')} z_t I_A] = \widetilde{P}(A) e^{-\frac{1}{2}|\eta|^2(t-s)}.$

Or,

$E_{\widetilde{P}}[e^{i\eta\cdot(w_t'-w_s')} \mid \Im_s] = e^{-\frac{1}{2}|\eta|^2(t-s)}.$

Therefore $w_t', 0 \leq t,$is a BM under probability $\hat{P}$. Now denote

$\widetilde{\theta_s^N} = \lambda_s \theta_s^N.$

Note that

$\triangle \widetilde{N_r'} = \triangle N_r, \ \triangle z_t^N = z_t^N - z_{t-}^N = z_{t-}^N \theta_t^N \triangle N_t.$

By Ito's formula for any $A \in \Im_s$

$d(\widetilde{N_t'} - \widetilde{N_s'}) z_t^N z_t^w I_A = I_A \{ (d\tilde{N}_t - \tilde{\theta}_t^N dt) z_{t-}^w z_{t-}^N + (\widetilde{N_t'} - \widetilde{N_s'}) z_t^N \theta_t^w \cdot z_t^w dw_t$
$+ (\widetilde{N_{t-}'} - \widetilde{N_s'}) z_{t-}^w \theta_t^N \cdot z_{t-}^N d\hat{N}_t + \sum_{r \in \{t\}} [(\widetilde{N_{r-}'} - \widetilde{N_s'} + \triangle \widetilde{N_r'}) z_{r-}^w (z_{r-}^N + \triangle z_r^N)$
$- (\widetilde{N_{r-}'} - \widetilde{N_s'}) z_{r-}^w z_{r-}^N - z_{r-}^w z_{r-}^N \triangle \widetilde{N_r'} - (\widetilde{N_{r-}'} - \widetilde{N_s'}) z_{r-}^w \triangle z_{r-}^N] \}$
$= I_A [dM_t - \widetilde{\theta_t^N} z_t dt + \theta_t^N z_{t-} \triangle N_t] = I_A [dM_t + \theta_t^N z_{t-} d\tilde{N}_t],$

where $s \leq t$

$M_t = \int_s^t z_{r-} d\tilde{N}_r + \int_s^t z_r (\widetilde{N_r'} - \widetilde{N_s'}) \theta_r^w \cdot dw_r + \int_s^t z_{r-} (\widetilde{N_{r-}'} - \widetilde{N_s'}) \theta_r^N d\tilde{N}_r.$

Let

$x_t = z_t (\widetilde{N_t'} - \widetilde{N_s'}) I_A.$

Then

$dx_t = I_A z_{t-} d\tilde{N}_t + x_t \theta_t^w \cdot dw_t + x_{t-} \theta_t^N d\tilde{N}_t + I_A \theta_t^N z_{t-} d\tilde{N}_t$
$= x_t \theta_t^w \cdot dw_t + (I_A z_{t-} + x_{t-} \theta_t^N + I_A \theta_t^N z_{t-}) d\tilde{N}_t.$

By Ito's formula

$|x_t|^2 = 2 \int_s^t x_{r-} (I_A z_{r-} + x_{r-} \theta_r^N + I_A \theta_r^N z_{r-}) d\tilde{N}_r$
$+ 2 \int_s^t |x_r|^2 \theta_r^w \cdot dw_r + \int_s^t |x_r \theta_r^w|^2 dr + \int_s^t \left| I_A z_{r-} + x_{r-} \theta_r^N + I_A \theta_r^N z_{r-} \right|^2 dN_r.$

Denote

$\overline{\tau}_n = \inf \{ t \geq s : |x_t| > n \}.$

By Lemma 3.9

$E |x_{t \wedge \overline{\tau}_n}|^2 \leq E \int_s^{t \wedge \overline{\tau}_n} k_0^2 |x_r|^2 dr + 3E \int_s^{t \wedge \overline{\tau}_n} (|z_r|^2 + k_0^2 |x_r|^2 + k_0^2 |z_r|^2) \lambda_r dr$
$\leq k_0' \int_s^t (E |x_{r \wedge \overline{\tau}_n}|^2 + E |z_r|^2)(1 + \lambda_r) dr.$

Hence, by Lemma 3.9, Gronwall's inequality and Fatou's lemma $\forall T \geq t \geq s$

$E |x_t|^2 \leq k_T e^{k_0' \int_s^t (1+\lambda_r) dr},$

where $0 \leq k_T$ is a constant depending on T and k_0 only, and $k_0' \geq 0$ is a constant depending on k_0 only. Therefore $x_t, s \leq t \leq T$, is a $P-$martingale. This shows that

$E[(\widetilde{N_t'} - \widetilde{N_s'}) z_t I_A] = 0.$

Therefore $\widetilde{N_t'}$ is a $\widetilde{P}-$martingale. Note that

$\widetilde{N_t'} = N_t - \int_0^t \lambda_s (1 + \theta_s^N) ds.$

Hence $\widetilde{N_t'}$ is a $\widetilde{P}-$martingale with compensator $\int_0^t \lambda_s (1 + \theta_s^N) ds.$

2): Now suppose that $\forall 0 \leq T < \infty$

$\int_0^T |\theta_t^w|^2 dt < \infty, \ \int_0^T |\theta_t^N|^2 dt < \infty, \ P - a.s.$

Let

$\theta_{n,t}^w = \theta_t^w I_{|\theta_t^w| < n}, \ \theta_{n,t}^N = \theta_t^N I_{|\theta_t^N| < n}, \ z_{n,t}^w = \exp[\int_0^t \theta_{n,s}^w \cdot dw_s - \frac{1}{2} \int_0^t |\theta_{n,s}^w|^2 ds],$

etc. Then it is obvious that as $n \to \infty$

$\int_0^T |\theta_t^w - \theta_{n,t}^w|^2 dt \to 0, \ \int_0^T |\theta_t^N - \theta_{n,t}^N|^2 \lambda_t dt \to 0.$

Hence, as $n \to \infty$

$\int_0^T |\theta_t^w - \theta_{n,t}^w|^2 dt \to 0, \ \int_0^T |\theta_t^N - \theta_{n,t}^N|^2 \lambda_t dt \to 0,$ in probability.

Denote

$$d\widetilde{P}_n|_{\Im_t} = z_{n,t}dP,$$
$$w_{n,t} = w_t - \int_0^t \theta^w_{n,s}ds,\ \widetilde{N'_{n,t}} = \tilde{N}_t - \int_0^t \lambda_s\theta^N_{n,s}ds = N_t - \int_0^t \lambda_s(1+\theta^N_{n,s})ds.$$
Then by above $w_{n,t}, t \in [0,T]$, is a BM under probability measure $\widetilde{P}_n$, and $\widetilde{N'_t}, t \in [0,T]$, is a $\widetilde{P}_n-$ martingale with compensator $\int_0^t \lambda_s(1+\theta^N_{n,s})ds$. Moreover, $\forall t \in [0,T]$, as $n \to \infty$
$$w_{n,t} \to w_t,\ \widetilde{N'_{n,t}} \to \tilde{N}_t, \text{ in probability.}$$
Now let us show that as $n \to \infty$
$$E[e^{i\eta\cdot(w'_{n,t}-w'_{n,s})}z_{n,t}I_A] \to E[e^{i\eta\cdot(w'_t-w'_s)}z_tI_A].\ (@)$$
At first, since $w_{n,t}$ converges to w_t in probability,
$$E[e^{i\eta\cdot(w'_{n,t}-w'_{n,s})}z_tI_A] \to E[e^{i\eta\cdot(w'_t-w'_s)}z_tI_A], \text{ as } n \to \infty.$$
Secondly, since by assumption $Ez_t = Ez_{n,t} = 1,\ \forall n, \forall 0 \le t \le T$,
$$E|z_t - z_{n,t}| = E(|z_t - z_{n,t}| + z_t - z_{n,t}).$$
However,
$$0 \le |z_t - z_{n,t}| + z_t - z_{n,t} \le 2z_t,$$
and as $n \to \infty$ for a fixed $t \in [s,T]$ (take a subsequence, if necessary)
$$|z_t - z_{n,t}| + z_t - z_{n,t} \to 0, \text{ in probability.}$$
(Indeed, as $t \in [0,T]$
$$P(\left|\int_0^t \theta^w_{n,s}\cdot dw_s - \int_0^t \theta^w_s \cdot dw_s\right| > \varepsilon) \le P(\int_0^T \left|\theta^w_t - \theta^w_{n,t}\right|^2 dt > \delta) + \varepsilon^{-2}\delta.$$
Hence, as $n \to \infty$.
$$\int_0^t \theta^w_{n,s}\cdot dw_s \to \int_0^t \theta^w_s \cdot dw_s, \text{ in probability.}$$
From this as $n \to \infty$
$$z^w_{n,t} \to z^w_t, \text{ in probability.}$$
On the other hand, since $z^N_{n,t} \to z^N_t$, as $n \to \infty$, $\forall\omega$. Therefore, for each fixed $t \in [0,T]$ one can choose a subsequence $\{n_k\}$, denote it by $\{n\}$ again, such that as $n \to \infty$
$$z_{n,t} = z^w_{n,t}z^N_{n,t} \to z^w_t z^N_t = z_t).$$
Applying Lebesgue's dominated convergence theorem one finds that for a fixed $t \in [s,T]$ as $n \to \infty$
$$E|z_t - z_{n,t}| = E(|z_t - z_{n,t}| + z_t - z_{n,t}) \to 0.$$
Therefore, $\forall A \in \Im_s$, for a fixed $t \in [s,T]$ by Lebesgue's dominated convergence theorem again
$$\left|Ee^{i\eta\cdot(w'_{n,t}-w'_{n,s})}z_{n,t}I_A - Ee^{i\eta\cdot(w'_t-w'_s)}z_tI_A\right| \le E|z_{n,t} - z_t|$$
$$+E(\left|e^{i\eta\cdot(w'_{n,t}-w'_{n,s})} - e^{i\eta\cdot(w'_t-w'_s)}\right| z_t) \to 0, \text{ as } n \to \infty.$$
(@) is proved. Recall that
$$Ee^{i\eta\cdot(w'_{n,t}-w'_{n,s})}z_{n,t}I_A = e^{-\frac{1}{2}|\eta|^2(t-s)}Ez_{n,t}I_A.$$
By letting $n \to \infty$ one finds that
$$Ee^{i\eta\cdot(w'_t-w'_s)}z_tI_A = e^{-\frac{1}{2}|\eta|^2(t-s)}Ez_tI_A = \widetilde{P}_T(A)e^{-\frac{1}{2}|\eta|^2(t-s)},$$
where we have used the result that
$$Ez_tI_A = E(z_tI_AE(z^t_T|\Im_t)) = Ez_TI_A = \widetilde{P}_T(A).$$
Therefore, $w'_t, t \in [0,T]$, is a $\widetilde{P}_T-$BM. Now let
$$\tau_k = \inf\{t \ge 0 : N'_t > k\}.$$

By $E[(\widetilde{N'_{n,t}} - \widetilde{N'_{n,s}})z_{n,t}I_A] = 0$ By letting $n \to \infty$ one has that

$E[(\widetilde{N'_{t\wedge\tau_k}} - \widetilde{N'_{s\wedge\tau_k}})z_{n,t}I_A] = 0.$

Therefore, $\widetilde{N'_t}, t \in [0,T]$, is a local $\widetilde{P}_T$ martingale. ■

3.11 Corollary. If $P - a.s.$ $\theta_t^N = 0$, $\forall t \geq 0$, and

$\int_0^T |\theta_t^w|^2 dt < \infty, \forall 0 \leq T < \infty,$

then $w'_t, t \geq 0$,is a $\widetilde{P}-$BM, and $\widetilde{N_t}, t \geq 0$,is a $\widetilde{P}-$Poisson process with the same compensator $\int_0^t \lambda_s ds$.

In the following, we always make the assumption of Corollary 3.11:

$\int_0^T |\theta_t^w|^2 dt < \infty, \forall 0 \leq T < \infty;$ $\theta_t^N = 0$, $\forall t \geq 0$.

By the same idea one can obtain the following

3.12 Theorem.(Girsanov type Theorem). Denote

(3.9) $d\widetilde{P}_t = z_t^w dP.$

Assume that $\hat{P}_T$ is a probability measure, and assume that $(\Omega, \Im)$ is a standard measurable space. Then there exists a probability measure $\widetilde{P}$ defined on $(\Omega, \Im)$ such that $\widetilde{P}\,|_{\Im_T} = \hat{P}_T$, for each $0 \leq T < \infty$; moreover,

1) $w'_t = w_t - \int_0^t \theta_s^w ds, 0 \leq t,$

is a BM under probability $\hat{P}$;

2) $q(dt,dz) = p(dt,dz) - \pi(dz)dt$

is still a Poisson random martingale measure with the same compensator $\pi(dz)dt$ under probability $\hat{P}$, where $p(dt,dz)$ is a Poisson counting measure with compensator $\pi(dz)dt$ under the original probability P, and $\pi(.)$ is a $\sigma-$finite measure on $(Z, \Re(Z))$.

Proof. The proof of 1) is the same as that of Theorem 3.10.

2): Now by Ito's formula again for all $A \in \Im_s, U_1, \cdots, U_m \in \Re(Z)$, where $\{U_i\}_{i=1}^m$ are disjointed such that $\pi(U_i) < \infty, i = 1, 2, \cdots, m$; and for $\lambda_1, \cdots, \lambda_m > 0, 0 \leq s < t$

$f(t)z_t^w - f(s)z_s^w$

$=\exp\{-\sum_{i=1}^m \lambda_i p((0,t],U_i)\}\, z_t^w - \exp\{-\sum_{i=1}^m \lambda_i p((0,s],U_i)\}\, z_s^w$

$= \int_s^t \int_Z f(r) z_r^w (e^{-\sum_{i=1}^m \lambda_i I_{U_i}(z)} - 1)\pi(dz)dr + M_t,\ P - a.s.,$

where

$M_t = \int_s^t f(r) z_r^w \theta_r^w \cdot dw_r + \int_s^t \int_Z f(r) z_r^w (e^{-\sum_{i=1}^m \lambda_i I_{U_i}(z)} - 1) q(ds,dz).$

Now assume that θ_r^w is bounded temporarily:

$|\theta_r^w| \leq k_0.$

Then by Lemma 3.9 $E\left|z_t^w\right|^2 \leq k_T$, as $t \leq T$. Note that $0 \leq f(r) \leq 1$, and by the Lagrange mean value formula

$\left|e^{-\sum_{i=1}^m \lambda_i I_{U_i}(z)} - 1\right| \leq \sum_{i=1}^m \lambda_i I_{U_i}(z).$

Hence in this case it is easily seen that $M_t, 0 \leq t \leq T$, is a $P-$martingale. It yields that

$Ez_t^w I_A[f(t)/f(s)] - Ez_s^w I_A = \sum_{i=1}^m (e^{-\lambda_i} - 1)\pi(U_i) \int_s^t Ez_r^w I_A[f(r)/f(s)]dr.$

Therefore

$E_{\widetilde{P}}\exp\{-\sum_{i=1}^m \lambda_i p((s,t],U_i)\} I_A = \widetilde{P}(A)\exp[(t-s)\sum_{i=1}^m (e^{-\lambda_i}-1)\pi(U_i)]$.
This means that $p(dt,dz)$ is still a Poisson counting measure with the same compensator $\pi(dz)dt$ under this new probability measure $\widetilde{P}$.
Now consider the general case: $\forall 0 \leq T < \infty$

$\int_0^T |\theta_t^w|^2 \, dt < \infty$.

Define $\theta_{n,t}^w$ and $z_{n,t}^w$ as that in the proof of 2) in Theorem 3.10. Denote

$d\widetilde{P}_n|_{\Im_t} = z_{n,t}^w dP$, $w_{n,t} = w_t - \int_0^t \theta_{n,s}^w ds$.

Then by above one has that as $0 \leq s \leq t \leq T$

$Ez_{n,t}^w I_A \exp\{-\sum_{i=1}^m \lambda_i p((s,t],U_i)\} = Ez_{n,t}^w I_A \exp[(t-s)\sum_{i=1}^m (e^{-\lambda_i}-1)\pi(U_i)]$.

Since as $n \to \infty$

$E\left|z_{n,t}^w - z_t^w\right| I_A exp\{-\sum_{i=1}^m \lambda_i p((s,t],U_i)\} \leq E\left|z_{n,t}^w - z_t^w\right| \to 0$,

which can be proved as that in Theorem 3.10. Hence letting $n \to \infty$ one obtains

$E_{\widetilde{P}}\exp\{-\sum_{i=1}^m \lambda_i p((s,t],U_i)\} I_A = \widetilde{P}(A)\exp[(t-s)\sum_{i=1}^m (e^{-\lambda_i}-1)\pi(U_i)]$.

Therefore, we have proved that for any given $0 \leq T < \infty$, w_t' and $q(dt,dz)$ on $t \in [0,T]$ is a $\widetilde{P}_T$–BM and $\widetilde{P}_T$ Poisson martingale measure with the same compensator $\pi(dz)dt$, respectively. However, $\widetilde{P}|_{\Im_T} = \widetilde{P}_T, \forall T \geq 0$. Hence the conclusion can be derived for $t \geq 0$. The proof is complete. ■

Theorem 3.12 is a powerful tool in the study of the existence of weak solutions for RSDE. For example, by using Theorem 3.1 and 3.12 we easily obtain the following

3.13 Theorem. Assume that for $t \geq 0, x \in \overline{\Theta}, z \in Z$

1° $b(t,x), \sigma(t,x), c(t,x,z)$ are jointly measurable, and there exists a constant $k_0 > 0$ such that

$|b|^2 + \|\sigma\|^2 + \int_Z |c|^2 \pi(dz) \leq k_0$,

2° $\sigma(t,x)$ is jointly continuous, and as $|x-y| \to 0, |t-s| \to 0$

$\int_0^t |c(t,x,z) - c(s,y,z)|^2 \pi(dz) \to 0$,

3° there exists a constant $\delta > 0$ such that as $(t,x) \in [0,T] \times \overline{\Theta}$

$(A\lambda, \lambda) \geq \delta |\lambda|^2$,

where $A = 2^{-1}\sigma\sigma^*$;

4° $\pi(dz) = dz/|z|^{d+1}$, $(Z = R^d - \{0\})$.

Then (1.1) with (1.2) has a weak solution. Furthermore, if the following conditions also hold:

5° $2(x-y)\cdot(b(t,x,\omega) - b(t,y,\omega)) + \|\sigma(t,x,\omega) - \sigma(t,y,\omega)\|^2$

$+ \int_0^t |c(t,x,z,\omega) - c(t,y,z,\omega)|^2 \pi(dz) \leq k_N(t)\rho_N(|x-y|^2)$,

as $|x|, |y| \leq N$, where $0 \leq k_N(t), \int_0^t k_N(t)ds < \infty$, and $\rho_N(u)$ is concave and strictly increasing in $u \geq 0$ such that $\rho_N(0) = 0$, and

$\int_{0+} du/\rho_N(u) = \infty$, for each $N = 1, 2, ...$,

6° $x + c(t,x,z,\omega) \in \overline{\Theta}$,

but if condition 1° is weaken to

1°′ $|b|^2 + \|\sigma\|^2 \leq k_0(1+|x|^2), \int_Z |c|^2 \pi(dz) \leq k_0$,

then (1.1) with (1.2) has a pathwise unique strong solution. Moreover, in this case $\phi(t)$ is continuous.

Proof. Case 1. Assume that

$$|b| + \|\sigma\|^2 + \int_Z |c|^2 \pi(dz) \leq k_0',$$

and assume that conditions $2° - 4°$ hold. By Theorem 3.1 there exists a weak solution (x_t, ϕ_t) defined on some probability space $(\Omega, \Im)$ (without losing any generality we may assume that $(\Omega, \Im)$ is a standard measurable space) satisfying the following RSDE

$$(3.10) \begin{cases} x_t = x_0 + \int_0^t \sigma(s, x_s) dw_s + \int_0^t \int_Z c(s, x_{s-}, z) q(ds, dz) + \phi_t, \\ \quad \text{and the other statements in (1.2) hold for } (x_t, \phi_t), \end{cases} \quad P - a.s.$$

Let $d\widetilde{P}_t = z_t dP$, where

$$z_t = \exp[\int_0^t \theta_s \cdot dw_s - \frac{1}{2} \int_0^t |\theta_s|^2 ds], \ \theta_t = \sigma^{-1}(s, x_s) b(t, x_s).$$

Then by Lemma 3.9, Theorem 3.12 there exists a probability measure $\widetilde{P}$ defined on $(\Omega, \Im)$ such that $\widetilde{P}|_{\Im_T} = \hat{P}_T$, for each $0 \leq T < \infty$; moreover,

1) $w_t' = w_t - \int_0^t \theta_s ds, 0 \leq t,$

is a BM under probability $\hat{P}$;

2) $q(dt, dz) = p(dt, dz) - \pi(dz)dt$

is still a Poisson random martingale measure with the same compensator $\pi(dz)dt$ under probability $\hat{P}$. Therefore by (3.10) we get that (x_t, ϕ_t) satisfies the following RSDE

$$\begin{cases} x_t = x_0 + \int_0^t b(s, x_s) ds + \int_0^t \sigma(s, x_s) dw_s' + \int_0^t \int_Z c(s, x_{s-}, z) q(ds, dz) + \phi_t, \\ \quad \text{and the other statements in (1.2) hold for } (x_t, \phi_t). \end{cases}$$

$\tilde{P} - a.s.$

Now assume that conditions $5°$ and $6°$ also hold. Then by the same proof as that of Theorem 3.7 we can get that (1.1) with (1.2) has a pathwise unique strong solution. Moreover, $\phi(t)$ is continuous.

Case 2. (General case). In the case that $1°'$ is satisfied, the proof can be completed just as in Theorem 3.8. ■

When using Theorem 3.12 the following theorem is also useful.

3.14 Theorem.(Girsanov type Theorem). Assume that (x_t, ϕ_t) solves the following RSDE

$$\begin{cases} x_t = x_0 + \int_0^t \sigma(s, x_s) dw_s + \int_0^t \int_Z c(s, x_{s-}, z) q(ds, dz) + \phi_t, \\ \quad \text{and the other statements in (1.2) hold for } (x_t, \phi_t), \end{cases}$$

moreover, assume that $\phi(t)$ is continuous, and

$$|b|^2 + \|\sigma\|^2 + \int_Z |c|^2 \pi(dz) \leq k_0(1 + |x|^2),$$

σ^{-1} exists such that

$$\|\sigma\|^{-1} \leq k_0.$$

Denote

$$z_t(\sigma^{-1}b) = \exp[\int_0^t (\sigma^{-1}b)(s, x_s) \cdot dw_s - \frac{1}{2} \int_0^t |(\sigma^{-1}b)(s, x_s)|^2 ds],$$

$$d\widetilde{P}_t = z_t dp.$$

Then $\widetilde{P}_t$ is a probability measure.

Proof. Let

$$b^N(t, x) = \begin{cases} b(t, x), & \text{as } |x| \leq N, \\ 0, & \text{otherwise.} \end{cases}$$

Then by Lemma 3.9

$d\widetilde{P}_t^N = z_t(\sigma^{-1}b^N)dp$

is a probability measure. Applying Theorem 3.12 for any given $0 \leq T < \infty$ as $0 \leq t \leq T$

1) $w'_t = w_t - \int_0^t (\sigma^{-1}b)(s, x_s)ds$,

is a BM under probability $\hat{P}_T^N$;

2) $q(dt, dz) = p(dt, dz) - \pi(dz)dt$

is still a Poisson random martingale measure with the same compensator $\pi(dz)dt$ under probability $\hat{P}_T^N$, where $p(dt, dz)$ is a Poisson counting measure with compensator $\pi(dz)dt$ under the original probability P. Therefore (x_t, ϕ_t) solves the following RSDE as $0 \leq t \leq T$

$$\begin{cases} x_t = x_0 + \int_0^t b(s, x_s)ds + \int_0^t \sigma(s, x_s)dw'_s + \int_0^t \int_Z c(s, x_{s-}, z)q(ds, dz) + \phi_t, \\ \qquad \text{and the other statements in (1.2) hold for } (x_t, \phi_t), \end{cases}$$

$\hat{P}_T^N - a.s.$

By Corollary 2.7

$E^{\tilde{P}_T^N}(\sup_{t \leq T} x_t^2) \leq k_T$.

Hence

$N^2 \hat{P}_T^N(\tau_N < T) \leq k_T$,

where

$\tau_N = \inf(t \in [0, T] : |x_t| > N)$, and $\tau_N = T$, for $\inf\{\phi\}$.

Therefore, as $N \to \infty$

$\tilde{P}_T^N(\tau_N < T) \to 0$.

It yields that

$Ez_T(\sigma^{-1}b) \geq Ez_T(\sigma^{-1}b^N)I_{\tau_N \geq T} = P^N(\tau_N \geq T) \to 1$, as $N \to \infty$.

On the other hand we always have that

$Ez_T(\sigma^{-1}b) \leq 1$.

The proof is complete. ■

3.15 Theorem. Assume that

1° $b(t, x), \sigma(t, x), c(t, x, z)$ are jointly measurable, and there exists a constant $k_0 > 0$ such that

$\|\sigma\|^2 + \int_Z |c|^2 \pi(dz) \leq k_0$, $|b|^2 \leq k_0(1 + |x|)$

2° $\sigma(t, x)$ is jointly continuous, and as $|x - y| \to 0, |t - s| \to 0$

$\int_0^t |c(t, x, z) - c(s, y, z)|^2 \pi(dz) \to 0$,

3° there exists a constant $\delta > 0$ such that as $(t, x) \in [0, T] \times R^d$

$(A\lambda, \lambda) \geq \delta |\lambda|^2$,

where $A = 2^{-1}\sigma\sigma^*$;

4° $\pi(dz) = dz/|z|^{d+1}$, $(Z = R^d - \{0\})$,

5° $x + c(t, x, z, \omega) \in \overline{\Theta}$.

Then (1.1) with (1.2) has a weak solution. Moreover, $\phi(t)$ is continuous. Furthermore, if

6° $2(x - y) \cdot (b(t, x, \omega) - b(t, y, \omega)) + \|\sigma(t, x, \omega) - \sigma(t, y, \omega)\|^2$
$+ \int_0^t |c(t, x, z, \omega) - c(t, y, z, \omega)|^2 \pi(dz) \leq k_N(t)\rho_N(|x - y|^2)$, as $|x|, |y| \leq N$,

where $0 \leq k_N(t), \int_0^t k_N(t)ds < \infty$, and $\rho_N(u)$ is concave and strictly increasing in $u \geq 0$ such that $\rho_N(0) = 0$, and

$\int_{0+} du/\ \rho_N(u) = \infty$, for each $N = 1, 2, ...,$

and in this case condition 1° can be weakened to

$1^{\circ\prime}$ $b(t,x), \sigma(t,x), c(t,x,z)$ are jointly measurable, and there exists a constant $k_0 > 0$ such that

$|b|^2 + \|\sigma\|^2 \leq k_0(1 + |x|), \int_Z |c|^2 \pi(dz) \leq k_0,$

then (1.1) with (1.2) has a pathwise unique strong solution.

Proof. The proof for the existence of a weak solution is a combination of the conclusions of Theorem 3.14 and 3.12. The proof for the second part is completely the same as that of the general case for Theorem 3.8. ■

3.3.3 Case for Coordinate Planes as a Boundary

Consider RSDE (1.1) with (1.2) in a special domain in this subsection, which is useful for the stochastic population model:

(3.11) $\Theta = \{x = (x^1, ...x^d) \in R^d : x^i > 0, 1 \leq i \leq d\} = R^d_+.$

Obviously, if we let $e = (\frac{1}{\sqrt{d}}, ..., \frac{1}{\sqrt{d}})$, then there exists a constant $c_0 > 0$ such that

$e \cdot n \geq c_0 > 0$, for all $n \in \cup_{x \in \partial\Theta} \aleph_x$.

Hence, Assumption 2.7 of Chapter 2 is satisfied. In this case, we can discuss the existence of solutions to RSDE with discontinuous coefficients. First, we introduce the following Krylov estimate.

3.16 Theorem. Assume that $b(t,x), \sigma(t,x)$ and $c(t,x,z)$ are non-random and jointly measurable, where b and c are R^d-valued, σ is $R^d \otimes R^d$-valued, which satisfy as $(t,x) \in [0,T] \times R^d$

$|b| + \|\sigma\|^2 + \int_Z |c|^2 \pi(dz) \leq k_0,$

and there exists a constant $\delta > 0$ such that as $(t,x) \in [0,T] \times R^d$

$(A\lambda, \lambda) \geq \delta |\lambda|^2,$

where $A = 2^{-1}\ \sigma\ \sigma^*$.Then for all $0 \leq f$- bounded measurable function and $p > d + 1$,

$E \int_0^T f(s, x_s)ds \leq k(k_0, \delta, d, p, T) \|f\|_{L_p([0,T] \times R^d)},$

where (x_t, ϕ_t) satisfies (1.1) (with (1.2)) on $[0,T]$, and k is a constant depending on k_0, δ, d, p, T only.

Proof. We only need to note that by Lemma 2.8 in Chapter 2 it is easily shown that

$E |\phi|^2_T \leq k_T < \infty,$

and

$E \sum_{0<s\leq t} |\Delta x_s|^2 \leq 2E \int_0^t \int_Z |c(s, x_s, z)|^2 \pi(dz) + 2E |\phi|_t \sum_{0<s\leq t} |\Delta\phi_s|$

$\leq 2(k_0 + E |\phi|^2_t) \leq \widetilde{k}_T < \infty.$

Hence, Corollary 1.7 applies. ■

The following theorem for Θ defined by (3.11) is with a weaker condition on c, since we have not placed any continuity condition on c.

3.17 Theorem. Suppose that the assumption of Theorem 3.16 holds on $(t,x) \in [0,\infty) \times R^d$ and suppose that

$\pi(dz) = dz/|z|^{d+1}$,

and for all $t \geq 0, x \in R^d, z \in Z = R^d - \{0\}$, $x_i + c_i(t,x,z) \geq 0, 1 \leq i \leq d$.

Then (1.1) with (1.2) (in $\overline{R^d_+}$) has a weak solution (x_t, ϕ_t) on $t \geq 0$. Moreover, in this case ϕ_t is continuous.

Proof. Let us smooth out b, σ, c to get b^n, σ^n, c^n as follows. Let for $y \in R^d$

$f_d(y) = c\exp[(1-|y|^2)^{-1}]$, as $|y| < 1$; $f_d(y) = 0$, otherwise;

where c is a constant such that

$\int_{R^d} f_d(y)dy = 1.$

For $n = 1, 2, ..., t \in [0, \infty)$ set

$b^n(t,x) = \int_{R^1 \times R^d} b(t - n^{-1}u, x - n^{-1}y) f_1(u) f_d(y) dy du,$

where we let $b(t,x) = 0$, as $t < 0$.

$\sigma^n(t,x)$ and $c^n(t,x,z)$ are defined similarly. Note that for all $1 \leq i \leq d$

$x_i + c_i^n(t,x,z)$

$= \int_{R^1 \times R^d} [x_i - n^{-1}y_i + c_i(t - n^{-1}u, x - n^{-1}y, z) + n^{-1}y_i] f_1(u) f_d(y) dy du$

$= \int_{R^1 \times R^d} [x_i - n^{-1}y_i + c_i(t - n^{-1}u, x - n^{-1}y, z)] f_1(u) f_d(y) dy du \geq 0.$

Hence by Theorem 3.8 there exists a pathwise unique strong solution (x_t^n, ϕ_t^n), for each $n = 1, 2, ...$, which satisfies (1.1) and (1.2) with the coefficients b^n, σ^n and c^n. Moreover, ϕ_t^n is continuous. Let us denote this RSDE by $(1.1)_n$ and $(1.2)_n$.

Now, similar to the proof of Theorem 3.1, we have that there exist a probability space $(\widetilde{\Omega}, \widetilde{\Im}, \widetilde{P})$ and cadlag processes $(\widetilde{x}_t^n, \widetilde{\phi}_t^n, \widetilde{w}_t^n, \widetilde{\zeta}_t^n)$ defined on it with the same finite-dimensional probability distributions as that of $(x_t^n, \phi_t^n, w_t, \zeta_t)$ and there exist $(\widetilde{x}_t^0, \widetilde{\phi}_t^0, \widetilde{w}_t^0, \widetilde{\zeta}_t^0)$, where $\widetilde{w}_t^0$ is a BM on $(\widetilde{\Omega}, \widetilde{\Im}, \widetilde{P})$, and there exists a subsequence $\{n_k\}$ of $\{n\}$, denote it by $\{n\}$ again, such that as $n \to \infty$

$\widetilde{\eta}_t^n \to \widetilde{\eta}_t^0$, in probability, as $\widetilde{\eta}_t^n = \widetilde{x}_t^n, \widetilde{\phi}_t^n, \widetilde{w}_t^n, \widetilde{\zeta}_t^n, n = 0, 1, 2, \cdots.$

Set for $n = 0, 1, 2, \cdots$

$\widetilde{p}^n(dt,dz) = \sum_{s \in dt} I_{(0 = \Delta \widetilde{\zeta}_s^n \in dz)}(s)$, $\widetilde{q}^n(dt,dz) = \widetilde{p}^n(dt,dz) - \pi(dz)dt.$

Then $\widetilde{p}^n(dt,dz)$ is a Poisson random point measure with compensator $\pi(dz)dt$ for each $n = 0, 1, 2, ...$, and it satisfies the condition

$\widetilde{\zeta}_t^n = \int_0^t \int_{|z| \leq 1} z \widetilde{q}^n(ds,dz) + \int_0^t \int_{|z|>1} z \widetilde{p}^n(ds,dz), n = 0, 1, 2, \cdots.$

By the coincidence of finite probability distributions one has that $(\widetilde{x}_t^n, \widetilde{\phi}_t^n)$ satisfies $(1.1)_n$ (with $(1.2)_n$) with $\widetilde{w}_t^n$ and $\widetilde{q}^n(dt,dz)$ on $(\widetilde{\Omega}, \widetilde{\Im}, \widetilde{P})$. Now let us show that as $n \to \infty$

$(3.11)_1$ $\left| \int_0^t (b^n(s, \widetilde{x}_s^n) - b(s, \widetilde{x}_s^0)) ds \right|^2 \to 0$, in probability,

$(3.11)_2$ $\int_0^t \sigma^n(s, \widetilde{x}_s^n) d\widetilde{w}_s^n \to \int_0^t \sigma(s, \widetilde{x}_s^0) d\widetilde{w}_s^0$, in probability,

$(3.11)_3$ $\int_0^t \int_Z c^n(s, \widetilde{x}_s^n, z) \widetilde{q}^n(ds,dz) \to \int_0^t \int_Z c(s, \widetilde{x}_s^0, z) \widetilde{q}^0(ds,dz)$, in probability,

$(3.11)_4$ $\int_0^t \int_Z \left| c^n(s, \widetilde{x}_{s-}^n, z) - c(s, \widetilde{x}_{s-}^0, z) \right|^2 \pi(dz) ds \to 0$, in probability.

Note that for each $n^0 = 1, 2, ...$

$$\int_0^t \int_Z \left|c^n(s, \widetilde{x}_s^n, z) - c(s, \widetilde{x}_s^0, z)\right|^2 \pi(dz)ds$$
$$\leq 3\int_0^t \int_Z \left|c^n(s, \widetilde{x}_s^n, z) - c^{n^0}(s, \widetilde{x}_s^n, z)\right|^2 \pi(dz)ds$$
$$+3\int_0^t \int_Z \left|c^{n^0}(s, \widetilde{x}_s^n, z) - c^{n^0}(s, \widetilde{x}_s^0, z)\right|^2 \pi(dz)ds$$
$$+3\int_0^t \int_Z \left|c^{n^0}(s, \widetilde{x}_s^0, z) - c(s, \widetilde{x}_s^0, z)\right|^2 \pi(dz)ds = 3(I_1^{n,n^0} + I_2^{n,n^0} + I_3^{n^0}).$$

However, for any $\varepsilon > 0$ by Theorem 3.16

$$\tilde{P}(I_1^{n,n^0} > \varepsilon) \leq \tilde{P}(\sup_{s\leq T} |\widetilde{x}_s^n| > N)$$
$$+\widetilde{k}\varepsilon^{-2} \left\|\int_Z \left|c^n - c^{n^0}\right|^2 \pi(dz)\right\|_{d+1,[0.T]\times S_N},$$
$$\tilde{P}(I_3^{n,n^0} > \varepsilon) \leq \tilde{P}(\sup_{s\leq T} |\widetilde{x}_s^0| > N)$$
$$+\widetilde{k}\varepsilon^{-2} \left\|\int_Z \left|c^{n^0} - c\right|^2 \pi(dz)\right\|_{d+1,[0.T]\times S_N}.$$

On the other hand, for each n^0 fixed, as $n \to \infty$

$$\int_Z \left|c^{n^0}(s, \widetilde{x}_s^n, z) - c^{n^0}(s, \widetilde{x}_s^0, z)\right|^2 \pi(dz) \leq k_{n^0} \left|\widetilde{x}_s^n - \widetilde{x}_s^0\right|^2 \to 0, \text{ in probability,}$$

and

$$\int_Z \left|c^{n^0}(s, \widetilde{x}_s^n, z) - c^{n^0}(s, \widetilde{x}_s^0, z)\right|^2 \pi(dz) \leq 2k_0.$$

Hence by Lebesgue's dominated convergence theorem, one has that for each n^0 fixed, as $n \to \infty$

$$\widetilde{E}\int_Z \left|c^{n^0}(s, \widetilde{x}_s^n, z) - c^{n^0}(s, \widetilde{x}_s^0, z)\right|^2 \pi(dz) \to 0.$$

Therefore, for each n^0 fixed, as $n \to \infty$

$$(3.12)\quad \widetilde{E}\int_0^t \int_Z \left|c^{n^0}(s, \widetilde{x}_s^n, z) - c^{n^0}(s, \widetilde{x}_s^0, z)\right|^2 \pi(dz)ds \to 0.$$

From these results it is obvious that as $n \to \infty$

$$\tilde{P}(\int_0^t \int_Z \left|c^n(s, \widetilde{x}_s^n, z) - c(s, \widetilde{x}_s^0, z)\right|^2 \pi(dz)ds > 3\varepsilon) \to 0.$$

$(3.11)_4$ is proved.

The proof of $(3.11)_3$ is similar. One only needs to show that for each fixed n^0 as $n \to \infty$

$$I_t^n = \tilde{P}(\left|\int_0^t \int_Z c^{n^0}(s, \widetilde{x}_s^n, z)\widetilde{q}^n(ds, dz) - \int_0^t \int_Z c^{n^0}(s, \widetilde{x}_s^0, z)\widetilde{q}^0(ds, dz)\right| > \varepsilon)$$
$$\to 0.$$

However,

$$I_t^n$$
$$\leq \tilde{P}(\left|\int_0^t \int_Z c^{n^0}(s, \widetilde{x}_s^n, z)\widetilde{q}^n(ds, dz) - \int_0^t \int_Z c^{n^0}(s, \widetilde{x}_s^0, z)\widetilde{q}^n(ds, dz)\right| > \varepsilon/2)$$
$$+\tilde{P}(\left|\int_0^t \int_Z c^{n^0}(s, \widetilde{x}_s^0, z)\widetilde{q}^n(ds, dz) - \int_0^t \int_Z c^{n^0}(s, \widetilde{x}_s^0, z)\widetilde{q}^0(ds, dz)\right| > \varepsilon/2)$$
$$= \widetilde{I}_t^{1n} + \widetilde{I}_t^{2n}.$$

Obviously, by (3.12) as $n \to \infty$

$$\widetilde{I}_t^{1n} \leq (2/\varepsilon)^2 E\int_0^t \int_Z \left|c^{n^0}(s, \widetilde{x}_s^n, z) - c^{n^0}(s, \widetilde{x}_s^0, z)\right|^2 \pi(dz)ds \to 0.$$

On the other hand,

$\widetilde{I}_t^{2n}$
$= \tilde{P}(\left|\int_0^t \int_Z c^{n^0}(s, \widetilde{x}_s^0, z)\widetilde{p}^n(ds, dz) - \int_0^t \int_Z c^{n^0}(s, \widetilde{x}_s^0, z)\widetilde{p}^0(ds, dz)\right| > \varepsilon/2).$
To show that for each n^0 fixed as $n \to \infty$
(3.13) $\widetilde{I}_t^{2n} \to 0$
we need the following

3.18 Lemma. For any given N and $T > 0, p > 1$
$\lim_{\varepsilon' \to 0} \int_0^T \int_{|x| \le N} \left|\int_{|z| \le \varepsilon'} |c(t, x, z)|^2 \pi(dz)\right|^p dxdt = 0.$

Recall that $Z = R^d - \{0\}$. Then Lemma 3.18 can be derived by Lebesgue's domination convergence theorem.

Now we have that by Theorem 3.16 and Lemma 3.18 as $p \ge d + 1$
$\widetilde{I}_t^{2n} \le \tilde{P}(\sup_{s \le T} |\widetilde{x}_s^0| > N)$
$+ 2(\frac{16}{\varepsilon})^2 k_{T,N} (\int_0^T \int_{|x| \le N} \left|\int_{|z| \le \varepsilon'} |c(t, x, z)|^2 \pi(dz)\right|^p dxdt)^{1/p}$
$+ \tilde{P}(| \int_0^t \int_{|z| > \varepsilon'} c^{n^0}(s, \widetilde{x}_s^0, z)\widetilde{p}^n(ds, dz)$
$- \sum_{i=0}^{2^k - 1} \int_{|z| > \varepsilon'} c^{n^0}(s_{it/2^k}, \widetilde{x}_{s_{it/2^k}}^0, z)\widetilde{p}^n(ds, dz) | > \varepsilon/16)$
$+ \tilde{P}(| \int_0^t \int_{|z| > \varepsilon'} c^{n^0}(s, \widetilde{x}_s^0, z)\widetilde{p}^0(ds, dz)$
$- \sum_{i=0}^{2^k - 1} \int_{|z| > \varepsilon'} c^{n^0}(s_{it/2^k}, \widetilde{x}_{s_{it/2^k}}^0, z)\widetilde{p}^0(ds, dz) | > \varepsilon/16)$
$+ \tilde{P}(| \sum_{i=1}^{2^k} \int_{|z| > \varepsilon'} c^{n^0}(s_{it/2^k}, \widetilde{x}_{s_{it/2^k}}^0, z)\widetilde{p}^n(ds, dz)$
$- \sum_{i=0}^{2^k - 1} \int_{|z| > \varepsilon'} c^{n^0}(s_{it/2^k}, \widetilde{x}_{s_{it/2^k}}^0, z)\widetilde{p}^0(ds, dz) | > \varepsilon/16).$

From this, applying Lemma 3.4 and so on, it is not difficult to show that (3.13) holds. Hence $(3.11)_3$ is true. The proof of $(3.11)_1$ and $(3.11)_2$ are similar and even simpler. Now by passing to the limit, one finds that $(\widetilde{x}_t^0, \widetilde{\phi}_t^0)$ satisfies
$\widetilde{x}_t^0 = x_0 + \int_0^t b(s, \widetilde{x}_s^0)ds + \int_0^t \sigma(s, \widetilde{x}_s^0)d\widetilde{w}_s^0$
$+ \int_0^t \int_Z c(s, \widetilde{x}_{s-}^0, z)\widetilde{q}^0(ds, dz) + \widetilde{\phi}_t^0 = x_0 + \widetilde{y}_t^0 + \widetilde{\phi}_t^0.$

Now by Theorem 2.9 in Chapter 2 there exists a unique (x_t, ϕ_t) satisfying the Skorohod problem (1.1) and (1.2) with x_0 and $\widetilde{y}_t^0$, i.e.
$x_t = x_0 + \widetilde{y}_t^0 + \phi_t,$
and (x_t, ϕ_t) satisfies the statements in (1.2).
If one can show that as $n \to \infty$ (or for a subsequence tending to ∞)
(3.14) $E \sup_{t \le T} |\widetilde{y}_t^n - \widetilde{y}_t^0|^2 \to 0,$
where
$\widetilde{y}_t^n = \int_0^t b^n(s, \widetilde{x}_s^n)ds + \int_0^t \sigma^n(s, \widetilde{x}_s^n)d\widetilde{w}_s^n + \int_0^t \int_Z c^n(s, \widetilde{x}_{s-}^n, z)\widetilde{q}^n(ds, dz),$
then by Corollary 2.10 in Chapter 2 as $n \to \infty$
$E \sup_{t \le T} |\widetilde{x}_t^n - x_t|^2 \to 0,$
and
$E \sup_{t \le T} \left|\widetilde{\phi}_t^n - \phi_t\right|^2 \to 0.$
By the uniqueness of the limit it follows that
$\widetilde{x}_t^0 = x_t, \ \widetilde{\phi}_t^0 = \phi_t.$

This means that $(\tilde{x}_t^0, \tilde{\phi}_t^0)$ is a weak solution of (1.1) with (1.2). Moreover, by 2) of Remark 2.5 $\tilde{\phi}_t^0$ is continuous. Therefore in what follows we are going to establish (3.14).

However, by using $(3.11)_3$, $\int_Z |c^n|^2 \pi(dz) \leq k_0, \forall n$, and Lemma 3.6 one has that as $n \to \infty$

$$\tilde{E}\left|\int_0^t \int_Z c^n(s, \tilde{x}_s^n, z)\tilde{q}^n(ds, dz) - \int_0^t \int_Z c(s, \tilde{x}_s^0, z)\tilde{q}^0(ds, dz)\right|^2 \to 0,$$

(cf. the proof after Lemma 3.6 for Theorem 3.1). Hence, by using similar discussion (3.14) is easily derived. The proof of Theorem 3.17 is now complete. ■.

Applying the Girsanov type theorem we can obtain a better result on the existence of weak solution for RSDE in this subsection.

3.19 Theorem. Assume that

1° $b(t,x), \sigma(t,x)$ and $c(t,x,z)$ are non-random, and jointly measurable, where b and c are R^d−valued, σ is $R^d \otimes R^d$-valued, which satisfy as $(t,x) \in [0,T] \times R^d$

$$\|\sigma\|^2 + \int_Z |c|^2 \pi(dz) \leq k_0, \ |b|^2 \leq k_0(1+|x|^2),$$

2° there exists a constant $\delta > 0$ such that as $(t,x) \in [0,T] \times R^d$

$$(A\lambda, \lambda) \geq \delta |\lambda|^2,$$

where $A = 2^{-1}\, \sigma\, \sigma^*$;

3° $\pi(dz) = dz/|z|^{d+1}$,

4° for all $t \geq 0, x \in R^d, z \in Z = R^d - \{0\}$, $x_i + c_i(t,x,z) \geq 0, 1 \leq i \leq d$.

Then (1.1) with (1.2) (in $\overline{R_+^d}$) has a weak solution (x_t, ϕ_t) on $t \geq 0$. Moreover, in this case ϕ_t is continuous. Furthermore, if we also have the condition

5° $2(x-y)\cdot(b(t,x,\omega) - b(t,y,\omega)) + \|\sigma(t,x,\omega) - \sigma(t,y,\omega)\|^2$

$$+ \int_0^t |c(t,x,z,\omega) - c(t,y,z,\omega)|^2 \pi(dz) \leq k_N(t)\rho_N(|x-y|^2), \text{ as } |x|, |y| \leq N,$$

where $0 \leq k_N(t)$, $\int_0^t k_N(t)ds < \infty$, and $\rho_N(u)$ is concave and strictly increasing in $u \geq 0$ such that $\rho_N(0) = 0$, and

$$\int_{0+} du/\ \rho_N(u) = \infty, \text{ for each } N = 1, 2, ...,$$

and if the condition 1° is weakened to

1°′ $b(t,x), \sigma(t,x)$ and $c(t,x,z)$ are non-random, and jointly measurable, where b and c are R^d−valued, σ is $R^d \otimes R^d$-valued, which satisfy as $(t,x) \in [0,T] \times R^d$

$$\|\sigma\|^2 + \int_Z |c|^2 \pi(dz) + |b|^2 \leq k_0(1+|x|^2)$$

then (1.1) with (1.2) has a pathwise unique strong solution, where $\phi(t)$ is continuous.

Proof. The proof for the existence of a weak solution under condition 1° is a combination of the conclusions of Theorem 3.17 and 3.14. We now discuss the case for condition 1°′. Let

$$b^N(t,x) = \begin{cases} b(t,x), & \text{as } |x| \leq N, \\ 0, & \text{otherwise.} \end{cases}$$

$\sigma^N(t,x)$ and $c^N(t,x,z)$ are similarly defined. Then by Theorem 3.17 (1.1) with (1.2) with coefficients (b^N, σ^N, c^N) (in $\overline{R_+^d}$) has a weak solution (x_t^N, ϕ_t^N), which satisfies $x_{it-}^N + c_i^N(t, x_{t-}^N, z) \geq 0$, $\forall 1 \leq i \leq d, \forall t \geq 0$. Hence, ϕ_t^N is continuous.

By condition 5° the pathwise uniqueness holds for (1.1) with (1.2) with coefficients (b^N, σ^N, c^N). Hence, (1.1) with (1.2) with coefficients (b^N, σ^N, c^N) has a pathwise unique strong solution (x_t^N, ϕ_t^N) by Theorem 1.3. Moreover, ϕ_t^N is continuous. Now define

$x_t = x_t^N, \phi_t = \phi_t^N$, as $t \in [0, \tau_N)$,

where $\tau_N = \inf(t \geq 0 : |x_t^N| > N)$.

Then the proof for the second part can be completed in the same manner as that for the general case for Theorem 3.8. ■

3.3.4 Existence of Solutions for RSDE with Functional Coefficients Depending on Reflection

In this subsection 3.3.4 we will discuss the existence of solution for RSDE (1.1) with (1.2) and with jumps with functional coefficients depending on reflection, i.e. the coefficients $b(t, x(.), \phi(.)), \sigma(t, x(.), \phi(.))$, $c(t, x(.), \phi(.), z)$and may depend on the whole trajectory of $x_s, s \leq t$, and $\phi_s, s \leq t$. Such a case is useful in the filtering problem and in the stochastic control problem.

3.20 Theorem. Assume that Assumption 2.7 in Chapter 2 holds and assume that

1° $b(t, x(.), \phi(.)), \sigma(t, x(.), \phi(.)), c(t, x(.), \phi(.), z)$ are jointly Borel measurable; for each t, $b(t, x(.), \phi(.)), \sigma(t, x(.), \phi(.))$ are $\sigma(x_s, s \leq t) \times \sigma(\phi_s, s \leq t)$- measurable, and $c(t, x(.), \phi(.), z)$ is $\sigma(x_s, s \leq t) \times \sigma(\phi_s, s \leq t) \times \Re(Z)$ - measurable; and there exists a constant $k_0 > 0$ such that

$|b|^2 + \|\sigma\|^2 + \int_Z |c|^2 \pi(dz) \leq k_0,$

2° $b(t, x(.), \phi(.)), \sigma(t, x(.), \phi(.))$ are Lipschitzian continuous as follows:

$$\left|b(t, x(.), \phi(.)) - b(t, y(.), \widetilde{\phi}(.))\right|^2 + \left|\sigma(t, x(.), \phi(.)) - \sigma(t, y(.), \widetilde{\phi}(.))\right|^2$$
$$+ \int_Z \left|c(t, x, \phi(.), z) - c(t, y, \widetilde{\phi}(.), z)\right|^2 \pi(dz) \leq k_0(|x_t - y_t|^2 + \left|\phi_t - \widetilde{\phi}_t\right|^2),$$
$$|b(t, x(.), \phi(.)) - b(s, x(.), \phi(.))|^2 + |\sigma(t, x(.), \phi(.)) - \sigma(s, x(.), \phi(.))|^2$$
$$+ \int_Z |c(t, x(.), \phi(.), z) - c(s, x(.), \phi(.), z)|^2 \pi(dz) \leq k_0 |t - s|^2,$$

3° $\pi(dz) = dz / |z|^{d+1}$, $(Z = R^d - \{0\})$.

Then (1.1) with (1.2) with coefficients

$(b, \sigma, c) = (b(t, x(.), \phi(.)), \sigma(t, x(.), \phi(.)), c(t, x, \phi(.), z))$

has a weak solution. Furthermore, if

4° $x_{t-} + c(t-, x(.), \phi(.)) \in \overline{\Theta}$, $\forall x(\cdot) \in D^d([0, \infty), \overline{\Theta})$,

then (1.1) with (1.2) with coefficients

$(b, \sigma, c) = (b(t, x(.), \phi(.)), \sigma(t, x(.), \phi(.)), c(t, x, \phi(.), z))$

has a pathwise unique strong solution (x_t, ϕ_t). Moreover, ϕ_t is continuous.

3.21 Remark. $b(t, x(.), \phi(.))$ is $\sigma(x_s, s \leq t) \times \sigma(\phi_s, s \leq t)$ - measurable, means that $b(t, x(.), \phi(.)) = b(t, x_s, \phi_s, s \leq t)$, where b is a Baire function from $[0, \infty) \times D([0, \infty); R^d) \times D([0, \infty); R^d)$ into R^d.

To prove Theorem 3.20 we need Theorem 3.22, which is similar to Theorems 2.1 and 2.3 for the moment estimates and pathwise uniqueness of solution to RSDE (1.1) and (1.2) with Markov-type coefficients:

3.22 Theorem. Assume that

1° $|b(t,x(.),\phi(.))|^2 + \|\sigma(t,x(.),\phi(.))\|^2 + \int_Z |c(t,x(.),\phi(.),z)|^2 \pi(dz)$
$\leq k_0(1+\sup_{s\leq t}|x_s|^2 + \sup_{s\leq t}|\phi_s|^2)$,

2° $x_{t-} + c(t-,x(.),\phi(.)) \in \overline{\Theta}$, $\forall x(\cdot) \in D^d([0,\infty),\overline{\Theta})$.

If (x_t,ϕ_t) satisfies (1.1) with (1.2) with coefficients

$(b,\sigma,c) = (b(t,x(.),\phi(.)),\sigma(t,x(.),\phi(.)),c(t,x,\phi(.),z))$

such that $E|x_0|^2 < \infty$, then as $0 \leq t \leq T$

$E(\sup_{s\leq t}|x_s|^2 + \sup_{s\leq t}|\phi_s|^2) \leq k_T$,

where $0 \leq k_T$ is a constant depending on T and $E|x_0|^2$ only.

Furthermore, assume that

3° $2(x-y)\cdot(b(t,x(.),\phi(.)) - b(t,y(.),\psi(.)))$
$+\|\sigma(t,x(.),\phi(.)) - \sigma(t,y(.),\psi(.))\|^2$
$+\int_Z |c(t,x(.),z,\phi(.)) - c(t,y(.),z,\psi(.))|^2 \pi(dz)$
$\leq k_N(t)\rho_N(\sup_{s\leq t}|x_s - y_s|^2 + \sup_{s\leq t}|\phi_s - \psi_s|^2)$,
as $\sup_{s\leq t}|x_s|, \sup_{s\leq t}|y_s| \leq N, \sup_{s\leq t}|\phi_s|, \sup_{s\leq t}|\psi_s| \leq N$,

where $0 \leq k_N(t)$, $\int_0^t k_N(t)ds < \infty$, and $\rho_N(u)$ is concave and strictly increasing in $u \geq 0$ such that $\rho_N(0) = 0$, and $\int_{0+} du/\ \rho_N(u) = \infty$, for each $N = 1,2,....$
Then the pathwise uniqueness of solution for (1.1) with (1.2) with coefficients $(b(t,x(.),\phi(.)),\sigma(t,x(.),\phi(.)),c(t,x,\phi(.),z))$ holds.

Proof. If (x_t,ϕ_t) satisfies (1.1) with (1.2) with coefficients

$(b,\sigma,c) = (b(t,x(.),\phi(.)),\sigma(t,x(.),\phi(.)),c(t,x,\phi(.),z))$

such that $E|x_0|^2 < \infty$, then by condition 2° applying 2) in Remark 2.5 ϕ_t is continuous. By Ito's formula, by 4') of Remark 1.1 and by condition 1° one easily gets that as $t \leq T$

$E(\sup_{s\leq t\wedge\tau_N}|x_s - x_0|^2) \leq k_T(1 + E\int_0^{t\wedge\tau_N}(\sup_{r\leq s}|x_r|^2 + \sup_{r\leq s}|\phi_r|^2)ds)$,

where $0 \leq k_T'$ is a constant depending on T and $E|x_0|^2$ only, and

$\tau_N = \inf\left\{t \geq 0 : |x_t|^2 + |\phi_t|^2 > N\right\}$,

(cf. the proof of Theorem 2.1). Hence

$E(\sup_{s\leq t\wedge\tau_N}|x_s|^2) \leq k_T'(1 + E\int_0^{t\wedge\tau_N}(\sup_{r\leq s}|x_r|^2 + \sup_{r\leq s}|\phi_r|^2)ds)$.

On the other hand,

$$E\sup_{s\leq t\wedge\tau_N}|\phi_s|^2 \leq 5E(|x_t|^2 + |x_0|^2 + \left|\int_0^{t\wedge\tau_N} bds\right|^2 + \left|\int_0^{t\wedge\tau_N}\sigma dw_s\right|^2$$
$$+\left|\int_0^{t\wedge\tau_N}\int_Z cq(ds,dz)\right|^2) \leq k_T''(1 + E\int_0^{t\wedge\tau_N}(\sup_{r\leq s}|x_r|^2 + \sup_{r\leq s}|\phi_r|^2)ds).$$

Hence

$E(\sup_{s\leq t\wedge\tau_N}|x_s|^2) + E\sup_{s\leq t\wedge\tau_N}|\phi_s|^2$
$\leq k_T'''(1 + E\int_0^{t\wedge\tau_N}(\sup_{r\leq s}|x_r|^2 + \sup_{r\leq s}|\phi_r|^2)ds)$.

From this it is easily shown that there exists a constant $0 \leq k_T$ depending on T and $E|x_0|^2$ only such that

$E(\sup_{s\leq t}|x_s - y_s|^2 + \sup_{s\leq t}|\phi_s - \psi_s|^2) \leq k_T.$

Assume now that $(x_t^i, \phi_t^i), i = 1, 2$, are two solutions on the same probability space with the same BM w_t and the same Poisson martingale measure $q(dt, dz)$. Then by condition 2° applying 2) in Remark 2.5 $\phi_t^i, i = 1, 2$, are continuous. Again by Ito's formula, but with condition 3°, one finds as above, that

$E(\sup_{s\leq t}|x_s^1 - x_s^2|^2 + \sup_{r\leq t}|\phi_r^1 - \phi_r^2|^2)$

$\leq k_0'(E\int_0^t k_N(s)\rho_N(\sup_{r\leq s}|x_r^1 - x_r^2|^2 + \sup_{r\leq s}|\phi_r^1 - \phi_r^2|^2)ds).$

From this it is also easily shown that the pathwise uniqueness conclusion is true. ■

Now we are in a position to show Theorem 3.20.

Proof. Denote $c(t-, x(.), \phi(.), z) = c(s, x_s, \phi_s, s < t, z)$.

Now, following the proof of Theorem 3.1, there exists a unique solution (x_t^n, ϕ_t^n) satisfying

$$(3.16)\begin{cases} x_t^n = x_0 + \int_0^t b^n(s, x^n(.), \phi^n(.))ds + \int_0^t \sigma^n(s, x^n(.), \phi^n(.))dw_s \\ \qquad + \int_0^t \int_Z c^n(s-, x^n(.), \phi^n(.), z)q(ds, dz) + \phi_t^n, \\ \quad \text{and the other statements in (1.2) hold for } (x_t^n, \phi_t^n), \end{cases}$$

$n = 1, 2, ..,$

where

$b^n(s, x^n(.), \phi^n(.)) = b(h_n(s), x^n(.), \phi^n(.)),$

$h_n(0) = 0$

$h_n(t) = (k-1)2^{-n}$, as $(k-1)2^{-n} < t \leq k2^{-n}$.

Now by Theorem 2.1, similar to the proof of Theorem 3.1, we still find that there exist a probability space $(\widetilde{\Omega}, \widetilde{\Im}, \widetilde{P})$ and cadlag processes $(\widetilde{x}_t^n, \widetilde{\phi}_t^n, \widetilde{w}_t^n, \widetilde{\zeta}_t^n)$ defined on it with the same finite-dimensional probability distributions as that of $(x_t^n, \phi_t^n, w_t, \zeta_t)$ and there exist $(\widetilde{x}_t^0, \widetilde{\phi}_t^0, \widetilde{w}_t^0, \widetilde{\zeta}_t^0)$, where $\widetilde{w}_t^0$ is a BM on $(\widetilde{\Omega}, \widetilde{\Im}, \widetilde{P})$, and there exists a subsequence $\{n_k\}$ of $\{n\}$, denote it by $\{n\}$ again, such that as $n \to \infty$

(3.17) $\widetilde{\eta}_t^n \to \widetilde{\eta}_t^0$, in probability, as $\widetilde{\eta}_t^n = \widetilde{x}_t^n, \widetilde{\phi}_t^n, \widetilde{w}_t^n, \widetilde{\zeta}_t^n$.

Moreover, for $n = 0, 1, 2, ...$ set

$\widetilde{p}^n(dt, dz) = \sum_{s\in dt} I_{(0=\Delta\widetilde{\zeta}_s^n\in dz)}(s)$, $\widetilde{q}^n(dt, dz) = \widetilde{p}^n(dt, dz) - \pi(dz)dt$,

Then $\widetilde{p}^n(dt, dz)$ is a Poisson random point measure with compensator $\pi(dz)dt$ for each $n = 0, 1, 2, ...$, and it verifies

$\widetilde{\zeta}_t^n = \int_0^t \int_{|z|\leq 1} z\widetilde{q}^n(ds, dz) + \int_0^t \int_{|z|>1} z\widetilde{p}^n(ds, dz), n = 0, 1, 2, ...$

By the coincidence of finite probability distributions one easily sees that $(\widetilde{x}_t^n, \widetilde{\phi}_t^n)$ satisfies (1.1) and (1.2) with $\widetilde{w}_t^n$ and $\widetilde{q}^n(dt, dz)$ on $(\widetilde{\Omega}, \widetilde{\Im}, \widetilde{P})$. Now by (3.17) and conditions 1° and 2° one still sees that as $n \to \infty$

$(3.18)_1$ $\left|\int_0^t (b^n(s, \widetilde{x}^n(.), \widetilde{\phi}^n(.)) - b(s, \widetilde{x}^0(.), \widetilde{\phi}^0(.)))ds\right| \to 0$, in probability.

$(3.18)_2$ $\int_0^t \sigma^n(s, \widetilde{x}^n(.), \widetilde{\phi}^n(.))d\widetilde{w}_s^n \to \int_0^t \sigma(s, \widetilde{x}^0(.), \widetilde{\phi}^0(.))d\widetilde{w}_s^0$, in probability,

$(3.18)_3$ $\int_0^t \int_Z c^n(s-, \widetilde{x}^n(.), \widetilde{\phi}^n(.), z)\widetilde{q}^n(ds, dz)$

$\to \int_0^t \int_Z c(s-,\widetilde{x}^0(.),\widetilde{\phi}^0(.),z)\widetilde{q}^0(ds,dz)$, in probability,

$(3.18)_4$ $\int_0^t \int_Z \left| c^n(s,\widetilde{x}^n(.),\widetilde{\phi}^n(.),z) - c(s,\widetilde{x}^0(.),\widetilde{\phi}^0(.),z)\right|^2 \pi(dz)ds \to 0,$

in probability.

Indeed, comparing the proof of $(3.2)_1$–$(3.2)_4$, one finds that Lemma 3.2 is not necessarily true here. However, one can still have Lemma 3.3. Moreover, one can still prove (3.3) directly without using Lemma 3.2, since now we have the stronger condition 2°. For example,

$$P(\int_0^T \int_Z \left| c^n(s,\widetilde{x}^n(.),\widetilde{\phi}^n(.),z) - c^{n^0}(s,\widetilde{x}^n(.),\widetilde{\phi}^n(.),z)\right|^2$$

$$\cdot I_{\sup_{s\leq t}|x_s|+\sup_{s\leq t}|\phi_s|\leq N} ds > \varepsilon) \leq \sup_n \sup_{s\in[0,T]} \widetilde{P}(\left|\widetilde{x}^n_{h_n(s)} - \widetilde{x}^n_{h_{n^0}(s)}\right| > \eta)$$

$$+ \sup_n \sup_{s\in[0,T]} \widetilde{P}(\left|\widetilde{\phi}^n_{h_n(s)} - \widetilde{\phi}^n_{h_{n^0}(s)}\right| > \eta) + k_0\varepsilon^{-1} \int_0^T (|h_n(s) - h_{n^0}(s)|^2 + 2\eta)ds.$$

Hence, from this and by Lemma 3.3 (3.3) still follows. Therefore $(3.18)_1$–$(3.18)_4$ can still be derived. Now, by using the same technique as this without using Lemma 3.2 one still goes through the whole proof as in Theorem 3.1. Hence, (1.1) with (1.2) with coefficients

$$(b,\sigma,c) = (b(t,x(.),\phi(.)),\sigma(t,x(.),\phi(.)),c(t,x,\phi(.),z))$$

has a weak solution (x_t,ϕ_t).

Now, suppose that 4° is satisfied. Then by 2) of Remark 2.5 ϕ_t is continuous. Hence applying the following Theorem 3.22 the solution is also pathwise unique. From the pathwise uniqueness applying Theorem 3.1 (Yamada-Watanabe type theorem) the existence of a strong solution is also derived. ■

Again, similar to Theorem 3.8, we can relax the bounded condition for coefficients to the one that is a less than linear growth condition.

3.23 Theorem. Under assumption 2° – 4° of Theorem 3.20, and assumption that

1°′ $b(t,x(.),\phi(.)),\sigma(t,x(.),\phi(.)),c(t,x(.),\phi(.),z)$ are jointly measurable, and there exists a constant $k_0 > 0$ such that

$$|b(t,x(.),\phi(.))|^2 + \|\sigma(t,x(.),\phi(.))\|^2 \leq k_0(1+\sup_{s\leq t}|x_s|^2 + \sup_{s\leq t}|\phi_s|^2),$$

$$\int_Z |c(t,x(.),\phi(.),z)|^2 \pi(dz) \leq k_0,$$

then (1.1) with (1.2) has a pathwise unique strong solution. Moreover, in this case $\phi(t)$ is continuous.

Proof. Let

$b^N(t,x(.),\phi(.)) = b(t,x(.),\phi(.)),$

as $\sup_{s\leq t}|x_s| \leq N$, and $\sup_{s\leq t}|\phi_s| \leq N$;

$b^N(t,x(.),\phi(.)) = b(t,Nx(.)/\sup_{s\leq t}|x_s|,\phi(.)),$

as $\sup_{s\leq t}|x_s| > N$, and $\sup_{s\leq t}|\phi_s| \leq N$;

$b^N(t,x(.),\phi(.)) = b(t,x(.),N\phi(.)/\sup_{s\leq t}|\phi_s|),$

as $\sup_{s\leq t}|\phi_s| > N$, and $\sup_{s\leq t}|x_s| \leq N$;

$b^N(t,x(.),\phi(.)) = b(t,Nx(.)/\sup_{s\leq t}|x_s|,N\phi(.)/\sup_{s\leq t}|\phi_s|),$

as $\sup_{s\leq t}|\phi_s| > N$, and $\sup_{s\leq t}|x_s| > N$;

σ^N is similarly defined. Then by Theorem 3.20 there exists a pathwise unique strong solution (x_t^N, ϕ_t^N), where ϕ_t^N is continuous, satisfying (1.1) and (1.2) with the coefficients b^N, σ^N and c. Set now

$$\tau_N = \inf\{t \geq 0 : |x_t^N| > N, |\phi_t^N| > N\}.$$

$$x_t = x_t^N,\ \phi_t = \phi_t^N, \text{ as} \leq t \leq \tau_N.$$

By the pathwise uniqueness the above definition is well posed, and

$$x_{t\wedge\tau_N} = x_0 + \int_0^{t\wedge\tau_N} b(s, x(.), \phi(.))ds + \int_0^{t\wedge\tau_N} \sigma(s, x(.), \phi(.))dw_s$$
$$+ \int_0^{t\wedge\tau_N} \int_Z c(s-, x(.), \phi(.), z)q(ds, dz) + \phi_{t\wedge\tau_N}, \text{ for all } t \geq 0.$$

(cf. the proof of general case for Theorem 2.8). Note that for any $T < \infty$

$$N^2 P(\tau_N \leq T) \leq E|x_{t\wedge\tau_N}|^2 \leq k_T, \text{ for all } N = 1, 2, ...$$

Hence, this shows that

$$P(\lim_{N\to\infty} \tau_N = \infty) = 1.$$

The proof is complete. ■

3.3.5 Existence of Solutions for RSDE with Discontinuous Diffusion Coefficients

So far, the RSDEs we discussed are all with continuous diffusion coefficients. In this section we will discuss the discontinuous case. The results are derived by the Krylov's estimate and smoothness technique. Actually, we have the following

3.24 Theorem. Consider (1.1) with (1.2). Assume that
$1°$ $b(t,x), \sigma(t,x), c(t,x,z)$ are jointly measurable, on $t \geq 0, x \in R^d, z \in Z$, and there exists a constant $k_0 > 0$ such that

$$|b|^2 \leq k_0(1 + |x|),$$
$$\|\sigma\|^2 + \int_Z |c|^2 \pi(dz) \leq k_0,$$

$2°$ $\sigma(t,x)$ is uniformly non-degenerate, i.e., there exists a $\delta_0 > 0$ such that for all $\lambda \in R^d$

$$\langle A(t,x)\lambda, \lambda\rangle \geq |\lambda|^2 \delta_0,$$

where $A = \frac{1}{2}\sigma\sigma^*$,
$3°$ as $|x - y| \to 0, |s - t| \to 0$

$$\int_Z |c(t,x,z) - c(s,y,z)|^2 \pi(dz) \to 0,$$

$4°$ $x \in \overline{\Theta} \Longrightarrow x + c(t,x,z) \in \overline{\Theta}$,
where Θ satisfies Assumption 2.7 in Chapter 2 (as that in the beginning of section §3.3),

$5°$ $\pi(dz) = dz/|z|^{d+1}$, $(Z = R^d - \{0\})$.

Then (1.1) with (1.2) has a weak solution (x_t, ϕ_t) and,moreover, ϕ_t is continuous.

Proof. Case 1. Assume that

$$|b|^2 + \|\sigma\|^2 + \int_Z |c|^2 \pi(dz) \leq k_0.$$

Then the proof of Theorem 3.24 is similar to that of Theorem 3.17 (in the proof here we only need to smooth out b and σ, but to keep c unchanged). So we omit the details.

Case 2. General case. Under condition 1° we first apply the result of case 1 to get the existence of a weak solution (x_t, ϕ_t) to (1.1) with (1.2) for the coefficient $(0, \sigma, c)$, where ϕ_t is continuous. Then, applying Theorem 3.14 and 3.12 (Girsanov type Theorem), we easily derive the conclusion. ■

3.4 Existence of Solutions with Jump Reflection in a Half Space

3.4.1 Existence of Strong Solutions for RSDE with Random Coefficients

In this section we will discuss RSDE (1.1) with (1.2)'. More precisely, we will consider the following RSDE with Poisson jumps and jump reflection in the half space of R^d as follows:

$$
(4.1)\begin{cases}
x_t^i = x_0^i + \int_0^t b^i(s, x_s, \omega)ds + \sum_{j=1}^d \int_0^t \sigma^{ij}(s, x_s, \omega)dw_s^j \\
\quad + \int_0^t \int_Z c^i(s, x_{s-}, z, \omega)q(ds, dz), \\
\quad i = 1, ..., d-1; \\
x_t^d = x_0^d + \int_0^t b^d(s, x_s, \omega)ds + \sum_{j=1}^d \int_0^t \sigma^{dj}(s, x_s, \omega)dw_s^j \\
\quad + \int_0^t \int_Z c^d(s, x_{s-}, z, \omega)q(ds, dz) + \phi_t, \\
x_t^d \geq 0, \text{ for all } t \geq 0, \ x_t \text{ is } R^d - \text{valued } \Im_t - \text{adapted and cadlag;} \\
\phi_t \text{ is a real } \Im_t - \text{adapted increasing process with } \phi_0 = 0 \text{ such that} \\
\int_0^t x_s^d d\phi_s^c = 0, \text{ for all } t \geq 0, \ \phi_t^c = \phi_t - \sum_{0<s\leq t} \Delta\phi_s, \\
\Delta\phi_t > 0 \Rightarrow \Delta\phi_t = \alpha_t x_t^d;
\end{cases}
$$

where $2 \leq \alpha_t$ is a given non-random real function on $t \geq 0$; and w_t is a d-dimensional standard Brownian Motion process (BM), $q(dt, dz)$ is a real Poisson random martingale measure such that

$q(dt, dz) = p(dt, dz) - \pi(dz)dt,$

where $p(dt, dz)$ is a Poisson random counting measure with the compensator $\pi(dz)dt$, and for simplicity let

$\pi(dz) = dz/ |z|^{d+1}, \ Z = R^d - \{0\}.$

4.1 Remark. In the case $\alpha_t = 2$, for all $t \geq 0$, (4.1) is reduced to the stochastic system considered in M. Chaleyat et al.(1980). In this whole section we make the following natural

4.2 Assumption. b, σ, c are all jointly measurable.

First, we have the following

4.3 Lemma. If (x_t, ϕ_t) and $(\hat{x}_t, \hat{\phi}_t)$ satisfies (4.1) with coefficients (b, σ, c) and $(\hat{b}, \hat{\sigma}, \hat{c})$, respectively, on the same probability space with the same w_t and $q(dt, dz)$, denote the dth component of (4.1) by

$x_t^d = y_t^d + \phi_t$, and $\hat{x}_t^d = \hat{y}_t^d + \hat{\phi}$, $(x_0^d = y_0^d, \ \hat{x}_0^d = \hat{y}_0^d)$,

respectively, then

(4.2) $|x_t - \hat{x}_t|^2 = |x_0 - \hat{x}_0|^2 + 2\int_0^t (x_s - \hat{x}_s)\cdot(b(s,x_s,\omega) - \hat{b}(s,\hat{x}_s,\omega))ds$
$+2\int_0^t (x_s - \hat{x}_s)\cdot(\sigma(s,x_s,\omega) - \widehat{\sigma}(s,\hat{x}_s,\omega))dw_s$
$+2\int_0^t\int_Z (x_{s-} - \hat{x}_{s-})\cdot(c(s,x_{s-},z,\omega) - \hat{c}(s,\hat{x}_{s-},z,\omega))q(ds,dz)$
$+\int_0^t \|\sigma(s,x_s,\omega) - \widehat{\sigma}(s,\hat{x}_s,\omega)\|^2 ds - 2\int_0^t x_s^d d\hat{\phi}_s^c - 2\int_0^t \hat{x}_s^d d\phi_s^c$
$+\int_0^t\int_Z |c(s,x_{s-},z,\omega) - \hat{c}(s,\hat{x}_{s-},z,\omega)|^2 p(ds,dz)$
$-2\sum_{0<s\le t}\alpha_s x_s^d \hat{x}_s^d (I_{\triangle\phi_s>0,\triangle\hat{\phi}_s=0} + I_{\triangle\phi_s=0,\triangle\hat{\phi}_s>0})$
$-\sum_{0<s\le t}\alpha_s(\alpha_s - 2)((x_s^d)^2 I_{\triangle\phi_s>0,\triangle\hat{\phi}_s=0} + (\hat{x}_s^d)^2 I_{\triangle\phi_s=0,\triangle\hat{\phi}_s>0})$
$-\sum_{0<s\le t}\alpha_s(\alpha_s - 2)(x_s^d - \hat{x}_s^d)^2 I_{\triangle\phi_s>0,\triangle\hat{\phi}_s>0}.$

Proof. By (4.1) we have

$$(4.3)\quad \begin{cases} -(\alpha_t - 1)x_t^d = x_{t-}^d + \triangle y_t^d, \text{ as } \triangle\phi_t > 0, \\ x_t^d = x_{t-}^d + \triangle y_t^d, \text{ as } \triangle\phi_t = 0. \end{cases}$$

For $(\hat{x}_t, \hat{\phi}_t)$ we also have the similar (4.3)'. In fact, as $\triangle\phi_t > 0$,

$x_t^d - x_{t-}^d + \triangle y_t^d = \triangle\phi_t = \alpha_t x_t^d.$

Hence, the 1st equality in (4.3) is true. The 2nd equality in (4.3) is evident. Now by Ito's formula

$|x_t - \hat{x}_t|^2 = \sum_{i=1}^7 I_t^i + \sum_{i=1}^{d-1}\int_0^t\int_Z |c^i(s,x_{s-},z,\omega) - \hat{c}^i(s,\hat{x}_{s-},z,\omega)|^2 p(ds,dz)$
$+2\sum_{0<s\le t}(x_{s-}^d - \hat{x}_{s-}^d)\triangle(\phi_s - \hat{\phi}_s) + \sum_{0<s\le t}\triangle(x_s^d - \hat{x}_s^d)^2 = \sum_{i=1}^{10} I_t^i,$

where I_t^i denotes the ith term in the right side of (4.2), $i = 1, 2, ..., 7$, respectively. Note that

$I_t^9 + I_t^{10} = \sum_{0<s\le t}(\triangle(y_s^d - \widehat{y}_s^d)^2 + \triangle(\phi_s - \hat{\phi}_s)^2$
$+2\triangle(y_s^d - \widehat{y}_s^d)\triangle(\phi_s - \hat{\phi}_s) + 2(x_{s-}^d - \hat{x}_{s-}^d)\triangle(\phi_s - \hat{\phi}_s)) = \sum_{0<s\le t} i(s).$

Hence, as $\triangle\phi_s = \triangle\hat{\phi}_s = 0$

$i(s) = \triangle(y_s^d - \widehat{y}_s^d)^2.$

In case s is such that

$\triangle\hat{\phi}_s = 0,\ \triangle\phi_s > 0,$

then by (4.1) and (4.3)

$i(s) = \triangle(y_s^d - \widehat{y}_s^d)^2 + \alpha_s^2(x_s^d)^2 - 2(\hat{x}_{s-}^d + \triangle\widehat{y}_s^d)\triangle\phi_s + 2(x_{s-}^d + \triangle y_s^d)\triangle\phi_s$
$= \triangle(y_s^d - \widehat{y}_s^d)^2 - \alpha_s(\alpha_s - 2)(x_s^d)^2 - 2\alpha_s x_s^d \hat{x}_s^d.$

A similar equality holds for $\triangle\hat{\phi}_s > 0,\ \triangle\phi_s = 0$. Suppose now that $\triangle\hat{\phi}_s > 0$, $\triangle\phi_s > 0$, then

$i(s) = \triangle(y_s^d - \widehat{y}_s^d)^2 + (1 - 2\alpha_s^{-1}(\alpha_s - 1))\triangle(\phi_s - \hat{\phi}_s)^2$
$= \triangle(y_s^d - \widehat{y}_s^d)^2 - \alpha_s(\alpha_s - 2)(x_s^d - \widehat{x}_s^d)^2.$

Therefore (4.2) follows. ■

4.4 Corollary. If (x_t, ϕ_t) satisfies

$$\begin{cases} x_t = y_t + \phi_t, \\ \text{and the other statements in (4.1) for } (x_t, \phi_t^d) \text{ hold,} \\ \text{where } \phi_t = (0, ..., 0, \phi_t^d);\ x_t, y_t, \phi_t \text{ are all cadlag, } \Im_t\text{-adapted,} \\ \text{and } y_t \text{ is a } \Im_t - \text{semi-martingale;} \end{cases}$$

and $(\hat{x}_t, \hat{\phi}_t)$ satisfies a similar system with $\widehat{y}_t$, then

(4.2)' $|x_t - \hat{x}_t|^2 = |x_0 - \hat{x}_0|^2 + 2\int_0^t (x_{s-} - \hat{x}_{s-})\cdot d(y_s - \widehat{y}_s) + [M - \widehat{M}]_t$

$-2\int_0^t x_s^d d\hat{\phi}_s^c - 2\int_0^t \hat{x}_s^d d\phi_s^c - 2\sum_{0<s\leq t}\alpha_s x_s^d \hat{x}_s^d (I_{\triangle\phi_s>0,\triangle\hat{\phi}_s=0} + I_{\triangle\phi_s=0,\triangle\hat{\phi}_s>0})$
$-\sum_{0<s\leq t}\alpha_s(\alpha_s - 2)((x_s^d)^2 I_{\triangle\phi_s>0,\triangle\hat{\phi}_s=0} + (\hat{x}_s^d)^2 I_{\triangle\phi_s=0,\triangle\hat{\phi}_s>0})$
$-\sum_{0<s\leq t}\alpha_s(\alpha_s - 2)(x_s^d - \hat{x}_s^d)^2 I_{\triangle\phi_s>0,\triangle\hat{\phi}_s>0}$,
where M_t and $\widehat{M}_t$ are the local martingale parts of y_t and $\widehat{y}_t$, respectively.

The proof of Corollary 4.4 is completely similar to that of Lemma 4.3.

4.5 Remark. Intuitively, if
$x_{t-}^d + \triangle y_t^d \geq 0$,
then at this point t no reflection is needed. Hence, it is natural to require that
$x_{t-}^d + \triangle y_t^d \geq 0 \Rightarrow \triangle\phi_t = 0$.
Or, equivalently, to assume that
$\triangle\phi_t > 0 \Rightarrow x_{t-}^d + \triangle y_t^d < 0$.
But in this case
$-(\alpha_t - 1)x_t^d < 0$.
Hence, it is natural to assume that
$\alpha_t \geq 1$, for all $t \geq 0$.
Moreover, observing the formula (4.2), one sees that if
$\alpha_t \geq 2$, for all $t \geq 0$,
then the last term in (4.2) is always less than zero. From this observation many effective estimates can be derived. That is why we always make the assumption $\alpha_t \geq 2$, for all $t \geq 0$.

We have the following estimate for the solution of (4.1).

4.6 Theorem. Assume that
1° $|b|^2 + \|\sigma\|^2 + \int_Z |c|^2 \pi(dz) \leq k_0(1 + |x|^2)$.

If (x_t, ϕ_t) satisfies (4.1) with x_0 - $\Im_0$-measurable and $E|x_0|^2 < \infty$, then for any $0 \leq T < \infty$
$E(\sup_{t\leq T}|x_t|^2 + \phi_T^2) \leq k_T$,
$E(\sup_{s\leq r\leq t}|x_r - x_s|^2 + (\phi_t - \phi_s)) \leq k_T(t-s)$, as $s \leq t \leq T < \infty$,
where $0 < k_T$ is a constant depending on k_0, T and $E|x_0|^2$ only.

4.7 Theorem. (Pathwise uniqueness). Assume that
$2(x-y)\cdot(b(t,x,\omega) - b(t,y,\omega)) + \|\sigma(t,x,\omega) - \sigma(t,y,\omega)\|^2$
$+\int_0^t |c(s,x,z,\omega) - c(s,y,z,\omega)|^2 \pi(dz) \leq k_N(t)\rho_N(|x-y|^2)$, as $|x|, |y| \leq N$,
where $0 \leq k_N(t)$, $\int_0^t k_N(s)ds < \infty$, and $\rho_N(u)$ is concave and strictly increasing in $u \geq 0$ such that $\rho_N(0) = 0$, and
$\int_{0+} du/\rho_N(u) = \infty$, for each $N = 1, 2, ...$

If $(x_t^i, \phi_t^i), i = 1, 2$, satisfy (4.2) with the same BM w_t and the same Poisson martingale measure $q(dt, dz)$, which has the compensator $\pi(dz)dt$ on the same probability space, then
$P(\sup_{t\geq 0}|x_t^1 - x_t^2| = 0) = 1, P(\sup_{t\geq 0}|\phi_t^1 - \phi_t^2| = 0) = 1$.

The proofs of Theorem 4.6 and 4.7 can be completed by using Lemma 4.3.

4.8 Theorem. Assume that

1° $b(t,x,\omega),\sigma(t,x,\omega),c(t,x,z,\omega)$ are jointly measurable, and as x and z are fixed, they are $\Im_t^{w,q}-$adapted, where c is $\Im_t^{w,q}-$predictable, moreover, there exists a constant $k_0>0$ such that

$|b|+\|\sigma\|^2+\int_Z |c|^2 \pi(dz) \leq k_0(1+|x|^2)$,

2° $|b(t,x,\omega)-b(t,y,\omega)|^2+\|\sigma(t,x,\omega)-\sigma(t,y,\omega)\|^2$
$+\int_Z |c(t,x,z,\omega)-c(t,y,z,\omega)|^2 \pi(dz) \leq k_N |x-y|^2$,
as $|x|,|y| \leq N$, $N=1,2,...$,

where k_N is a constant depending on N only.

Then (4.1) has a pathwise unique strong solution (x_t,ϕ_t).

Proof. Case 1. Assume that $k_N=k_0$ is independent of N, and $|b|+\|\sigma\|^2+\int_Z |c|^2 \pi(dz) \leq k_0$. As the proof of Theorem 2.8 by means of Theorem 3.2 in Chapter 2 one can use the iteration technique to show that there exists a unique strong solution (x_t^n,ϕ_t^n) satisfying

$$\begin{cases} x_t^n = y_t^{n-1}+\phi_t^n, \\ \text{and all the other statements in (4.1) hold for } (x_t^n,\phi_t^n), \end{cases}$$

where

$$y_t^{n-1}=x_0+\int_0^t b(s,x_s^{n-1},\omega)ds+\int_0^t \sigma(s,x_s^{n-1},\omega)dw_s$$
$$+\int_0^t \int_Z c(s,x_{s-}^{n-1},z,\omega)q(ds,dz),$$

and in the following we always denote

$$\phi_t^n=(0,...,0,\phi_t^{nd}).$$

Then by Corollary 4.4 and Borel-Cantelli lemma there exists a (x_t,ϕ_t) such that $P-a.s$

$$x_t=\lim_{n\to\infty} x_t^n,\ \phi_t=\lim_{n\to\infty}\phi_t^n, \text{ uniformly in } t\in[0,T].$$

On the other hand, for given

$$y_t=x_0+\int_0^t b(s,x_s,\omega)ds+\int_0^t \sigma(s,x_s,\omega)dw_s+\int_0^t\int_Z c(s,x_{s-},z,\omega)q(ds,dz)$$

by Theorem 3.2 in Chapter 2 there exists a unique $\Im_t^{w,q}-$adapted solution $(\tilde{x}_t,\tilde{\phi}_t)$ satisfying

$$\begin{cases} \tilde{x}_t = y_t+\tilde{\phi}_t, \\ \text{and all the other statements in (4.1) hold for } (\tilde{x}_t,\tilde{\phi}_t). \end{cases}$$

By assumption 2° (the continuity property) and the assumption of case 1 (the bounded assumption) applying Corollary 4.4 one easily shows that as $n\to\infty$

$$E\sup_{s\leq t}|x_s^n-\tilde{x}_s|^2\to 0.$$

In particular, as $n\to\infty$

$$E|x_t^n-\tilde{x}_t|\to 0.$$

Hence by the uniqueness of limit one has

$$\tilde{x}_t=x_t,\ \tilde{\phi}_t=\phi_t,$$

i.e. (x_t,ϕ_t) is a solution of (4.1). The uniqueness of solution to (4.1) can be derived by Theorem 4.7.

For the general case, the proof can be completed in a similar fashion to that of Theorem 2.8. ■

Now we are going to discuss the solution for RSDE (4.1) with non-Lipschitzian coefficients.

4.9 Theorem. Assume that

1° $b(t,x,\omega),\sigma(t,x,\omega),c(t,x,z,\omega)$ are jointly measurable, and as x and z are fixed, they are $\Im_t^{w,q}-$adapted, where c is $\Im_t^{w,q}-$predictable, moreover, there exists a constant $k_0>0$ such that

$|b|^2+\|\sigma\|^2+\int_Z |c|^2\,\pi(dz)\leq k_0(1+|x|^2)$,

2° $b(t,x,\omega),\sigma(t,x,\omega)$ are continuous in x, and as $|x-y|\to 0$,

$\int_0^t |c(s,x,z,\omega)-c(s,y,z,\omega)|^2\,\pi(dz)\to 0$,

3° $2(x-y)\cdot(b(t,x,\omega)-b(t,y,\omega))+\|\sigma(t,x,\omega)-\sigma(t,y,\omega)\|^2$

$+\int_0^t |c(s,x,z,\omega)-c(s,y,z,\omega)|^2\,\pi(dz)\leq k_N(t)\rho_N(|x-y|^2)$, as $|x|,|y|\leq N$,

where $0\leq k_N(t)$, $\int_0^t k_N(s)ds<\infty$, and $\rho_N(u)$ is concave and strictly increasing in $u\geq 0$ such that $\rho_N(0)=0$, and

$\int_{0+} du/\,\rho_N(u)=\infty$, for each $N=1,2,...$

Then (4.1) has a pathwise unique strong solution.

Proof. Case 1. Assume that $k_N(t)$ and $\rho_N(u)$ in 3° are independent of N, i.e. $k_N(t)=k(t),\rho_N(u)=\rho(u)$, for all $N=1,2,...$,and assume that

$|b|+\|\sigma\|^2+\int_Z |c|^2\,\pi(dz)\leq k_0$.

Let for $y\in R^d$

$f(y)=c\exp[(1-|y|^2)^{-1}]$, as $|y|<1$; $f(y)=0$, otherwise;

where c is a constant such that

$\int_{R^d} f(y)dy=1$.

For $n=1,2,...,(t,\omega)\in[0,\infty)\times\Omega$ set

$b^n(t,x,\omega)=\int_{R^d} b(t,x-n^{-1}y,\omega)f(y)dy$.

$\sigma^n(t,x,\omega)$ and $c^n(t,x,z,\omega)$ is defined similarly. Then by Theorem 4.8 there exists a pathwise unique strong solution (x_t^n,ϕ_t^n) satisfying

$$\begin{cases} x_t^n=x_0+\int_0^t b^n(s,x_s^n,\omega)ds+\int_0^t \sigma^n(s,x_s^n,\omega)dw_s \\ \quad +\int_0^t\int_Z c^n(s,x_{s-}^n,z,\omega)q(ds,dz)+\phi_t^n, \\ \text{and the other statements in (4.1) hold for } (x_t^n,\phi_t^{nd}), \end{cases}$$

where $\phi_t^n=(0,...,0,\phi_t^{nd})$.

By Lemma 4.3 one verifies that

$E|x_t^n-x_t^m|^2\leq\int_0^t\int_{R^d} k(s)\rho(E|x_s^n-x_s^m-n^{-1}y+m^{-1}y|^2)f(y)dyds$

$+\bar{k}_T(n^{-1}+m^{-1})$.

Now by Theorem 4.6 we argue as in the proof of Theorem 2.9 that there exists a $\Im_t-$adapted process $x_t\in L^2([0,T]\times\Omega,dt\times dP)$ such that as $n\to\infty$ (if necessary, take a subsequence)

$x_t^n(\omega)\to x_t(\omega)$, in R^d, $dt\times dP-a.e.$ $(t,\omega)\in[0,T]\times\Omega$.

We also have

$\lim_{n\to\infty}E\left|\int_0^t\int_Z(c^n(s,x_{s-}^n,z,\omega)-c(s,x_{s-},z,\omega))q(ds,dz)\right|^2$

$=\lim_{n\to\infty}E\int_0^t\int_Z\left|c^n(s,x_{s-}^n,z,\omega)-c(s,x_{s-},z,\omega)\right|^2\pi(dz)ds=0$,

etc. Denote

$\phi_t=x_t-(x_0+\int_0^t b(s,x_s,\omega)ds+\int_0^t\sigma(s,x_s,\omega)dw_s$

$+\int_0^t\int_Z c(s,x_{s-},z,\omega)q(ds,dz))$.

Then as $n \to \infty$

$$E|\phi_t^n - \phi_t| \to 0.$$

Now denote

$$y_t = x_0 + \int_0^t b(s, x_s, \omega)ds + \int_0^t \sigma(s, x_s, \omega)dw_s + \int_0^t \int_Z c(s, x_{s-}, z, \omega)q(ds, dz).$$

Similar to the proof of Theorem 4.8, one easily finds that (x_t, ϕ_t) satisfies (4.1) on $[0, T]$. By Theorem 4.7 the pathwise uniqueness of solution to (4.1) holds. Since $0 < T < \infty$ is arbitrarily given, hence we find a pathwise unique strong solution (x_t, ϕ_t) satisfying (4.1).

General case. The general case can be reduced to the special case 1 by using Lemma 2.10 and the following fact:

Let $w^N(x) \in c_0^\infty(R^d)$ such that

$$w^N(x) = \begin{cases} 1, \text{as } |x| \leq N+2, \\ 0, \text{ as } |x| \geq N+3, \end{cases}$$

and $0 \leq w^N(x) \leq 1$. Set

$$c^N(t, x, z, \omega) = c(t, x, z, \omega)w^N(x).$$

Then

$$c^N(t, x, z, \omega) = \begin{cases} c(t, x, z, \omega), \text{as } |x| \leq N+2, \\ \quad 0, \text{ as } |x| \geq N+3. \end{cases}$$

Obviously, c^N satisfies that

$$\int_Z |c^N|^2 \pi(dz) \leq \int_Z |c|^2 \pi(dz) \wedge (k_0(1 + (N+3)^2)).$$

Let us take b^N and σ_{ij}^N from Lemma 2.11 and c^N here. Then we argue, as in the proof of Lemma 2.11, that $\forall x, y \in R^d$

$$2(x-y) \cdot (b^N(t, x, \omega) - b^N(t, y, \omega)) + \left\|\sigma^N(t, x, \omega) - \sigma^N(t, y, \omega)\right\|^2$$
$$+ \int_0^t \left|c^N(s, x, z, \omega) - c^N(s, y, z, \omega)\right|^2 \pi(dz) \leq \overline{k}_N(t)\overline{\rho}_N(|x-y|^2), \ \forall x, y \in R^d,$$

where $0 \leq \overline{k}_N(t)$, $\int_0^t \overline{k}_N(s)ds < \infty$, and $\overline{\rho}_N(u)$ is concave and strictly increasing in $u \geq 0$ such that $\overline{\rho}_N(0) = 0$, and

$$\int_{0+} du/\, \overline{\rho}_N(u) = \infty, \text{ for each } N = 1, 2, ...$$

and such $\overline{k}_N(t), \overline{\rho}_N(u)$ exist.

Now, for coefficients (b^N, σ^N, c^N) case 1 can be applied. Then we argue, similar to the proof of Theorem 2.8 for the general case, that (4.1) has a strong solution. The pathwise uniqueness of solution to (4.1) is derived from Theorem 4.7. ■

3.4.2 Existence of Weak and Strong Solutions for RSDEs with Non-random Non-Lipschitzian Coefficients

Now let us show a Krylov type estimate as follows:

4.10 Theorem. Assume that $b(t, x), \sigma(t, x)$ and $c(t, , z)$ are non-random and jointly measurable, where b and c are R^d-valued, σ is $R^d \otimes R^d$-valued, which satisfy as $(t, x) \in [0, T] \times R^d$

$$|b|^2 + \|\sigma\|^2 + \int_Z |c|^2 \pi(dz) \leq k_0.$$

1) If there exists a constant $\delta > 0$ such that as $(t, x) \in [0, T] \times R^d$

$$(A\lambda, \lambda) \geq \delta |\lambda|^2,$$

where $A = 2^{-1}\sigma\sigma^*$, then for all $0 \leq f$- bounded measurable function and $p \geq d+1$,

$$E\int_0^T f(s,x_s)ds \leq k(k_0,\delta,d,p,T)\,\|f\|_{L_p([0,T]\times R^d)},$$

where (x_t,ϕ_t) satisfies (4.1) on $[0,T]$, and k is a constant depending on k_0,δ,d,p,T only.

2) If for each $N = 1,2,\cdots$there exists a $\delta_N > 0$ such that for all $\mu \in R^d$

$$\langle A(t,x)\mu,\mu\rangle \geq |\mu|^2\,\delta_N, \text{ as } |x| \leq N,$$

let

$$\tau_N = \inf\{t \geq 0 : |x_t| > .N\},$$

then

$$E\int_0^{\tau_N} f(s,x_s)ds \leq k(p,k_0,d,\delta_N,T)\,\|f\|_{p,[0,\infty)\times[-N,N]},$$

where (x_t,ϕ_t) satisfies (4.1) on $[0,T]$, and k is a constant depending on p, k_0, d, δ_N and T only.

Proof. We only need to note that by Theorem 4.6

$$E\,|\phi|_T^2 = E\,|\phi_T|^2 \leq k_T < \infty,$$

and

$$E\sum_{0<s\leq t}|\Delta x_s|^2 \leq 2E\int_0^t\int_Z |c(s,x_s,z)|^2\,\pi(dz) + 2E(\phi_t\sum_{0<s\leq t}\Delta\phi_s)$$
$$\leq 2(k_0 + k_T) < \infty.$$

Hence, Corollary 1.7 and 1.8 are applicable. ■

Now we are going to prove an existence theorem for a weak solution to (4.1) with discontinuous drift, diffusion, and jump coefficients.

4.11 Theorem. Suppose that assumption of Theorem 4.10 holds on $(t,x) \in [0,\infty)\times R^d$ and suppose that

$$\pi(dz) = dz/\,|z|^{d+1}.$$

Then (4.1) has a weak solution on $t \geq 0$.

Proof. Let us smooth out b,σ,c to get b^n,σ^n,c^n as follows. For each $n = 1,2,...$take c^n - smooth vector function with compact support on $[0,n]\times S_n \times \{\varepsilon_n \leq |z| \leq 1/\varepsilon_n\}$, where $S_n = \{x \in R^d : |x| \leq n\}$, such that

$$\int_Z |c^n|^2\,\pi(dz) \leq 2k_0,$$
$$\left\|\int_Z |c - c^n|^2\,\pi(dz)\right\|_{d+1,[0,n]\times S_n} \leq 1/2^n,$$

where $\varepsilon_n \downarrow 0$, as $n \uparrow \infty$, and

$$\|g\|_{d+1,[0,n]\times S_n}^{d+1} = \int_0^n\int_{|x|\leq n}|g(t,x)|^{d+1}\,dxdt.$$

Take b^n- smooth vector function with compact support on $[0,n]\times S_n$ such that

$$|b^n|^2 \leq k_0,$$
$$\||b - b^n|\|_{d+1,[0,n]\times S_n} \leq 1/2^n.$$

Take σ^n- smooth matrix function with compact support such that as $(t,x) \in [0,n]\times S_n$

$$|\sigma^n|^2 \leq k_0, (A^n\lambda,\lambda) \geq \delta\,|\lambda|^2,$$
$$\|\|\sigma^n - \sigma\|\|_{d+1,[0,n]\times S_n} \leq 1/2^n.$$

where $A^n = 2^{-1}\sigma^n\sigma^{n*}$.

Then by Theorem 4.8 there exists a pathwise unique strong solution (x_t^n, ϕ_t^n) for each $n = 1, 2, ...$, which satisfies (4.1) with coefficients b^n, σ^n and c^n. Let us denote this RSDE by $(4.1)_n$.

Now, similar to the proof of Theorem 3.17, we have that there exist a probability space $(\widetilde{\Omega}, \widetilde{\Im}, \widetilde{P})$ and cadlag processes $(\widetilde{x}_t^n, \widetilde{\phi}_t^n, \widetilde{w}_t^n, \widetilde{\zeta}_t^n)$ defined on it with the same finite-dimensional probability distributions as those of $(x_t^n, \phi_t^n, w_t, \zeta_t)$, and there exist $(\widetilde{x}_t^0, \widetilde{\phi}_t^0, \widetilde{w}_t^0, \widetilde{\zeta}_t^0)$, where $\widetilde{w}_t^0$ is a BM on $(\widetilde{\Omega}, \widetilde{\Im}, \widetilde{P})$, and there also exists a subsequence $\{n_k\}$ of $\{n\}$, denote it by $\{n\}$ again, such that as $n \to \infty$

$\widetilde{\eta}_t^n \to \widetilde{\eta}_t^0$, in probability, as $\widetilde{\eta}_t^n = \widetilde{x}_t^n, \widetilde{\phi}_t^n, \widetilde{w}_t^n, \widetilde{\zeta}_t^n, n = 0, 1, 2, ...$

Set

$\widetilde{p}^n(dt, dz) = \sum_{s \in dt} I_{(0 = \triangle \widetilde{\zeta}_s^n \in dz)}(s)$, $\widetilde{q}^n(dt, dz) = \widetilde{p}^n(dt, dz) - \pi(dz)dt$, $n = 0, 1, 2, ...$

Then $\widetilde{p}^n(dt, dz)$ is a Poisson random point measure with compensator $\pi(dz)dt$ for each $n = 0, 1, 2, ...$, and it verifies

$\widetilde{\zeta}_t^n = \int_0^t \int_{|z| \le 1} z \widetilde{q}^n(ds, dz) + \int_0^t \int_{|z| > 1} z \widetilde{p}^n(ds, dz), n = 0, 1, 2, ...$

By the coincidence of finite probability distributions one has that $(\widetilde{x}_t^n, \widetilde{\phi}_t^n)$ satisfies $(4.1)_n$ with $\widetilde{w}_t^n$ and $\widetilde{q}^n(dt, dz)$ on $(\widetilde{\Omega}, \widetilde{\Im}, \widetilde{P})$. Now we follow completely the same approach as the proof of Theorem 3.17. As a result we find that as $n \to \infty$

$(4.4)_1$ $\left| \int_0^t (b^n(s, \widetilde{x}_s^n) - b(s, \widetilde{x}_s^0)) ds \right|^2 \to 0$, in probability,

$(4.4)_2$ $\int_0^t \sigma^n(s, \widetilde{x}_s^n) d\widetilde{w}_s^n \to \int_0^t \sigma(s, \widetilde{x}_s^0) d\widetilde{w}_s^0$, in probability,

$(4.4)_3$ $\int_0^t \int_Z c^n(s, \widetilde{x}_s^n, z) \widetilde{q}^n(ds, dz) \to \int_0^t \int_Z c(s, \widetilde{x}_s^0, z) \widetilde{q}^0(ds, dz)$, in probability,

$(4.4)_4$ $\int_0^t \int_Z \left| c^n(s, \widetilde{x}_{s-}^n, z) - c(s, \widetilde{x}_{s-}^0, z) \right|^2 \pi(dz) ds \to 0$, in probability.

Hence, by passing to the limit, one gets that $(\widetilde{x}_t^0, \widetilde{\phi}_t^0)$ satisfies

$\widetilde{x}_t^0 = x_0 + \int_0^t b(s, \widetilde{x}_s^0) ds + \int_0^t \sigma(s, \widetilde{x}_s^0) d\widetilde{w}_s^0$
$+ \int_0^t \int_Z c(s, \widetilde{x}_{s-}^0, z) \widetilde{q}^0(ds, dz) + \widetilde{\phi}_t^0 = x_0 + \widetilde{y}_t^0 + \widetilde{\phi}_t^0,$

where $\widetilde{\phi}_t^0 = (0, ..., 0, \widetilde{\phi}_t^{0d})$. Let us show that the other statements in (4.1) also hold for $(\widetilde{x}_t^0, \widetilde{\phi}_t^0)$. Indeed, by Theorem 3.2 in Chapter 2 there exists a unique (x_t^d, ϕ_t^d) satisfying the Skorohod problem (4.1) with x_0^d and $\widetilde{y}_t^{0d}$, i.e.,

$x_t^d = x_0^d + \widetilde{y}_t^{0d} + \phi_t^d,$

and (x_t^d, ϕ_t^d) satisfies the other statements in (4.1).

Now denote

$x_t = (\widetilde{x}_t^{01}, ..., \widetilde{x}_t^{0(d-1)}, x_t^d)$, $\phi_t = (0, 0, ..., 0, \phi_t^d)$.

Then (x_t, ϕ_t) satisfies

$x_t = x_0 + \widetilde{y}_t^0 + \phi_t,$

and the other statements hold for (x_t, ϕ_t) in (4.1).

Now let us prepare one more lemma to show that actually, $\widetilde{x}_t^0 = x_t$, $\widetilde{\phi}_t^0 = \phi_t$.

4.12 Lemma. In Corollary 4.4 if $(\widehat{x}_t, \widehat{y}_t, \widehat{\phi}_t) = (x_t^n, y_t^n, \phi_t^n)$,

$y_t = x_0 + A_t + M_t$, $y_t^n = x_0^n + A_t^n + M_t^n$,

where A_t^n and A_t are continuous finite variational processes and, M_t^n and M_t are local square integrable martingales, then

$\lim_{n\to\infty} E|x_0 - x_0^n|^2 = 0$, $\lim_{n\to\infty} E|A - A^n|_t^2 = 0$,
$\lim_{n\to\infty} E[M - M^n]_t = 0$
implies that
$\lim_{n\to\infty} E\sup_{s\leq t}|x_s^n - x_s|^2 = 0$,
$\lim_{n\to\infty} E\sup_{s\leq t}|\phi_s^n - \phi_s|^2 = 0$,
where $|A|_t$ is the total variation of A on $[0,t]$,
$[M]_t = \langle M^c\rangle_t + \sum_{0<s\leq t}(\Delta M_s)^2$.

Proof. By (4.2)' in Corollary 4.4

$$|x_t^n - x_t|^2 \leq |x_0 - x_0^n|^2 + 2\int_0^t (x_s^n - x_s)\cdot d(A_s^n - A_s)$$
$$+2\int_0^t (x_{s-}^n - x_{s-})\cdot d(M_s^n - M_s) + [M^n - M]_t.$$

Hence,

$$E\sup_{s\leq t}|x_s^n - x_s|^2 \leq k_0'(E|x_0 - x_0^n|^2 + E|A - A^n|_t^2 + E[M^n - M]_t).$$

Now the conclusion is easily derived. ■

By Lemma 4.12 if one can show that

$$(4.5)\begin{cases} \lim_{n\to\infty} E\int_0^t |b^n(s,\tilde{x}_s^n) - b(s,\tilde{x}_s^0)|^2 ds = 0, \\ \lim_{n\to\infty} E[M - M^n]_t = 0 \end{cases}$$

where

$$M_t^n = \int_0^t \sigma^n(s,\tilde{x}_s^n)d\tilde{w}_s^n + \int_0^t\int_Z c^n(s,\tilde{x}_{s-}^n,z)\tilde{q}^n(ds,dz),$$

and M_t is similarly defined, then by Lemma 4.12 and by the uniqueness of a limit this shows that

$$\tilde{x}_t^0 = x_t,\ \tilde{\phi}_t^0 = \phi_t.$$

This means that $(\tilde{x}_t^0, \tilde{\phi}_t^0)$ is a weak solution of (4.1). Therefore, in what follows, we are going to show (4.5). For example, let us show that as $n\to\infty$

$$E\left|\int_0^t\int_Z c^n(s,\tilde{x}_{s-}^n,z)\tilde{q}^n(ds,dz) - \int_0^t\int_Z c(s,\tilde{x}_{s-}^0,z)\tilde{q}^0(ds,dz)\right|^2 \to 0.$$

In fact, by $(4.4)_3$ for any given $t\geq 0$ there exists a subsequence $\{n_k\}$, denote it by $\{n\}$ again, such that as $n\to\infty$

$$\int_0^t\int_Z c^n(s,\tilde{x}_{s-}^n,z)\tilde{q}^n(ds,dz) \to \int_0^t\int_Z c(s,\tilde{x}_{s-}^0,z)\tilde{q}^0(ds,dz),\ P-a.s.$$

Since by $(4.4)_4$ one has that as $n\to\infty$

$$E\int_0^t\int_Z |c^n(s,\tilde{x}_s^n,z) - c(s,\tilde{x}_s^0,z)|^2\pi(dz)ds \to 0.$$

Hence as $n\to\infty$

$$E\left|\int_0^t\int_Z c^n(s,\tilde{x}_{s-}^n,z)\tilde{q}^n(ds,dz)\right|^2 = E\int_0^t\int_Z |c^n(s,\tilde{x}_s^n,z)|^2\pi(dz)ds$$
$$\to E\int_0^t\int_Z |c(s,\tilde{x}_s^0,z)|^2\pi(dz)ds = E\left|\int_0^t\int_Z c(s,\tilde{x}_{s-}^0,z)\tilde{q}^0(ds,dz)\right|^2.$$

Applying Lemma 3.6, one has that as $n\to\infty$

$$E\left|\int_0^t\int_Z c^n(s,\tilde{x}_{s-}^n,z)\tilde{q}^n(ds,dz) - \int_0^t\int_Z c(s,\tilde{x}_{s-}^0,z)\tilde{q}^0(ds,dz)\right|^2 \to 0.$$

Now the 2nd statement in (4.5) can be shown similarly. The first one is even easier to show by Lebesgue's dominated convergence theorem, since b is bounded, and $(4.4)_1$ holds. The proof of Theorem 4.11 is now completed. ■

Applying Theorem 4.11, Theorem 4.7 and Theorem 1.3, the following theorem immediately follows.

4.13 Theorem. Assume that $b(t,x), \sigma(t,x)$ and $c(t,x,z)$ are non-random, and jointly measurable, where b and c are R^d−valued, σ is $R^d \otimes R^d$-valued, which satisfy as $(t,x) \in [0,\infty) \times R^d$

$$|b| + \|\sigma\|^2 + \int_Z |c|^2 \pi(dz) \leq k_0,$$

and there exists a constant $\delta > 0$ such that as $(t,x) \in [0,\infty) \times R^d$

$$(A\lambda, \lambda) \geq \delta |\lambda|^2,$$

where $A = 2^{-1}\sigma\sigma^*$; and assume that

$$\pi(dz) = dz/|z|^{d+1},$$

$$2(x-y)\cdot(b(t,x)-b(t,y)) + \|\sigma(t,x)-\sigma(t,y)\|^2$$

$$+ \int_0^t |c(s,x,z)-c(s,y,z)|^2 \pi(dz) \leq k_N(t)\rho_N(|x-y|^2), \text{ as } |x|, |y| \leq N,$$

where $0 \leq k_N(t), \int_0^t k_N(s)ds < \infty$, and $\rho_N(u)$ is concave and strictly increasing in $u \geq 0$ such that $\rho_N(0) = 0$, and

$$\int_{0+} du/\rho_N(u) = \infty, \text{ for each } N = 1, 2, ...$$

Then (4.1) has a pathwise unique strong solution.

For the existence of a weak solution with degenerate diffusion coefficient σ we have the following

4.14 Theorem. Assume that

1° $b(t,x), \sigma(t,x), c(t,x,z)$ are jointly measurable, and there exists a constant $k_0 > 0$ such that

$$|b| + \|\sigma\|^2 + \int_Z |c|^2 \pi(dz) \leq k_0,$$

2° $b(t,x), \sigma(t,x)$ are continuous in x, and as $|x-y| \to 0$,

$$\int_0^t |c(s,x,z)-c(s,y,z)|^2 \pi(dz) \to 0,$$

3° $\pi(dz) = dz/|z|^{d+1}$

Then (4.1) has a weak solution.

Proof. The proof is a modification of that for Theorem 4.11. However, instead of using Krylov's estimate, here we must use the continuity of b, σ and c because σ can also be degenerate.

As in the proof of Theorem 4.9 one can smooth out b, σ and c to get b^n, σ^n, and c^n, and show that there exists a pathwise unique strong solution (x_t^n, ϕ_t^n) satisfying

$$\begin{cases} x_t^n = x_0 + \int_0^t b^n(s,x_s^n)ds + \int_0^t \sigma^n(s,x_s^n)dw_s \\ \qquad + \int_0^t \int_Z c^n(s,x_{s-}^n,z)q(ds,dz) + \phi_t^n, \\ \text{and the other statements in (4.1) hold for } (x_t^n, \phi_t^{nd}), \\ \qquad \text{where } \phi_t^n = (0,...,0,\phi_t^{nd}). \end{cases}$$

Again, similar to the proof of Theorem 4.11, we have that there exists a probability space $(\widetilde{\Omega}, \widetilde{\Im}, \widetilde{P})$ and cadlag processes $(\widetilde{x}_t^n, \widetilde{\phi}_t^n, \widetilde{w}_t^n, \widetilde{\zeta}_t^n)$ defined on it with the same finite probability distributions as those of $(x_t^n, \phi_t^n, w_t, \zeta_t)$, and there exist $(\widetilde{x}_t^0, \widetilde{\phi}_t^0, \widetilde{w}_t^0, \widetilde{\zeta}_t^0)$, where $\widetilde{w}_t^0$ is a BM on $(\widetilde{\Omega}, \widetilde{\Im}, \widetilde{P})$, and there also exists a subsequence $\{n_k\}$ of $\{n\}$, denote it by $\{n\}$ again, such that as $n \to \infty$

$$\widetilde{\eta}_t^n \to \widetilde{\eta}_t^0, \text{ in probability, as } \widetilde{\eta}_t^n = \widetilde{x}_t^n, \widetilde{\phi}_t^n, \widetilde{w}_t^n, \widetilde{\zeta}_t^n.$$

For $n = 0, 1, 2, ...$ set

$$\widetilde{p}^n(dt,dz) = \sum_{s \in dt} I_{(0 = \Delta\widetilde{\zeta}_s^n \in dz)}(s), \ \widetilde{q}^n(dt,dz) = \widetilde{p}^n(dt,dz) - \pi(dz)dt.$$

Then $\widetilde{p}^n(dt,dz)$ is a Poisson random point measure with compensator $\pi(dz)dt$ for each $n = 0, 1, 2, ...$, and it verifies

$\widetilde{\zeta}_t^n = \int_0^t \int_{|z|\leq 1} z\widetilde{q}^n(ds,dz) + \int_0^t \int_{|z|>1} z\widetilde{p}^n(ds,dz), n = 0, 1, 2, ...$

By the coincidence of finite probability distributions we find that $(\widetilde{x}_t^n, \widetilde{\phi}_t^n)$ satisfies (4.1) with $\widetilde{w}_t^n$ and $\widetilde{q}^n(dt,dz)$ on $(\widetilde{\Omega}, \widetilde{\Im}, \widetilde{P})$. Now let us show that as $n \to \infty$

$(4.6)_1$ $E\left|\int_0^t (b^n(s,\widetilde{x}_s^n) - b(s,\widetilde{x}_s^0))ds\right|^2 \to 0,$

$(4.6)_2$ $\int_0^t \sigma^n(s,\widetilde{x}_s^n)d\widetilde{w}_s^n \to \int_0^t \sigma(s,\widetilde{x}_s^0)d\widetilde{w}_s^0$, in probability,

$(4.6)_3$ $\int_0^t \int_Z c^n(s,\widetilde{x}_{s-}^n,z)\widetilde{q}^n(ds,dz) \to \int_0^t \int_Z c(s,\widetilde{x}_{s-}^0,z)\widetilde{q}^0(ds,dz)$,
in probability,

$(4.6)_4$ $E\int_0^t \int_Z \left|c^n(s,\widetilde{x}_s^n,z) - c(s,\widetilde{x}_s^0,z)\right|^2 \pi(dz)ds \to 0.$

Note that for each $n^0 = 1, 2, \cdots$

$\int_0^t \int_Z \left|c^n(s,\widetilde{x}_s^n,z) - c(s,\widetilde{x}_s^0,z)\right|^2 \pi(dz)ds$

$\leq 3\int_0^t \int_Z \left|c^n(s,\widetilde{x}_s^n,z) - c^{n^0}(s,\widetilde{x}_s^n,z)\right|^2 \pi(dz)ds$

$+3\int_0^t \int_Z \left|c^{n^0}(s,\widetilde{x}_s^n,z) - c^{n^0}(s,\widetilde{x}_s^0,z)\right|^2 \pi(dz)ds$

$+3\int_0^t \int_Z \left|c^{n^0}(s,\widetilde{x}_s^0,z) - c(s,\widetilde{x}_s^0,z)\right|^2 \pi(dz)ds = 3(I_1^{n,n^0} + I_2^{n,n^0} + I_3^{n^0}).$

We must now be careful, because we cannot use Krylov's estimate. However,

$I_1^{n,n^0} \leq \int_0^t \int_Z | \int_{|y|\leq 1}(c(s,\widetilde{x}_s^n - n^{-1}y,z)$

$-c(s,\widetilde{x}_s^n - (n^0)^{-1}y,z))e^{-1/(1-|y|^2)}\widetilde{k}_0 dy |^2 \pi(dz)ds$

$\leq \int_{|y|\leq 1}[\int_0^t \int_Z | c(s,\widetilde{x}_s^n - n^{-1}y,z)$

$-c(s,\widetilde{x}_s^n - (n^0)^{-1}y,z) |^2 \pi(dz)ds]e^{-1/(1-|y|^2)}\widetilde{k}_0 dy$

By the continuity assumption 2° applying Lebesgue's dominated convergence theorem, one finds that as $n, n^0 \to \infty$

$I_1^{n,n^0} \to 0.$

Similarly, one has that as $n^0 \to \infty$

$I_3^{n^0} \to 0.$

Now for each fixed n^0 as $n \to \infty$

$\int_Z \left|c^{n^0}(s,\widetilde{x}_s^n,z) - c^{n^0}(s,\widetilde{x}_s^0,z)\right|^2 \pi(dz) \leq k_{n^0}\left|\widetilde{x}_s^n - \widetilde{x}_s^0\right|^2 \to 0$, in probability.

Moreover,

$\int_Z \left|c^{n^0}(s,\widetilde{x}_s^n,z) - c^{n^0}(s,\widetilde{x}_s^0,z)\right|^2 \pi(dz) \leq 2k_0.$

Hence, by Lebesgue's dominated convergence theorem for each fixed n^0 as $n \to \infty$

$\tilde{E}\int_Z \left|c^{n^0}(s,\widetilde{x}_s^n,z) - c^{n^0}(s,\widetilde{x}_s^0,z)\right|^2 \pi(dz) \to 0.$

Applying Lebesgue's dominated convergence theorem again, one also has that for each fixed n^0 as $n \to \infty$

$\tilde{E}I_2^{n,n^0} = \int_0^t E\int_Z \left|c^{n^0}(s,\widetilde{x}_s^n,z) - c^{n^0}(s,\widetilde{x}_s^0,z)\right|^2 \pi(dz)ds \to 0.(\#)$

Similarly, as $n, n^0 \to \infty$

$(4.7)_1$ $\tilde{E}I_1^{n,n^0} \to 0$,
and as $n^0 \to \infty$
$(4.7)_2$ $\hat{E}I_3^{n^0} \to 0$.
Therefore $(4.6)_4$ is true. To show $(4.6)_3$ one only needs to prove that for each fixed n^0 as $n \to \infty$

$$I_t^n = \check{P}(\left|\int_0^t \int_Z c^{n^0}(s, \tilde{x}_{s-}^n, z)\tilde{q}^n(ds, dz) - \int_0^t \int_Z c^{n^0}(s, \tilde{x}_{s-}^0, z)\tilde{q}^0(ds, dz)\right| > \varepsilon) \to 0. \quad (@)$$

However,

$$I_t^n \leq (2/\varepsilon)^2 \tilde{E} \int_0^t \int_Z \left|c^{n^0}(s, \tilde{x}_s^n, z) - c^{n^0}(s, \tilde{x}_s^0, z)\right|^2 \pi(dz)ds$$
$$+\tilde{P}(\left|\int_0^t \int_Z c^{n^0}(s, \tilde{x}_{s-}^0, z)\tilde{q}^n(ds, dz) - \int_0^t \int_Z c^{n^0}(s, \tilde{x}_{s-}^0, z)\tilde{q}^0(ds, dz)\right| > \varepsilon/2)$$
$$= 2(I_t^{1n} + I_t^{2n}).$$

By (@) for each fixed n^0 as $n \to \infty$

$$I_t^{1n} \to 0.$$

Note that as $t \leq T$

$$\tilde{E} \int_0^t \int_Z \left|c^{n^0}(s, \tilde{x}_s^0, z)\right|^2 \pi(dz)ds \leq k_0 T < \infty.$$

Hence for any given $\delta > 0$ one can take a $\varepsilon' > 0$ such that

$$(4.8) \quad (4/\varepsilon)^2 \tilde{E} \int_0^t \int_{|z| \leq \varepsilon'} \left|c^{n^0}(s, \tilde{x}_s^0, z)\right|^2 \pi(dz)ds \leq \delta/8.$$

Obviously,

$$I_t^{2n} \leq \delta/2$$
$$+\check{P}(\left|\int_0^t \int_{|z|>\varepsilon'} c^{n^0}(s, \tilde{x}_s^0, z)\tilde{q}^n(ds, dz) - \int_0^t \int_{|z|>\varepsilon'} c^{n^0}(s, \tilde{x}_s^0, z)\tilde{q}^0(ds, dz)\right| > \varepsilon/4).$$

However, by Lemma 3.4 for each fixed n^0 as $n \to \infty$

$$\int_0^t \int_{|z|>\varepsilon'} c^{n^0}(s, \tilde{x}_{s-}^0, z)\tilde{q}^n(ds, dz) \to \int_0^t \int_{|z|>\varepsilon'} c^{n^0}(s, \tilde{x}_{s-}^0, z)\tilde{q}^0(ds, dz),$$

in probability.

Hence, (@) is proved. $(4.6)_1$ and $(4.6)_2$ are proved similarly. Now, in the rest as the proof in Theorem 4.11, one only needs to show that

$$\lim_{n\to\infty} E[M^n - M^0]_t = 0,$$

where

$$M^n = \int_0^t \sigma^n(s, \tilde{x}_s^n) d\tilde{w}_s^n + \int_0^t \int_Z c^n(s, \tilde{x}_{s-}^n, z)\tilde{q}^n(ds, dz), \text{ etc.}$$

(The proof of $\lim_{n\to\infty} E \int_0^t \left|b^n(s, \tilde{x}_s^n) - b(s, \tilde{x}_s^0)\right|^2 ds = 0$ is obvious.) However, this can be completed just as the last part in the proof of Theorem 4.11. The proof of Theorem 4.14 is now complete. ∎

3.4.3 Existence of Weak Solutions for RSDE with Random Coefficients

By using the Girsanov theorem we can even get a weak solution on some probability space for some RSDE with random coefficients.

4.15 Theorem. Assume that there are a BM $w_t, t \geq 0$, and a Poisson martingale measure $q(dt, dz)$ with the compensator $\pi(dz)dt$, $\pi(dz) = dz/|z|^{d+1}$, given on a probability space $(\Omega, \Im, (\Im_t), P)$, and assume that

1° $b(t,x,\omega), \sigma(t,x,\omega), c(t,x,z,\omega)$ are jointly measurable, and as x and z are fixed, they are $\Im_t$−adapted, where c is $\Im_t$−predictable, moreover, there exists a constant $k_0 > 0$ such that

$$|b|^2 + \|\sigma\|^2 + \int_Z |c|^2 \pi(dz) \leq k_0(1+|x|^2),$$

2° $\sigma(t,x,\omega)$ is continuous in x, and as $|x-y| \to 0$,

$$\int_0^t |c(s,x,z,\omega) - c(s,y,z,\omega)|^2 \pi(dz) \to 0.$$

3° $\|\sigma(t,x,\omega) - \sigma(t,y,\omega)\|^2 + \int_0^t |c(s,x,z,\omega) - c(s,y,z,\omega)|^2 \pi(dz)$

$$\leq k_N(t)\rho_N(|x-y|^2), \text{ as } |x|, |y| \leq N,$$

where $0 \leq k_N(t), \int_0^t k_N(s)ds < \infty$, and $\rho_N(u)$ is concave and strictly increasing in $u \geq 0$ such that $\rho_N(0) = 0$, and

$$\int_{0+} du/\rho_N(u) = \infty, \text{ for each } N = 1, 2, ...,$$

4° there exists a constant $\delta > 0$ such that as $(t,x) \in [0,\infty) \times R^d$

$$(A\lambda, \lambda) \geq \delta |\lambda|^2,$$

where $A = 2^{-1}\sigma\sigma^*$.

Then there exist a probability measure $\widetilde{P}$, two $\Im_t$− adapted cadlag processes (x_t, ϕ_t), a BM $\hat{w}_t, t \geq 0$, and a Poisson martingale measure $\tilde{q}(dt,dz)$ with the compensator $\pi(dz)dt$, $\pi(dz) = dz/|z|^{d+1}$ defined on the probability space $(\Omega, \Im, (\Im_t), \tilde{P})$ such that (4.1) is satisfied by $(x_t, \phi_t, \hat{w}_t, \tilde{q}(dt,dz))$.

Proof. The proof is a combination of Theorem 4.9 and Theorem 4.16 below. Indeed, by Theorem 4.9, RSDE (4.9) below has a pathwise unique strong solution (x_t, ϕ_t). Applying the following Theorem 4.16 $(x_t, \phi_t, \widetilde{w}_t, q(dt,dz))$ satisfies (4.1) on probability space $(\Omega, \Im, (\Im_t), \tilde{P})$, where $\widetilde{w}_t, q(dt,dz)$ and $\tilde{P}$ are given in the following Theorem 4.16. ■

4.16 Theorem. (Girsanov type Theorem). Assume that (x_t, ϕ_t) satisfies

$$(4.9)\begin{cases} x_t^i = x_0^i + \sum_{j=1}^d \int_0^t \sigma^{ij}(s,x_s,\omega)dw_s^j + \int_0^t \int_Z c^i(s,x_{s-},z,\omega)q(ds,dz), \\ \qquad i = 1, ..., d-1; \\ x_t^d = x_0^d + \sum_{j=1}^d \int_0^t \sigma^{dj}(s,x_s,\omega)dw_s^j + \int_0^t \int_Z c^d(s,x_{s-},z,\omega)q(ds,dz) \\ \qquad +\phi_t, \\ x_t^d \geq 0, \text{ for all } t \geq 0, \ x_t \text{ is } R^d - \text{valued } \Im_t - \text{adapted and cadlag;} \\ \phi_t \text{ is a real } \Im_t - \text{adapted increasing process with } \phi_0 = 0 \text{ such that} \\ \qquad \int_0^t x_s^d d\phi_s^c = 0, \text{ for all } t \geq 0, \ \phi_t^c = \phi_t - \sum_{0<s\leq t} \triangle\phi_s, \\ \qquad \triangle\phi_t > 0 \Rightarrow \triangle\phi_t = \alpha_t x_t^d; \end{cases}$$

and

$$|b(t,x,\omega)| \leq k_0(1+|x|),$$

and assume that there exists a constant $\delta > 0$ such that

as $(t,x) \in [0,\infty) \times R^d$

$$(A\lambda, \lambda) \geq \delta |\lambda|^2,$$

where $A = 2^{-1}\sigma\sigma^*$.

Then

$$Ez_t = 1, \text{ for all } t \geq 0,$$

where

$z_t = \exp(\int_0^t (\sigma^{-1}b)(s, x_s, \omega) \cdot dw_s - \frac{1}{2}\int_0^t \left|(\sigma^{-1}b)(s, x_s, \omega)\right|^2 ds)$.
Moreover, there exists a probability measure $\tilde{P}$, which satisfies $\tilde{P}\ |_{\Im_t} = P_t$, where $d\tilde{P}_t = z_t dP$, (here without loss of generality we can assume that $(\Omega, \Im)$ is a standard measurable space), such that
1) $\widetilde{w}_t = w_t - \int_0^t (\sigma^{-1}b)(s, x_s, \omega)ds$, $t \geq 0$,
is a BM under probability measure $\tilde{P}$;
2) $q(dt, dz) = p(dt, dz) - \pi(dz)dt$
is still a $\tilde{P}$-martingale measure with the same compensator $\pi(dz)dt$.
Hence $(x_t, \phi_t, \widetilde{w}_t, q(dt, dz))$ solves (4.1) on probability space $(\Omega, \Im, \tilde{P})$.
Proof. The proof is completely the same as that of Theorems 3.12 and 3.14. ■

Chapter 4

Properties of Solutions to RSDE with Jumps

In this chapter we will discuss the convergence, stability, comparison and uniqueness of solutions to RSDE with jumps in d-dimensional space. As an application we also show how to use the comparison theorem to get the existence of a strong solution for an RSDE with Poisson jumps and with discontinuous drift and degenerate diffusion coefficients in 1-dimensional space. It should be noted that some results are derived only in some special refection domains.

4.1 Convergence Theorems for Solutions

In this section we will discuss the convergence theorems for solutions to RSDE (1.1) and (1.2) in Chapter 3. More precisely, let us consider the RSDEs as follows:

$$(1.1)\quad \begin{cases} dx_t^n = b^n(t,x_t^n)dt + \sigma^n(t,x_t^n)dw_t + \int_Z c^n(t,x_{t-}^n,z)q(dt,dz) + d\phi_t^n, \\ \qquad x_0^n = x_0^n \in \overline{\Theta}, \\ \text{and } (x_t^n,\phi_t^n),\ t\geq 0 \text{ satisfies the statements in (1.2) of Chapter 3;} \\ \qquad \text{moreover, } \phi_t^n \text{ is continuous;} \end{cases}$$

$n = 0,1,2,.....$

We have

1.1 Theorem. Assume that

$1°$ there exists a constant $k_0 \geq 0$, and for each $N = 1,2,...$ there exists a $\delta_N > 0$ such that

$|b^n(t,x)|^2 + \|\sigma^n(t,x)\|^2 + \int_Z |c^n(t,x,z)|^2 \pi(dz) \leq k_0,$

$\langle A^n\lambda,\lambda\rangle \geq \delta_N |\lambda|^2$, for all $|x| \leq N, \lambda \in R^d, n = 0,1,2,...$

where $N = 1,2,...$, and $A^n = \frac{1}{2}\sigma^n\sigma^{n*}$;

$2°$ $2(x-y)\cdot(b^0(t,x) - b^0(t,y)) \leq k(t)\rho(|x-y|^2),$

$$\left\|\sigma^0(t,x)-\sigma^0(t,y)\right\|^2 \leq k(t)\rho(|x-y|^2),$$
$$\int_Z \left|c^0(t,x,z)-c^0(t,y,z)\right|^2 \pi(dz) \leq k(t)\rho(|x-y|^2),$$
where $\rho(u)$ is continuous, strictly increasing, concave, $\rho(0)=0$, $k(t)\geq 0$,and they are such that for any $0<T<\infty$
$$\int_0^T k(t)dt<\infty,\ \int_{0+} du/\ \rho(u)=\infty;$$
3° as $n\to\infty$
$$\left\|\left|b^n-b^0\right|^2\right\|_{p,[0,\infty)\times[-N,N]} \to 0,\ \left\|\left\|\sigma^n-\sigma^0\right\|^2\right\|_{p,[0,\infty)\times[-N,N]} \to 0,$$
$$\left\|\int_Z \left|c^n-c^0\right|^2 \pi(dz)\right\|_{p,[0,\infty)\times[-N,N]} \to 0,$$
for some $p\geq d+1$, and for each $N=1,2,...$;

4° $E\left|x^n-x^0\right|^2 \to 0$, as $n\to\infty$,

Then for all $t\geq 0$, as $n\to\infty$
$$E\sup_{s\leq t}\left|x_s^n-x_s^0\right|^2 \to 0.$$
In addition, if
$$\lim_{y\to x}\sup_{s\leq t}\left|b^0(s,x)-b^0(s,y)\right|^2=0,$$
then for all $t\geq 0$, as $n\to\infty$
$$E\sup_{s\leq t}\left|\phi_s^n-\phi_s^n\right|^2 \to 0.$$
Proof. By Ito's formula
$$(1.2)\ \left|x_t^n-x_t^0\right|^2=\left|x_0^n-x_0^0\right|^2+\int_0^t[2(x_s^n-x_s^0)\cdot(b^n(s,x_s^n)-b^0(s,x_s^0))$$
$$+\left\|\sigma^n(s,x_s^n)-\sigma^0(s,x_s^0)\right\|^2]ds+2\int_0^t(x_s^n-x_s^0)\cdot(\sigma^n(s,x_s^n)-\sigma^0(s,x_s^0))dw_s$$
$$+2\int_0^t(x_s^n-x_s^0)\cdot\int_Z(c^n(s,x_{s-}^n,z)-c^0(s,x_{s-}^0,z))q(ds,dz)$$
$$+2\int_0^t(x_s^n-x_s^0)\cdot d(\phi_s^n-\phi_s^0)+\int_0^t\int_Z\left|c^n(s,x_{s-}^n,z)-c^0(s,x_{s-}^0,z)\right|^2 p(ds,dz)$$
$$\leq\left|x_0^n-x_0^0\right|^2+\int_0^t[\left|b^n(s,x_s^n)-b^0(s,x_s^n)\right|^2+2\left\|\sigma^n(s,x_s^n)-\sigma^0(s,x_s^n)\right\|^2]ds$$
$$+2\int_0^t(x_s^n-x_s^0)\cdot(\sigma^n(s,x_s^n)-\sigma^0(s,x_s^0))dw_s$$
$$+2\int_0^t(x_s^n-x_s^0)\cdot\int_Z(c^n(s,x_{s-}^n,z)-c^0(s,x_{s-}^0,z))q(ds,dz)$$
$$+2\int_0^t\int_Z\left|c^n(s,x_{s-}^n,z)-c^0(s,x_{s-}^n,z)\right|^2 p(ds,dz)+3\int_0^t k(s)\rho(\left|x_s^n-x_s^0\right|^2)ds$$
$$+\int_0^t\left|x_s^n-x_s^0\right|^2 ds+2\int_0^t\int_Z\left|c^0(s,x_{s-}^n,z)-c^0(s,x_{s-}^0,z)\right|^2 p(ds,dz),$$
where we have applied Remark 2.2 in Chapter 3
$$\int_0^t(x_s^n-x_s^0)\cdot d(\phi_s^n-\phi_s^0)\leq 0.$$
However, by condition 1° applying Theorem 2.1 in Chapter 3, one finds that
$$E(\sup_{0\leq t\leq T}|x_t^n|^2+\sup_{0\leq t\leq T}|\phi_t^n|^2)\leq k_T,\text{ for all } n=1,2,...$$
Hence by
$$E\int_0^T\left|b^n(t,x_s^n)-b^0(t,x_s^n)\right|^2 ds\leq k_T'\sup_n P(\sup_{0\leq t\leq T}|x_t^n|^2>N)$$
$$+k_{T,N}'\left\|\left|b^n-b^0\right|^2\right\|_{p,[0,\infty)\times[-N,N]}$$
so as $n\to\infty$
$$E\int_0^T\left|b^n(t,x_s^n)-b^0(t,x_s^n)\right|^2 ds\to 0.$$
Therefore by the martingale inequality and a similar technique to the one above we find that
$$(1.3)\ \overline{\lim}_{n\to\infty}E\sup_{s\leq t}\left|x_s^n-x_s^0\right|^2$$
$$\leq k_0'[\int_0^t k(r)\rho(\overline{\lim}_{n\to\infty}E\sup_{s\leq r}\left|x_s^n-x_s^0\right|^2)dr$$

$+ \int_0^t \overline{\lim}_{n\to\infty} E\sup_{s\le r} \left|x_s^n - x_s^0\right|^2 dr].$

Hence

(1.4) $\overline{\lim}_{n\to\infty} E\sup_{s\le t} \left|x_s^n - x_s^0\right|^2 = 0.$

Now

(1.5) $|\phi_t^n - \phi_t^n|^2 \le 5(\left|x_t^n - x_t^0\right|^2 + \left|x_0^n - x_0^0\right|^2$

$+ \left|\int_0^t (b^n(s,x_s^n) - b^0(s,x_s^n))ds\right|^2 + \left|\int_0^t (\sigma^n(s,x_s^n) - \sigma^0(s,x_s^0))dw_s\right|^2$

$+ \left|\int_0^t \int_Z (c^n(s,x_{s-}^n,z) - c^0(s,x_{s-}^0,z))q(ds,dz)\right|^2) = 5\sum_{i=1}^5 I_t^i.$

Note that as $t \le T$

$EI_t^3 \le k_T(E\int_0^t \left|b^n(s,x_s^n) - b^0(s,x_s^n)\right|^2 ds$

$+E\int_0^t \left|b^0(s,x_s^n) - b^0(s,x_s^0)\right|^2 ds) = I_t^{31} + I_t^{32}.$

However, for any $\varepsilon' > 0$

$EI_t^{32} \le E\int_0^t I_{|x_s^n - x_s^0| < \varepsilon'} \left|b^0(s,x_s^n) - b^0(s,x_s^0)\right|^2 ds$

$+k_T' P(\sup_{0\le s\le t} \left|x_s^n - x_s^0\right|^2 > \varepsilon').$

Hence, by assumption and the facts just proved,

$\lim_{n\to\infty} EI_t^3 = 0.$

Therefore, by the martingale inequality as $n \to \infty$

$E\sup_{s\le t} \left|\phi_s^n - \phi_s^0\right|^2 \to 0.$ ■

If $2°$ is a local condition we have

1.2 Theorem. Under the assumption of Theorem 1.1, but in place of $2°$, it is assuming that

$2°'$ $2(x-y)\cdot(b^0(t,x) - b^0(t,y)) \le k_N(t)\rho_N(|x-y|^2),$

$\left\|\sigma^0(t,x) - \sigma^0(t,y)\right\|^2 + \int_Z \left|c^0(t,x,z) - c^0(t,y,z)\right|^2 \pi(dz)$

$\le k_N(t)\rho_N(|x-y|^2)$, as $|x|, |y| \le N$,

where for each $N = 1, 2, ...,$ $k_N(t) \ge 0$, $\rho_N(u)$ is continuous, strictly increasing, concave, $\rho_N(0) = 0$, and they are such that for any $0 < T < \infty$

$\int_0^T k_N(t)dt < \infty$, $\int_{0+} du/\rho_N(u) = \infty$;

then for any $\varepsilon > 0$ and $t \ge 0$

$\lim_{n\to\infty} P(\sup_{s\le t} \left|x_s^n - x_s^0\right| > \varepsilon) = 0.$

In addition, if

$\lim_{y\to x} \sup_{s\le t} \left|b^0(s,x) - b^0(s,y)\right| = 0,$

then for all $t \ge 0$, as $n \to \infty$

$P(\sup_{s\le t} \left|\phi_s^n - \phi_s^0\right| > \varepsilon) \to 0.$

Proof. Set

$\tau^N = \inf\left\{t : \left|x_t^0\right| > N\right\},$

$\tau^{n,\varepsilon} = \inf\left\{t : \left|x_t^n - x_t^0\right| > \varepsilon\right\},$

$\tau_\varepsilon^N(n) = \tau^N \wedge \tau^{n,\varepsilon}.$

Then by Ito's formula, as above, we have

$\lim_{n\to\infty} E\sup_{s\le t} \left|x_{s\wedge\tau_\varepsilon^N(n)}^n - x_{s\wedge\tau_\varepsilon^N(n)}^0\right|^2 = 0.$

However, for all $t \ge 0$

$$E\left|x^n_{t\wedge\tau^N_\varepsilon(n)} - x^0_{t\wedge\tau^N_\varepsilon(n)}\right|^2 \geq \varepsilon^2 P(\tau^{n,\varepsilon} \leq \tau^N \wedge t).$$

Hence,

$$\lim_{n\to\infty} P(\tau^{n,\varepsilon} \leq \tau^N \wedge t) = 0.$$

We have that

$$P(\sup_{s\leq\tau^N\wedge t} \left|x^n_s - x^0_s\right| > 2\varepsilon) \leq P(\tau^{n,\varepsilon} \leq \tau^N \wedge t) \to 0, \text{ as } n \to \infty.$$

Note that

$$P(\sup_{s\leq T} \left|x^n_s - x^0_s\right| > \varepsilon) \leq P(\sup_{s\leq\tau^N\wedge T} \left|x^n_s - x^0_s\right| > \varepsilon)$$
$$+P(\sup_{\tau^N\wedge T<s\leq T} \left|x^n_s - x^0_s\right| > \varepsilon) \leq P(\sup_{s\leq\tau^N\wedge T} \left|x^n_s - x^0_s\right| > \varepsilon)$$
$$+P(\tau^N \wedge T < T).$$

Since

$$\lim_{N\to\infty} \tau^N = \infty.$$

Hence,

$$\lim_{N\to\infty} P(\tau^N \wedge T < T) = 0.$$

Therefore the first conclusion is derived. Now by (1.1)

$$P(\sup_{s\leq t} \left|\phi^n_s - \phi^0_s\right| > 4\varepsilon) \leq P(\sup_{s\leq t} \left|x^n_s - x^0_s\right| > \varepsilon)$$
$$+P(\sup_{r\leq t} \left|\int_0^r b^n(s, x^n_s)ds - \int_0^r b^0(s, x^0_s)ds\right| > \varepsilon)$$
$$+P(\sup_{r\leq t} \left|\int_0^r \sigma^n(s, x^n_s)dw_s - \int_0^r \sigma^0(s, x^0_s)dw_s\right| > \varepsilon)$$
$$+P(\sup_{r\leq t} \left|\int_0^r \int_Z c^n(s, x^n_{s-}, z)q(ds, dz) - \int_0^r \int_Z c^0(s, x^0_{s-}, z)q(ds, dz)\right| > \varepsilon)$$
$$= \sum_{i=1}^4 I_i.$$

Now it is known that as $n \to \infty$

$$I_1 \to 0.$$

Let us show that as $n \to \infty$

$$I_4 \to 0.$$

In fact, by the martingale inequality

$$I_4 \leq k_t\varepsilon^{-1}E(\int_0^t k_N(s)\rho_N(\sup_{r\leq s} \left|x^n_r - x^0_r\right|^2)I_{|x^n_s|\leq N, |x^0_s|\leq N}ds)^{1/2}$$
$$+[\sup_n P(\sup_{0\leq s\leq t} \left|x^n_s\right|^2 > N) + P(\sup_{0\leq s\leq t} \left|x^0_s\right|^2 > N)]$$
$$+k'_{t,N} \left\|\int_Z \left|c^n - c^0\right|^2 \pi(dz)\right\|^{1/2}_{p,[0,\infty)\times[-N,N]} = \sum_{i=1}^3 I^{n,N}_{4i}.$$

Obviously, for arbitrary given $\tilde{\varepsilon} > 0$, one can take an N_0 large enough that

$$I^{n,N_0}_{42} < \tilde{\varepsilon}/3, \text{ for all } n = 1, 2, ...$$

However,

$$I^{n,N_0}_{41} \leq k_t\varepsilon^{-1}E(\int_0^t k_{N_0}(s)ds)^{1/2}\rho_{N_0}(\overline{\delta}^2)^{1/2}$$
$$+k_t\varepsilon^{-1}P(\sup_{0\leq s\leq t} \left|x^n_s - x^0_s\right| > \overline{\delta})(\rho_{N_0}(4N_0^2) \int_0^t k_{N_0}(s)ds)^{1/2}.$$

From these it is easily seen that as $n \to \infty$

$$I_4 \to 0.$$

The proof of $I_3, I_2 \to 0$,as $n \to \infty$ is obtained in a similar fashion. Therefore, we arrive at the second conclusion. ■

Remark 1.3. By the proof of Theorem 1.2 one sees that in Theorem 1.2 the bounded assumption on b^n, σ^n,and $\int_Z |c|^2 \pi(dz)$ can be weakened to

$1^{\circ\prime}$ $|b^n(t,x)|^2 + \|\sigma^n(t,x)\|^2 + \int_Z |c^n(t,x,z)|^2 \pi(dz) \leq k_0(1+|x|^2), n = 0, 1, 2, ...$

Under the weaker condition $1^{\circ\prime}$, if one still wants to obtain the conclusion of Theorem 1.1, one has to strengthen assumption 3°. In fact, one can have the following

Theorem 1.4. In Theorem 1.1 if one weakens the whole 1° to the above assumption $1^{\circ\prime}$, strengthens 3° to

$3^{\circ\prime}$ $\lim_{n\to\infty} \sup_{x\in R^d} \int_0^t [|b^n(s,x) - b^0(s,x)|^2$
$+ \|\sigma^n(s,x) - \sigma^0(s,x)\|^2 + \int_Z |c^n(s,x,z) - c^0(s,x,z)|^2 \pi(dz)]ds = 0,$
for all $t > 0$;

and retains the other assumptions 2° and 4°, then the conclusion of Theorem 1.1 still holds true.

Proof. By Ito's formula we still get (1.2). Applying condition $3^{\circ\prime}$, we have

$$\lim_{n\to\infty} E\int_0^T |b^n(s,x_s^n) - b^0(s,x_s^n)|^2 ds = 0.$$

Therefore, a similar discussion as the one in the proof of Theorem 1.1 yields the conclusion. ■

We can also have another kind of conditions for the convergence theorem on solutions to RSDEs (1.1).

Theorem 1.5. Assume that $1^{\circ\prime}$ of theorem 1.4 and 4° of Theorem 1.1 hold, and

$2^{\circ\prime}$ $2(x-y)\cdot(b^n(t,x) - b^n(t,y)) \le k(t)\rho(|x-y|^2),$
$\|\sigma^n(t,x) - \sigma^n(t,y)\|^2 + \int_Z |c^n(t,x,z) - c^n(t,y,z)|^2 \pi(dz)$
$\le k(t)\rho(|x-y|^2)$, for all $n = 1, 2, ...$

where $k(t) \ge 0, \rho(u)$ is continuous, strictly increasing, concave, $\rho(0) = 0$, and they are such that for any $0 < T < \infty$

$$\int_0^T k(t)dt < \infty, \int_{0+} du/\ \rho(u) = \infty;$$

$3^{\circ\prime\prime}$ $\lim_{n\to\infty}[|b^n(s,x) - b^0(s,x)|^2$
$+ \|\sigma^n(s,x) - \sigma^0(s,x)\|^2 + \int_Z |c^n(s,x,z) - c^0(s,x,z)|^2 \pi(dz)]ds = 0,$
for all $s \ge 0, x \in R^d$;

4° is the same as 4° in Theorem 1.1.

Then, for all $t \ge 0$, as $n \to \infty$

$$E\sup_{s\le t} |x_s^n - x_s^0|^2 \to 0.$$

In addition, if

$$|b^n(t,x) - b^n(t,y)|^2 \le k(t)\rho(|x-y|^2),$$

where $k(t) \ge 0, \rho(u)$ have the same property as the above;
then for all $t \ge 0$, as $n \to \infty$

$$E\sup_{s\le t} |\phi_s^n - \phi_s^n|^2 \to 0.$$

Proof. By Ito's formula as (1.2) one gets that

$$|x_t^n - x_t^0|^2 \le |x_0^n - x_0^0|^2 + E\int_0^t [3k(s)\rho(|x_s^n - x_s^0|^2)$$
$$+ |x_s^n - x_s^0|^2 + |b^n(s,x_s^0) - b^0(s,x_s^0)|^2$$
$$+2\|\sigma^n(s,x_s^0) - \sigma^0(s,x_s^0)\|^2]ds + 2\int_0^t (x_s^n - x_s^0)\cdot(\sigma^n(s,x_s^n) - \sigma^0(s,x_s^0))dw_s$$
$$+2\int_0^t (x_s^n - x_s^0)\cdot\int_Z (c^n(s,x_{s-}^n,z) - c^0(s,x_{s-}^0,z))q(ds,dz)$$

$$+2\int_0^t \int_Z \left|c^n(s,x^0_{s-},z) - c^0(s,x^0_{s-},z)\right|^2 p(ds,dz)$$
$$+2\int_0^t \int_Z \left|c^n(s,x^n_{s-},z) - c^n(s,x^0_{s-},z)\right|^2 p(ds,dz).$$

Note that

$$E\int_0^t \left|b^n(s,x^0_s) - b^0(s,x^0_s)\right|^2 ds \leq 2k_0 E\int_0^t (1+\left|x^0_s\right|^2)ds < \infty.$$

Hence, applying Lebesgue's dominated convergence theorem, one finds that as $n \to \infty$

$$E\int_0^t \left|b^n(s,x^0_s) - b^0(s,x^0_s)\right|^2 ds \to 0.$$

By a similar technique and the martingale inequality one obtains (1.3), and then (1.4).

Now by (1.5)

$$|\phi^n_t - \phi^n_t|^2 \leq 5(\left|x^n_t - x^0_t\right|^2 + \left|x^n_0 - x^0_0\right|^2 + 2t\int_0^t k(s)\rho(\left|x^n_s - x^0_s\right|^2)ds$$
$$+2\left|\int_0^t (b^n(s,x^0_s) - b^0(s,x^0_s))ds\right|^2 + \left|\int_0^t (\sigma^n(s,x^n_s) - \sigma^0(s,x^0_s))dw_s\right|^2$$
$$+\left|\int_0^t \int_Z (c^n(s,x^n_{s-},z) - c^0(s,x^0_{s-},z))q(ds,dz)\right|^2).$$

Therefore, applying the martingale inequality and a similar technique to the one above we again find that as $n \to \infty$

$$E\sup_{s\leq t} |\phi^n_s - \phi^n_s|^2 \to 0. \blacksquare$$

4.2 Stability of Solutions

In this section we will discuss the stability of solutions to (1.1) with $n = 0$ and with coordinate planes as a boundary, i.e.

$$\Theta = R^d_+ = \left\{x = (x^1, x^2, ..., x^d) \in R^d : x^i > 0, 1 \leq i \leq d\right\},$$

because this case is important for the application to stochastic population control.

For simplicity, we omit the superscript 0 for $n = 0$ in (1.1) and denote this RSDE by $(1.1)_0$.

2.1 Remark. For domain $\Theta = R^d_+$ we have that as $x \in R^d_+$

$$\aleph_x = \phi.$$

If $x \in \partial R^d_+$, then $n = (n_1, n_2, ..., n_d) \in \aleph_x$, iff, as $x_i = 0, x_j = 0, j = i$,

$$n_i = 1,\ n_j = 0.j = i;$$

as $x_i = x_j = 0,\ x_k = 0, k = i, j,$

$$\begin{cases} n_i = \cos\theta, n_j = \sin\theta, 0 \leq \theta \leq \pi/2; \\ n_k = 0,\ \text{for all } k = i, j; \end{cases}$$

and as $x_{i_1} = x_{i_2} = \cdots = 0,\ x_j = 0, j = i_1, i_2, \cdots, i_k,$

$$\begin{cases} n_{i_1} = \sin\theta\ \cos\phi_1 \cdots \cos\phi_{k-2}, \\ n_{i_2} = \sin\theta\ \cos\phi_1 \cdots \cos\phi_{k-3}\sin\phi_{k-2}, \\ n_{i_3} = \sin\theta\ \cos\phi_1 \cdots \cos\phi_{k-4}\sin\phi_{k-3}, \\ \cdots\cdots\cdots \\ n_{i_{k-1}} = \sin\theta\sin\phi_1, \\ n_{i_k} = \cos\theta,\ 0 \leq \theta \leq \pi/2,\ 0 \leq \phi_1.\phi_2, \cdots, \phi_k \leq \pi/2, \\ n_j = 0,\ \text{as } j = i_1, i_2, \cdots, i_k; k \geq 3. \end{cases}$$

Since all $n_i \geq 0, 1 \leq i \leq d$; it follows that, if (x_t, ϕ_t) is a solution of $(1.1)_0$, then all the $\phi_i(t), 1 \leq i \leq d$, are increasing, where $\phi_t = \phi(t) = (\phi_1(t), \phi_2(t), \cdots, \phi_d(t))$. Moreover, $|\phi_t|$ is also increasing.

2.2 Theorem. Assume that coefficients b, σ and c in (1.1) satisfy all conditions in Theorem 3.8 (or Theorem 3.15) in Chapter 3, and

$$b(t,0) = 0, \sigma\ (t,0) = 0, c(t,0,z) = 0.$$

Then

1) $(0,0)$ is a pathwise unique strong solution of $(1.1)_0$ with initial condition $x_0 = 0$;

2) for any given $x_0 \in \bar{R}^d_+$,which is a $\Im_0-$measurable random variable, $(1.1)_0$ has a pathwise unique strong solution $(x_t, \phi_t), t \geq 0$;

3) furthermore, if there exists a positive constant $k_1 \geq 0$ such that

$$2x \cdot b(t,x) + \|\sigma(t,x)\|^2 + \int_Z |c(t,x,z)|^2 \pi(dz) \leq -k_1 |x|^2,$$

then

$$E|x_t|^2 \leq E|x_0|^2 e^{-k_1 t}, \text{ for all } t \geq 0.$$

In addition, if

$$|b(t,x)| \leq k_3 |x|,$$

then there exists a constant k_4 such that

$$E|\phi_t|^2 \leq k_4 E|x_0|^2 (1 + e^{-k_1 t}), \text{ for all } t \geq 0.$$

Hence, if one writes that $\phi_\infty = \lim_{n\to\infty} \phi_t$, then

$$E|\phi_\infty|^2 \leq k_4 E|x_0|^2.$$

Furthermore,

$$E(|\phi_\infty|^2 - |\phi_t|^2) \leq k''' E|x_0|^2 e^{-k_1 t}.$$

4) In the case that there exists a positive constant $k_5 > 0$ such that

$$2x \cdot b(t,x) + \|\sigma(t,x)\|^2 + \int_Z |c(t,x,z)|^2 \pi(dz) \geq k_5 |x|^2,$$

and

$$|b(t,x)|^2 + \|\sigma(t,x)\|^2 + \int_Z |c(t,x,z)|^2 \pi(dz) \leq k_0(1 + |x|^2),$$

then

$$E|x_t|^2 \geq E|x_0|^2 e^{k_5 t}, \text{ for all } t \geq 0.$$

Hence,

$$\lim_{t\to\infty} E|x_t|^2 = \infty.$$

Proof. By Theorem 3.8 or Theorem 3.15 there exists a pathwise unique strong solution (x_t, ϕ_t) and $(0,0)$ for the given initial condition $x_0 = x_0$ and $x_0 = 0$, respectively. Let us show 3): By Ito's formula

$$|x_t|^2 = |x_0|^2 + \int_0^t (2x_s \cdot b(s,x_s) + \|\sigma(s,x_s)\|^2) ds + 2\int_0^t x_s \cdot \sigma(s,x_s) dw_s$$
$$+2\int_0^t \int_Z x_s \cdot c(s, x_{s-}, z) q(ds,dz) + \int_0^t \int_Z |c(s, x_{s-}, z|^2 p(ds,dz).$$

Denote $\tau_N = \inf\{t \geq 0 : |x|_t > N\}$. Then

$$E|x_{t\wedge\tau_N}|^2 = E|x_0|^2 + E\int_0^t (2x_s \cdot b(s,x_s) + \|\sigma(s,x_s)\|^2$$
$$+ \int_Z |c(s,x_s,z|^2 \pi(dz)) I_{s\leq\tau_N} ds.$$

Hence,

$$\tfrac{d}{dt} E|x_{t\wedge\tau_N}|^2 = E(2x_t \cdot b(t,x_t) + \|\sigma(t,x_t)\|^2 + \int_Z |c(t,x_t,z|^2 \pi(dz)) I_{t\leq\tau_N}$$
$$\leq -k_1 E|x_{t\wedge\tau_N}|^2, \ a.e.\ t \geq 0.$$

Therefore,

$E|x_{t\wedge\tau_N}|^2 \leq E|x_0|^2 e^{-k_1 t}$, for all $t \geq 0$.

By Fatou's lemma the first conclusion of 3) is derived. On the other hand, by Ito's formula

$|\phi_T|^2 - |\phi_t|^2 = 2\int_t^T \phi_s \cdot d\phi_s = 2\int_t^T \phi_s \cdot (dx_s - b(s,x_s)ds - \sigma(s,x_s)dw_s$
$- \int_Z c(s,x_{s-},z)q(ds,dz)) = 2(\phi_T \cdot x_T - \phi_t \cdot x_t)$
$-2\int_t^T \phi_s \cdot (b(s,x_s)ds + \sigma(s,x_s)dw_s + \int_Z c(s,x_{s-},z)q(ds,dz))$,

where we have used the result that $\int_t^T x_s \cdot d\phi_s = 0$. Hence by assumption $|b(t,x)| \leq k_3|x|$, and we get

$E(|\phi_T|^2 - |\phi_t|^2) \leq k''(E|x_t|^2 + \int_t^T E|x_s|^2 ds + E|x_T|^2) \leq k'E|x_0|^2 (e^{-k_1 t} + e^{-k_1 T})$.

From this result the second and third conclusions of 3) follow. Since ϕ_t^i is increasing in t for each i, ϕ_∞ exists, and

$E(|\phi_\infty|^2 - |\phi_t|^2) \leq k'''E|x_0|^2 e^{-k_1 t}$.

Now let us show 4). In the above discussion, and by assumption we have

$\frac{d}{dt}E|x_t|^2 = E(2x_t \cdot b(t,x_t) + \|\sigma(t,x_t)\|^2 + \int_Z |c(t,x_t,z|^2 \pi(dz))ds \geq k_5 E|x_t|^2$.

Hence,

$E|x_t|^2 \geq E|x_0|^2 e^{k_5 t}$, for all $t \geq 0$. ■

4.3 Comparison of Solutions

In this section, we will discuss the comparison theorems for solutions to (1.1). More precisely, let us consider RSDEs

$$(3.1)\quad \begin{cases} dx_t^n = b^n(t,x_t^n)dt + \sigma(t,x_t^n)dw_t + \int_Z c(t,x_{t-}^n,z)q(dt,dz) + d\phi_t^n, \\ x_0^n = x_0^n \in \overline{R_+^d} = \overline{\Theta}, \\ \text{and } (x_t^n,\phi_t^n),\ t \geq 0 \text{ satisfies the statements in (1.2) of Chapter 3;} \\ \text{moreover, } \phi_t^n \text{ is continuous;} \end{cases}$$

$n = 1, 2$.

Before discussing the comparison theorem, we need some preparation. We first recall a Tanaka type formula stated in Chapter 3 as Theorem 1.4, and improve it as follows:

3.1 Theorem. (Tanaka type formula). Assume that

1° b^1, b^2, σ and $\int_Z |c|^2 \pi(dz)$ are locally bounded, i.e. for each $r = 1, 2, ...$

$|h(t,x)| \leq k_r$, as $|x| \leq r$, $h = b^1, b^2, \sigma$ and $\int_Z |c|^2 \pi(dz)$;

where $0 \leq k_r$ is a constant depending on r only;

2° $\|\sigma_{ik}(t,x) - \sigma_{ik}(t,y)\|^2 \leq k_N(t)\rho_N(|x_i - y_i|^2)$,as $|x|, |y| \leq N$, $\forall i,k$;

where $0 \leq k_N(t)$, for any $0 \leq T < \infty$,

$\int_0^T k_N(t)dt < \infty$,

and $\rho_N(u) > 0$,as $u > 0$; $\rho_N(0) = 0$; and $\rho_N(u)$ is strictly increasing in u and such that

$\int_{0+} du/\rho_N(u) = \infty$, for $N = 1, 2, ...$

If (x_t^n, ϕ_t^n) satisfies (3.1), $n = 1, 2$, then

$$(x_i^1(t)-x_i^2(t))^+ = (x_i^1(0)-x_i^2(0))^+ + \int_0^t I_{(x_i^1(s-)>x_i^2(s-)}d(x_i^1(s)-x_i^2(s))+J_i(t),$$

where for $i = 1, 2, ..., d$

$$J_i(t) = \int_0^t \int_Z [(x_i^1(s-) - x_i^2(s-) + c_i(s, x^1(s-), z) - c_i(s, x^2(s-), z))^+$$
$$-(x_i^1(s-) - x_i^2(s-))^+$$
$$-I_{(x_i^1(s-)>x_i^2(s-)}(c_i(s, x^1(s-), z) - c_i(s, x^2(s-), z))]p(ds, dz).$$

Furthermore, if

3° $x_i \geq y_i \Rightarrow x_i + c_i(t, x, z) \geq y_i + c_i(t, y, z)$, for all $1 \leq i \leq d$,

then for all $t \geq 0$, all $1 \leq i \leq d$

$$J_i(t) = 0.$$

More precisely, under condition 1°, 2° and 3°

$$(x_i^1(t) - x_i^2(t))^+ = (x_i^1(0) - x_i^2(0))^+$$
$$+ \int_0^t I_{(x_i^1(s)>x_i^2(s)}(b_i^1(s, x^1(s)) - b_i^2(s, x^2(s)))ds$$
$$+ \int_0^t I_{(x_i^1(s)>x_i^2(s)} \sum_{k=1}^d (\sigma_{ik}^1(s, x^1(s)) - \sigma_{ik}^2(s, x^2(s)))dw_k(s)$$
$$+ \int_0^t \int_Z I_{(x_i^1(s-)>x_i^2(s-)}(c_i(s, x^1(s-), z) - c_i(s, x^2(s-), z))q(ds, dz)$$
$$- \int_0^t I_{(x_i^1(s)>x_i^2(s))}d\phi_i^2(s),$$

where for $i = 1, 2, ..., d$.

Proof. By Theorem 1.4 of Chapter 3 we only need to show the last formula. By the following Lemma 3.2

$$\int_0^t I_{(x_i^1(s)>x_i^2(s))}d\phi_i^1(s) = 0.$$

Hence, the conclusion is derived. ■

3.2 Lemma. For $i = 1, 2, \cdots, d$

$$\phi_i(t) = \int_0^t I_{(x_i(s)=0)}d\phi_i(s),$$

where $(x(t), \phi(t))$ satisfies (3.1), and

$$x(t) = (x_1(t), ..., x_d(t)), \phi(t) = (\phi_1(t), \cdots, \phi_d(t)).$$

Proof. Since

$$\phi_i(t) = \int_0^t n_i(s)d\left|\phi\right|(s),$$

by Remark 2.1 $\phi_i(t)$ is increasing in t, and

$$n_i(t) = 0, \text{ iff } x_i(t) = 0.$$

Hence, the conclusion follows directly. ■

Now we are in a position to derive some comparison theorems for solutions to (3.1) as follows:

3.3 Theorem. Assume that all conditions 1° – 3° in Theorem 3.1 hold, and there exists a $k_0, 1 \leq k_0 \leq d$, such that

1° for any $x = (x_1, ..., x_d), y = (y_1, ..., y_d) \in \overline{R}_+^d$ satisfying

$x_j \leq y_j$, as $1 \leq j \leq k_0$,

$x_j \geq y_j$, as $k_0 < j \leq d$;

if $x_i = y_i$, then

$b_i^1(t, x) < b_i^2(t, x)$, as $1 \leq i \leq k_0$,

or

$b_i^1(t, x) > b_i^2(t, x)$,as $k_0 < i \leq d$.

If $(x^n(t),\phi^n(t))$ satisfies (3.1), $n=1,2$, then

$$\begin{cases} x_j^1(0) \leq x_j^2(0), \text{as } 1 \leq j \leq k_0, \\ x_j^1(0) \geq x_j^2(0), \text{ as } k_0 < j \leq d; \end{cases}$$

implies that

$$\begin{cases} x_j^1(t) \leq x_j^2(t), \text{as } 1 \leq j \leq k_0, \\ x_j^1(t) \geq x_j^2(t), \text{ as } k_0 < j \leq d; \end{cases}$$

for all $t \geq 0$.

Proof. Assume that for some i

(3.2) $x_i^1(0) = x_i^2(0)$.

In the case that $1 \leq i \leq k_0$ set

$$\tau = \inf\left\{t \geq 0 : b_i^1(t,x^1(t)) > b_i^2(t,x^2(t))\right\},$$
$$\tau_N = \inf\left\{t \geq 0 : \left|x^1(t)\right| + \left|x^2(t)\right| > N\right\}.$$

Then by Theorem 3.1

$$E(x_i^1(t\wedge\tau_N\wedge\tau) - x_i^2(t\wedge\tau_N\wedge\tau))^+$$
$$\leq E\int_0^{t\wedge\tau_N\wedge\tau} I_{(x_i^1(s)>x_i^2(s))}(b_i^1(s,x^1(s)) - b_i^2(s,x^2(s)))ds \leq 0,$$

where we have used the result

$$\int_0^t I_{(x_i^1(s)>x_i^2(s))} d(\phi_s^1 - \phi_s^2) = -\int_0^t I_{(x_i^1(s)>x_i^2(s)=0)} d\phi_s^2 \leq 0.$$

From this and by Fatou's lemma etc., it is easily shown that $P-a.s.$

$$x_i^1(t) \leq x_i^2(t), \text{ as } t \leq \tau.$$

For case that $k_0 < i \leq d$ under (3.2) it is similarly shown that $P-a.s.$

$$x_i^1(t) \geq x_i^2(t), \text{ as } t \leq \overline{\tau},$$

where

$$\overline{\tau} = \inf\left\{t \geq 0 : b_i^1(t,x^1(t)) < b_i^2(t,x^2(t))\right\}.$$

Now set

$$\tau_i = \begin{cases} \sup(t \geq 0 : x_i^1(s) \leq x_i^2(s), \text{ foe all } s \in [0,t]), \text{ as } 1 \leq i \leq k_0; \\ \sup(t \geq 0 : x_i^1(s) \geq x_i^2(s), \text{ foe all } s \in [0,t]), \text{ as } k_0 < i \leq d; \end{cases}$$
$$\widetilde{\tau} = \tau_1 \wedge \tau_2 \wedge \ldots \wedge \tau_d.$$

Let us show that

(3.3) $\widetilde{\tau} = \infty$.

If not, there exists a $\Lambda \in \Im, P(\Lambda) > 0$, such that as $\omega \in \Lambda$

$$\widetilde{\tau}(\omega) < \infty.$$

Therefore there exists a $i_0, 1 \leq i_0 \leq d$, and there exists a $\Lambda_1 \in \Im, \Lambda_1 \subset \Lambda$ such that $P(\Lambda_1) > 0$ and as $\omega \in \Lambda_1$

$$\tau_{i_0}(\omega) < \infty, \tau_{i_0}(\omega) = \widetilde{\tau}(\omega).$$

Now set

$$(\widetilde{\Omega}, \widetilde{\Im}, (\widetilde{\Im}_t), \widetilde{P}) = (\Lambda_1, \Lambda_1 \cap \Im, (\Lambda_1 \cap \Im_t), \widetilde{P}),$$
$$\widetilde{P}(A) = P(A)/P(\Lambda_1), \text{ as } A \in \Lambda_1 \cap \Im,$$

and set

$$y^n(t) = x^n(t + \tau_{i_0}), n = 1,2.$$

Then $\widetilde{P}-a.s.$ on $\widetilde{\Omega}$, $n=1,2$,

$$\begin{cases} y^n(t) = y^n(0) + \int_0^t b^n(s+\tau_{i_0}, y^n(s))ds + \int_0^t \overline{\sigma}(s+\tau_{i_0}, y^n(s))d\overline{w}(s) \\ \qquad + \int_0^t \int_Z c(s+\tau_{i_0}, y^n(s-), z)q(ds,dz) + \overline{\phi}^n(t), \\ \text{and the other statements in (3.1) hold for } (y^n(t), \overline{\phi}^n(t)), t \geq 0; \end{cases}$$

where $\overline{\sigma} = \sigma/P(\Lambda_1)^{1/2}, \overline{\phi}^n(t) = \phi^n(t+\tau_{i_0}) - \phi^n(\tau_{i_0})$; and

$$\overline{w}(t) = P(\Lambda_1)^{1/2}(w(t+\tau_{i_0}) - w(\tau_{i_0}))$$

is a $\widetilde{P}-$BM., and $q(dt,dz)$ is a $\widetilde{P}-$Poisson martingale measure with compensator $\pi(dz)dt/P(\Lambda_1)$ such that $\widetilde{P}-a.s.$ on $\widetilde{\Omega}$

$$q(dt,dz) = p(dt,dz) - \pi(dz)dt/P(\Lambda_1).$$

Obviously,

$$y^1_{i_0}(0) = y^2_{i_0}(0), \text{ as } \omega \in \widetilde{\Omega}.$$

Now assume that $1 \leq i_0 \leq k_0$. Then, by assumption and by the facts just proved above it yields that $\widetilde{P}-a.s.$ on $\widetilde{\Omega}$

$$y^1_{i_0}(t) \leq y^2_{i_0}(t), \text{ as } 0 \leq t \leq \widetilde{\tau}_{i_0},$$

where

$$\widetilde{\tau}_{i_0} = \inf\left\{t \geq 0 : b^1_i(t+\tau_{i_0}, y^1(t)) > b^2_i(t+\tau_{i_0}, y^2(t))\right\}.$$

However, by assumption $\widetilde{P}-a.s.$ on $\widetilde{\Omega}$

$$\widetilde{\tau}_{i_0} > 0.$$

This is in contradiction with the definition of τ_{i_0}. Hence, (3.3) holds. In the case that $d \geq i_0 > k_0$, the proof is the same. ■

To weaken 1° in Theorem 3.3 we have the following

3.4 Theorem. Assume that all assumptions of Theorem 3.3 except 1° hold, and assume that

$$|b^1|^2 + |b^2|^2 + \|\sigma\|^2 + \int_Z |c|^2 \pi(dz) \leq \overline{k}_0(1+|x|^2),$$

$$\int_Z |c(t,x,z) - c(s,y,z)|^2 \pi(dz) \to 0, \text{ as } |x-y| \to 0, |t-s| \to 0;$$

$$\pi(dz) = dz/|z|^{d+1},$$

$$\int_Z |c(t,x,z) - c(t,y,z)|^2 \pi(dz) \leq k_N(t)\rho_N(|x-y|^2), \text{as } |x|, |y| \leq N,$$

where $k_N(t)$ and $\rho_N(u)$ satisfy the same condition as that in Theorem 3.1,

$$x + c(t,x,z) \in \overline{R}^d_+,$$

and there exists a $k_0, 1 \leq k_0 \leq d$, such that for any $x, y \in \overline{R}^d_+, x = (x_1, ..., x_d), y = (y_1, ..., y_d)$ satisfying

$$x_i \leq y_i, \text{ as } 1 \leq i \leq k_0,$$

$$x_i \geq y_i, \text{ as } k_0 < i \leq d,$$

one has that if $x_i = y_i$, then

$$b^1_i(t,x) \leq b^2_i(t,x), \text{ as } 1 \leq i \leq k_0,$$

or

$$b^1_i(t,x) \geq b^2_i(t,x), \text{as } k_0 < i \leq d;$$

moreover, assume that $b^1(t,x)$ and $b^2(t,x)$ are jointly continuous, and the pathwise uniqueness holds for (3.1) as $n = 1$ (or $n = 2$). Then the conclusion of Theorem 3.3 still holds.

Proof. Let us assume that the pathwise uniqueness holds for $n = 2$. Since there exist $\overline{b}^n_i(t,x), n = 1, 2, ...$ which are Lipschitzian continuous such that

$$b^1_i(t,x) \leq b^2_i(t,x) < b^2_i(t,x) + \tfrac{1}{n+1} < \overline{b}^n_i(t,x) < b^2_i(t,x) + \tfrac{1}{n}, \text{ as } 1 \leq i \leq k_0,$$

$$b^1_i(t,x) \geq b^2_i(t,x) > b^2_i(t,x) - \tfrac{1}{n+1} > \overline{b}^n_i(t,x) > b^2_i(t,x) - \tfrac{1}{n}, \text{ as } k_0 < i \leq d.$$

Hence, by Theorem 3.8 or Theorem 3.15 in Chapter 3 there exists a pathwise unique strong solution $(\overline{x}^n_t, \overline{\phi}^n_t)$, for each $n = 1, 2, ...$, satisfying

$$
(3.4)\begin{cases} d\overline{x}_t^n = \overline{b}^n(t,\overline{x}_t^n)dt + \sigma(t,\overline{x}_t^n)dw_t + \int_Z c(t,\overline{x}_{t-}^n,z)q(dt,dz) + d\overline{\phi}_t^n, \\ \overline{x}_0^n = x_0^2 \in \overline{R}_+^d, \\ \text{and } (\overline{x}_t^n,\overline{\phi}_t^n),\ t \geq 0 \text{ satisfies the statements in (1.2) of Chapter 3;} \\ \text{moreover, } \overline{\phi}_t^n \text{ is continuous;} \end{cases}
$$

$n = 1, 2, ...$

By Theorem 3.3

$x_i^1(t) \leq \overline{x}_i^n(t) \leq \overline{x}_i^{n+1}(t)$,as $1 \leq i \leq k_0$,

$x_i^1(t) \geq \overline{x}_i^n(t) \geq \overline{x}_i^{n+1}(t)$,as $k_0 < i \leq d$.

Hence

$\widetilde{x}_t^2 = \lim_{n\to\infty} \overline{x}_t^n$

exists and it is finite, $P-a.s.$ Set

$$
d\overline{y}_t^n = \overline{b}^n(t,\widetilde{x}_t^2)dt + \sigma(t,\widetilde{x}_t^2)dw_t + \int_Z c(t,\widetilde{x}_{t-}^2,z)q(dt,dz),
$$
$$
\overline{y}_0^n = x_0^2 \in \overline{R}_+^d,
$$
$$
d\overline{y}_t = b^2(t,\widetilde{x}_t^2)dt + \sigma(t,\widetilde{x}_t^2)dw_t + \int_Z c(t,\widetilde{x}_{t-}^2,z)q(dt,dz),
$$
$$
\overline{y}_0 = x_0^2 \in \overline{R}_+^d.
$$

Then it is not difficult to show that there exist an $n_k \uparrow \infty$, as $k \uparrow \infty$; such that

$\lim_{k\to\infty} \sup_{s\leq t} |\overline{y}_s^{n_k} - \overline{y}_s| = 0.$

Note that by Theorem 2.9 of Chapter 2 there exists a unique solution $(\widetilde{x}_t, \widetilde{\phi}_t)$ satisfying

$$
\begin{cases} \widetilde{x}_t = \overline{y}_t + \widetilde{\phi}_t \in \overline{R}_+^d, \\ \text{and the other statements in (2.1) of Chapter 2 hold for } (\widetilde{x}_t, \widetilde{\phi}_t). \end{cases}
$$

Applying Corollary 2.10 of Chapter 2, one has that

$\lim_{k\to\infty} \sup_{s\leq t} |\overline{x}_s^{n_k} - \widetilde{x}_s| = 0,$

$\lim_{k\to\infty} \sup_{s\leq t} \left|\overline{\phi}_s^{n_k} - \widetilde{\phi}_s\right| = 0$, for all $t \geq 0$.

By the uniqueness of a limit

$\widetilde{x}_t = \widetilde{x}_t^2.$

Hence, $(\widetilde{x}_t^2, \widetilde{\phi}_t)$ satisfies

$$
\begin{cases} d\widetilde{x}_t^2 = b^2(t,\widetilde{x}_t^2)dt + \sigma(t,\widetilde{x}_t^2)dw_t + \int_Z c(t,\widetilde{x}_{t-}^2,z)q(dt,dz) + d\widetilde{\phi}_t, \\ \widetilde{x}_0^2 = x_0^2 \in \overline{R}_+^d. \end{cases}
$$

Since the pathwise uniqueness holds for (3.1) as $n = 2$, therefore

$(\widetilde{x}_t^2, \widetilde{\phi}_t) = (x_t^2, \phi_t^2).$

It follows that $P-a.s.$

$x_i^1(t) \leq \widetilde{x}_i^2(t) = x_i^2(t)$,as $1 \leq i \leq k_0$,

$x_i^1(t) \geq \widetilde{x}_i^2(t) = x_i^2(t)$,as $k_0 < i \leq d$,

for all $t \geq 0$. ■

For the comparison of $\phi_t^n, n = 1, 2$, we need more conditions.

3.5 Theorem. Under conditions of Theorem 3.3 or 3.4 if $\sigma(t) = \sigma(t, x)$, $c(t, z) = c(t, x, z)$ do not depend on x, and $b^i(t, x), i = 1, 2$, are jointly continuous, then for all $t \geq 0$

$d\phi_i^1(t) \geq d\phi_i^2(t)$,as $1 \leq i \leq k_0$,

$d\phi_i^1(t) \leq d\phi_i^2(t)$,as $k_0 < i \leq d$.

To prove Theorem 3.5 we need the following

3.6 Lemma. If (x_t, ϕ_t) satisfies (3.1) with coefficients (b, σ, c), where c is such that

$x + c(t, x, z) \in \bar{R}_+^d$,

and the pathwise uniqueness of solutions to (3.1) with (b, σ, c) holds, then (x_t, ϕ_t) is such that

$\phi_i(t) = \sup_{0 \leq s \leq t}((-y_i(s)) \vee 0)$,

$x_i(t) = y_i(t) + \phi_i(t)$, $y_i(0) \geq 0$,

for all $t \geq 0$, where

$y_i(t) = x_i(0) + \int_0^t b_i(s, x(s))ds + \sum_{k=1}^d \int_0^t \sigma_{ik}(s, x(s))dw_k(s)$

$+ \int_0^t \int_Z c_i(s, x(s-), z)q(ds, dz)$, $i = 1, 2, ...d$.

Proof. If (x_t, ϕ_t) solves (3.1), set

$\overline{\phi}_i(t) = \sup_{0 \leq s \leq t}((-y_i(s)) \vee 0)$,

$\overline{x}_i(t) = y_i(t) + \overline{\phi}_i(t)$,

then it is obvious that $0 \leq \overline{\phi}_i(t)$ is increasing, continuous, $\overline{\phi}_i(0) = 0$, and $\overline{x}_i(t) \geq 0$, for all $i = 1, 2, ..., d$. Moreover,

$\overline{\phi}_i(t) = \int_0^t I_{(\overline{x}_i(s)=0)} d\overline{\phi}_i(s)$.

Hence, $(\overline{x}_i(t), \overline{\phi}_i(t))$ satisfies

$$(3.5)\begin{cases} \overline{x}_i(t) = y_i(t) + \overline{\phi}_i(t), \\ \overline{x}_i(t) \geq 0, \text{ for all } t \geq 0, \\ 0 \leq \overline{\phi}_i(t) \text{ is increasing, continuous}, \overline{\phi}_i(0) = 0, \text{and} \\ \overline{\phi}_i(t) = \int_0^t I_{(\overline{x}_i(s)=0)} d\overline{\phi}_i(s), \end{cases}$$

i.e. it is a solution of the Skorohod's problem (3.5) in 1-dimensional space. Since $(x_i(t), \phi_i(t))$ also satisfies (3.5); by the uniqueness of solution to the Skorohod problem we have

$(x_i(t), \phi_i(t)) = (\overline{x}_i(t), \overline{\phi}_i(t))$. ■

Now we are in a position to prove Theorem 3.5.

Proof. First, let us assume that the assumptions of Theorem 3.3 hold. For any $t \geq 0$ and $1 \leq i \leq k_0$ if

$x_i^1(t) = x_i^2(t)$,

then by the right continuity of $x_i^1(s)$ and $x_i^2(s)$, and the continuity of b^1 and b^2 there exists a $\delta > 0$ such that as $0 \leq s \leq \delta$

$b_i^1(t + s, x^1(t + s)) < b_i^2(t + s, x^2(t + s))$.

Note that

$x^2(t + s) = x^2(t) + \int_t^{t+s} b^2(r, x^2(r))dr + \int_t^{t+s} \sigma(r)dw(r)$

$+ \int_t^{t+s} \int_Z c(r, z)q(dr, dz) + \phi^2(t + s) - \phi^2(t)$.

So

$$(3.6)\begin{cases} x^2(t + s) = x^2(t) + \int_0^s b^2(t + r, x^2(t + r))dr + \int_0^s \sigma(t + r)d\overline{w}(r) \\ \quad + \int_0^s \int_Z c(t + r, z)\overline{q}(dr, dz) + \phi^2(t + s) - \phi^2(t), s \geq 0, \\ \text{and the other statements in (3.1) hold for} \\ (x^2(t + s), \phi^2(t + s) - \phi^2(t)), \end{cases}$$

where

$\overline{w}(r) = w(t+r) - w(t)$,

$\overline{q}((0,r],dz) = q((t,t+r],dz)$.

This means that $(x^2(t+s), \phi^2(t+s) - \phi^2(t))$ solves RSDE (3.6) for the given fixed t with the BM $\overline{w}(s)$ and the Poisson martingale measure $\overline{q}(ds,dz)$. Similar facts hold for $(x^1(t+s), \phi^1(t+s) - \phi^1(t))$. Applying Lemma 3.6 and Theorem 3.3, one gets that as $0 \leq s \leq \delta$,

$\phi_i^2(t+s) - \phi_i^2(t) = \sup_{0 \leq r \leq s}[(-x_i^2(t) - \int_0^r b_i^2(t+u, x^2(t+u))du - z_r) \vee 0]$

$\leq \sup_{0 \leq r \leq s}[(-x_i^1(t) - \int_0^r b_i^1(t+u, x^1(t+u))du - z_r) \vee 0] = \phi_i^1(t+s) - \phi_i^1(t)$,

where

$z_r = \sum_{k=1}^d \int_0^r \sigma_{ik}(t+u)d\overline{w}_k(u) + \int_0^r \int_Z c_i(t+u,z)\overline{q}(du,dz)$.

Hence,

$d\phi_i^2(t) \leq d\phi_i^1(t)$.

For $t \geq 0$ and $1 \leq i \leq k_0$ if

$(0 \leq) x_i^1(t) < x_i^2(t)$,

then it is evident that

$d\phi_i^2(t) = 0 \leq d\phi_i^1(t)$.

For any $t \geq 0$ and $k_0 < i \leq d$ a similar argument gives

$d\phi_i^2(t) \geq d\phi_i^1(t)$.

Now assume that the assumptions of Theorem 3.4 hold. From the proof of Theorem 3.4 and the facts just proved this shows that for any $t \geq 0$

$d\phi_i^n(t) \leq d\phi_i^1(t)$, as $1 \leq i \leq k_0$,

$d\phi_i^n(t) \geq d\phi_i^1(t)$, as $k_0 < i \leq d$.

where $(\overline{x}_t^n, \overline{\phi}_t^n)$, for each $n = 1, 2, \ldots$, satisfies (3.4). Since

$\lim_{k \to \infty} \phi^{n_k}(t) = \phi^2(t)$, for all $t \geq 0$.

Hence, the conclusion follows. ■

The comparison theorem for solutions is a useful tool in discussing the existence of strong solutions under weaker condition.

3.7 Theorem. Assume that the coefficients (b, σ, c) satisfy the conditions in Theorem 3.4 (e.g.

$|b|^2 + \|\sigma\|^2 + \int_Z |c|^2 \pi(dz) \leq k_0(1 + |x|^2)$, and $b(t,x)$ is jointly continuous, etc.).

Then (3.1) with coefficients (b, σ, c) has a strong solution (x_t, ϕ_t).

Furthermore, if

$2(x-y) \cdot (b(t,x) - b(t,y)) \leq k_N(t)\rho_N(|x-y|^2)$,as $|x|, |y| \leq N$,

where $k_N(t)$ and $\rho_N(u)$ satisfy the same conditions as that in 2° of Theorem 3.1 and, moreover, $\rho_N(u)$ is concave, then the solution (x_t, ϕ_t) is also pathwise unique.

Proof. The proof is similar to that of Theorem 3.4. Take jointly Lipschitzian continuous functions $b_i^n(t,x)$ such that for $n = 0, 1, \cdots; 1 \leq i \leq d$

$b_i(t,x) - 1 < b_i^0(t,x) \leq b_i(t,x) < b_i(t,x) + 1/(n+1) < b_i^n(t,x) < b_i(t,x) + 1/n$.

Then by Theorem 3.8 of Chapter 3 there exists a pathwise unique strong solution (x_t^n, ϕ_t^n) satisfying (3.4) with coefficients (b^n, σ, c) for each $n = 0, 1, 2, \cdots$. By Theorem 3.3

$x_i^0(t) \leq x_i^n(t) \leq x_i^{n+1}(t)$, as $1 \leq i \leq d$.

Hence,

$x_t = \lim_{n\to\infty} x_t^n$

exists and it is finite, $P - a.s.$ Set

$$dy_t = b(t, x_t)dt + \sigma(t, x_t)dw_t + \int_Z c(t, x_{t-}, z)q(dt, dz),$$
$$y_0 = x_0 \in \overline{R}_+^d.$$

By Theorem 2.9 of Chapter 2 there exists a unique solution $(\widetilde{x}_t, \widetilde{\phi}_t)$ satisfying

$$\begin{cases} \widetilde{x}_t = y_t + \widetilde{\phi}_t \in \overline{R}_+^d, \\ \text{and the other statements in (2.1) of Chapter 2 hold for } (\widetilde{x}_t, \widetilde{\phi}_t). \end{cases}$$

Following the same approach as in the proof of Theorem 3.4, it follows that $(\widetilde{x}_t, \widetilde{\phi}_t)$ satisfies

$$\begin{cases} d\widetilde{x}_t = b(t, \widetilde{x}_t)dt + \sigma(t, \widetilde{x}_t)dw_t + \int_Z c(t, \widetilde{x}_{t-}, z)q(dt, dz) + d\widetilde{\phi}_t, \\ \widetilde{x}_0^2 = x_0^2 \in \overline{R}_+^d, \\ \text{and the other statements in (1.2) of Chapter 3 hold for } (\widetilde{x}_t, \widetilde{\phi}_t). \end{cases}$$

The pathwise uniqueness of solutions to (3.1) with coefficients (b, σ, c) now follows by Theorem 2.3 of Chapter 3. ■

4.4 Applications of Comparison Theorem to 1-Dimensional RSDEs

Now let us discuss the comparison theorems for solutions of 1-dimensional RSDEs and their applications. More precisely, consider

$$(4.1)\quad \begin{cases} dx_t^i = b^i(t, x_t^i, \omega)dt + \sigma(t, x_t^i, \omega)dw_t + \int_Z c(t, x_{t-}^i, z, \omega)q(dt, dz) \\ \qquad + d\phi_t^i - d\eta_t^i, \\ x_t^i \in \overline{\Theta} = [r_1, r_2], \forall t \geq 0; \\ \text{where } r_1 < r_2 \text{ are constants, } 0 \leq \phi_t^i,\ \eta_t^i, \text{ with } \phi_0^i = \eta_0^i = 0, \\ \phi_t^i \text{ and } \eta_t^i \text{ are continuous and increasing such that} \\ \int_0^t (x_s^i - r_1)d\phi_s^i = \int_0^t (r_2 - x_s^i)d\eta_s^i = 0, \forall t \geq 0, \end{cases}$$

$i = 1, 2,$

where all random processes in (4.1) are R^1- valued and, actually (4.1) can be obtained from (1.1) and (1.2) in Chapter 3 by taking the dimension $d = 1$, i.e. (4.1) can be rewritten as

$$(4.1)'\quad \begin{cases} dx_t^i = b^i(t, x_t^i, \omega)dt + \sigma(t, x_t^i, \omega)dw_t + \int_Z c(t, x_{t-}^i, z, \omega)q(dt, dz) \\ \qquad + d\widetilde{\phi}_t^i, \\ x_t^i \in \overline{\Theta} = [r_1, r_2], \forall t \geq 0; \\ \text{where } r_1 < r_2 \text{ are constants, and } \left(x_t^i, \widetilde{\phi}_t^i\right) \text{ satisfies} \\ \text{statements (1.2) in Chapter 3.} \end{cases}$$

Indeed, if (4.1)' holds, let

$\phi_t^i = \int_0^t I_{(x_s=r_1)} d\left|\widetilde{\phi}\right|_s$, $\eta_t^i = \int_0^t I_{(x_s=r_2)} d\left|\widetilde{\phi}\right|_s$,

then (4.1) follows. The inverse is obvious. Beware that (4.1) is not a special case of (3.1) above. For the comparison of solutions to (4.1) we have

4.1 Theorem. Assume that

1° b^1, b^2, σ and $\int_Z |c|^2 \pi(dz)$ are locally bounded, i.e. for each $r = 1, 2, ...$

$|h(t,x,\omega)| \leq k_r$, as $|x| \leq r$, $h = b^1, b^2, \sigma$ and $\int_Z |c|^2 \pi(dz)$;

where $0 \leq k_r$ is a constant depending on r only;

2° $|\sigma(t,x+y,\omega) - \sigma(t,x,\omega)|^2 \leq \rho_N(|y|)^2$

$\cdot(k_N(t,\omega) + g_N(x) |\sigma(t,x,\omega)|^2)$, as $|x|, |y| \leq N$,

where $0 \leq k_N(t,\omega)$ and $0 \leq g_N(u)$ satisfy conditions: $\forall 0 \leq T < \infty$,

$\int_0^T k_N(t,\omega)dt < \infty$, $\int_{[-N,N]} g_N(u)du < \infty, \forall N$,

and $\rho_N(u) > 0$,as $u > 0$; $\rho_N(0) = 0$; $\rho_N(u)$ is strictly increasing in u and such that

$\int_{0+} du/\rho_N(u) = \infty$, for $N = 1, 2, ...$

3° $x \geq y \Rightarrow x + c(t,x,z,\omega) \geq y + c(t,y,z,\omega)$,

4° $b^1(t,x,\omega) \geq b^2(t,x,\omega)$,

5° as $i = 1$ (or $i = 2$)

$sgn(x-y) \cdot (b^1(t,x,\omega) - b^1(t,y,\omega)) \leq \overline{k}_N(t,\omega)\overline{\rho}_N(|x-y|)$, as $|x|, |y| \leq N$,

where $\overline{k}_N(t,\omega)$ and $\overline{\rho}_N(u)$ have the same property as that in 2° above, moreover, $\overline{\rho}_N(u)$ is concave, $\forall N = 1, 2, \cdots$.

Then

$x_0^1 \geq x_0^2$

implies that $P-a.s.$

$x_t^1 \geq x_t^2, \forall t \geq 0.$

Proof. By Theorem 10.8 in Chapter 1 one has that

$$E(x_{t\wedge\tau_n}^1 - x_{t\wedge\tau_n}^2)^+ = E(x_0^1 - x_0^2)^+ - E\int_0^{t\wedge\tau_n} I_{(x_s^1 > x_s^2)} d\phi_s^2 - E\int_0^{t\wedge\tau_n} I_{(x_s^1 > x_s^2)} d\eta_s^1$$

$$+E\int_0^{t\wedge\tau_n} I_{(x_s^1 > .x_s^2)} (b^1(s,x_s^1,\omega) - b^2(s,x_s^2,\omega))ds$$

$$\leq E\int_0^{t\wedge\tau_n} \overline{k}_N(s,\omega)\overline{\rho}_N((x_s^1 - x_s^2)^+)ds,$$

where we have applied that

$$\int_0^{t\wedge\tau_n} I_{(x_s^1 > x_s^2)} d\phi_s^1 = \int_0^{t\wedge\tau_n} I_{(x_s^1 > x_s^2 \geq r_1)} d\phi_s^1 = 0,$$

$$\int_0^{t\wedge\tau_n} I_{(x_s^1 > x_s^2)} d\eta_s^2 = \int_0^{t\wedge\tau_n} I_{(r_2 \geq x_s^1 > x_s^2)} d\eta_s^2 = 0.$$

Let

$$A_t = \int_0^t \overline{k}_N(s,\omega) + t,$$

$$T_t = A_t^{-1}.$$

Then, since A_t and T_t are strictly increasing, one has

$$E(x_{T_t\wedge\tau_n}^1 - x_{T_t\wedge\tau_n}^2)^+ \leq E\int_0^{T_t\wedge\tau_n} \overline{\rho}_N((x_s^1 - x_s^2)^+)dA_s$$

$$\leq E\int_0^{t\wedge A(\tau_n)} \overline{\rho}_N((x_{T_s}^1 - x_{T_s}^2)^+)ds \leq E\int_0^t \overline{\rho}_N((x_{T_s\wedge\tau_n}^1 - x_{T_s\wedge\tau_n}^2)^+)ds$$

$$\leq \int_0^t \overline{\rho}_N(E(x_{T_s\wedge\tau_n}^1 - x_{T_s\wedge\tau_n}^2)^+)ds.$$

From this one easily derives the conclusion. ■

To remove condition 5° we have to strengthen condition 4°. Actually, we have the following

4.2 Theorem. Assume that all coefficients in RSDE (4.1) are non-random, i.e. $b^i, i = 1, 2; \sigma, c$ all do not depend on ω. and assume that for $t \geq 0, x \in \overline{\Theta}, z \in Z$
$1^{o\prime}$ $b^i(t,x), \sigma(t,x), c(t,x,z)$ are jointly measurable, and there exists a constant $k_0 > 0$ such that

$$|b^i|^2 + \|\sigma\|^2 + \int_Z |c|^2 \pi(dz) \leq k_0(1 + |x|^2),\ i = 1, 2,$$

$2^{o\prime}$ $\sigma(t,x)$ is jointly continuous, and as $|x - y| \to 0, |t - s| \to 0$

$$\int_Z |c(t,x,z) - c(s,y,z)|^2 \pi(dz) \to 0,$$

$3^{o\prime}$ $\pi(dz) = dz/|z|^{d+1}$, $(Z = R^d - \{0\})$; $x \geq y \Rightarrow x + c(t,x,z) \geq y + c(t,y,z)$,
$4^{o\prime}$ $\|\sigma(t,x) - \sigma(t,y)\|^2$

$$+ \int_Z |c(t,x,z) - c(t,y,z)|^2 \pi(dz) \leq k_N(t)\rho_N(|x - y|^2),\ \text{as } |x|, |y| \leq N,$$

where $0 \leq k_N(t), \int_0^t k_N(t)ds < \infty$, and $\rho_N(u)$ is concave and strictly increasing in $u \geq 0$ such that $\rho_N(0) = 0$, and

$$\int_{0+} du/\ \rho_N(u) = \infty, \text{ for each } N = 1, 2, ...,$$

$5^{o\prime}$ $x + c(t,x,z) \in \overline{\Theta}$,
$6^{o\prime}$ $b^1(t,x) > b^2(t,x)$,
where $b^i(t,x), i = 1, 2$, are jointly continuous.
Then the conclusion of Theorem 4.1 still holds.
Proof. Take a jointly Lipschitzian continuous function $\overline{b}(t,x)$ such that

$$b^2(t,x) < \overline{b}(t,x) < b^1(t,x).$$

Then by Theorem 3.8 of Chapter 3 there exists a pathwise unique strong solution $(\overline{x}_t, \phi_t, \eta_t)$ satisfying (4.1) with coefficients $(\overline{b}, \sigma, c)$. Moreover, ϕ_t and η_t are continuous. By Theorem 4.1 $P - a.s.$

$$x_t^1 \geq \overline{x}_t \geq x_t^2, \forall t \geq 0. \blacksquare$$

To weaken the condition $6^{o\prime}$ in Theorem 4.2 we have the following

4.3 Theorem. Under conditions $1^{o\prime} - 5^{o\prime}$ of Theorem 4.2 if
$6^{o\prime\prime}$ $b^1(t,x) \geq b^2(t,x)$,
where $b^i(t,x), i = 1, 2$, are jointly continuous,
$7^{o\prime\prime}$ the pathwise uniqueness holds for (4.1) as $i = 1$ (or $i = 2$);
then the conclusion of Theorem 4.2 still holds.

The proof of Theorem 4.3 is almost completely the same as that of Theorem 3.4 by using Theorem 4.2. We omit it here.

For the comparison of $\phi_t^i, i = 1, 2$, we need more conditions.

4.4 Theorem. Under conditions of Theorem 4.3 if $\sigma(t) = \sigma(t,x), c(t,z) = c(t,x,z)$ do not depend on x, and $7^{o\prime\prime}$ in Theorem 4.3 holds for $i = 1$, 2, then $\forall t \geq 0$

$$d\phi_t^1 \leq d\phi_t^2,\ d\eta_t^1 \geq d\eta_t^2.$$

To show Theorem 4.4 we need the following

4.5 Lemma. If (x_t, ϕ_t, η_t) satisfies (4.1), and coefficient c satisfies that

$$x + c(t,x,z) \in [r_1, r_2],$$

then it satisfies

$(4.2)_1$ $\phi_t = \sup_{0\le s\le t}((r_1 - y_s + \eta_s) \vee 0), \eta_t = \sup_{0\le s\le t}((y_s + \phi_s - r_2) \vee 0),$

$\phi_0 = \eta_0 = 0,$

$(4.2)_2$ $x_t = y_t + \phi_t - \eta_t, \forall t \ge 0,$

where

$$y_t = x_0 + \int_0^t b(s, x_s)ds + \int_0^t \sigma(s, x_s)dw_s + \int_0^t \int_Z c(s, x_{s-}, z)q(ds, dz).$$

Proof. Obviously, ϕ_t and η_t by $(4.2)_1$ are increasing, continuous, and x_t by $(4.2)_2$ takes values in $[r_1, r_2]$. Moreover, this (x_t, ϕ_t, η_t) satisfies (4.1). By the uniqueness theorem this is the unique solution of (4.1). ■

Now we are in a position to show Theorem 4.4.

Proof. Let $t \ge 0$ be arbitrary given.

(i) If $r_2 > x_t^1 \ge x_t^2 > r_1$, then $d\phi_t^1 = d\phi_t^2 = d\eta_t^1 = d\eta_t^2 = 0$.

(ii) If $r_2 = x_t^1 > x_t^2 > r_1$, then $d\phi_t^1 = d\phi_t^2 = 0$, $d\eta_t^1 \ge 0 = d\eta_t^2$.

(iii) If $r_2 > x_t^1 > x_t^2 = r_1$, then $d\phi_t^1 = 0 \le d\phi_t^2$, $d\eta_t^1 = d\eta_t^2 = 0$.

(iv) For case $x_t^1 = x_t^2$ first we assume that $\forall(s, x)$

$$b^1(s, x) > b^2(s, x).$$

By the continuity of $b^i(t, x), i = 1, 2$, and the right continuity of $x_t^i, i = 1, 2$, $\exists\delta > 0$, as $0 < s < \delta$

$$b^1(t + s, x_{t+s}^1) > b^2(t + s, x_{t+s}^2).$$

Note that

$$x_{t+s}^1 = x_t^1 + \int_0^s b^1(t + r, x_{t+r}^1)dr + \int_0^s \sigma(t + r)d\overline{w}_r$$
$$+ \int_0^t \int_Z c(t + r, z)\overline{q}(dr, dz) + \phi_{t+s}^1 - \phi_t^1 - (\eta_{t+s}^1 - \eta_t^1),$$

where

$$\overline{w}_r = w_{t+r} - w_t,$$
$$\overline{q}((0, r], dz) = q((t, t + r], dz).$$

Hence $(x_{t+s}^1, \phi_{t+s}^1 - \phi_t^1, \eta_{t+s}^1 - \eta_t^1)$ solves RSDE (4.1) as $i = 1$ for the given fixed t with the initial value x_t^1 and with the BM $\overline{w}(s)$ and the Poisson martingale measure $\overline{q}(ds, dz)$. Similar facts hold for $(x_{t+s}^2, \phi_{t+s}^2 - \phi_t^2, \eta_{t+s}^2 - \eta_t^2)$. Now suppose that $x_t^1 = x_t^2 = r_1$. By Lemma 4.5

$$\phi_{t+s}^i - \phi_t^i = \sup_{0\le r\le s}[(-x_t^i + r_1 - \int_0^r b^i(t + u, x_{t+u}^i)du + \eta_{t+r}^i - \eta_t^i - z_r) \vee 0],$$

$i = 1, 2,$

where

$$z_r = \int_0^r \sigma(t + u)d\overline{w}(u) + \int_0^r \int_Z c(t + u, z)\overline{q}(du, dz).$$

However, as $0 < s < \delta_1 < \delta$ one has that $\eta_{t+r}^i - \eta_t^i = 0$, $i = 1, 2$. Therefore, as $0 < s < \delta_1 < \delta$

$$\phi_{t+s}^1 - \phi_t^1 = \sup_{0\le r\le s}[(-x_t^1 + r_1 - \int_0^r b^1(t + u, x_{t+u}^1)du - z_r) \vee 0]$$
$$\le \sup_{0\le r\le s}[(-x_t^2 + r_1 - \int_0^r b^2(t + u, x_{t+u}^2)du - z_r) \vee 0] = \phi_{t+s}^2 - \phi_t^2.$$

Hence

$$d\phi_i^1(t) \le d\phi_i^2(t).$$

Similarly, one finds that

$$d\eta_t^1 \ge d\eta_t^2.$$

Next, for the general case one can replace $b^1(t, x)$ by $b^1(t, x) + 1/n$, and then let $n \to \infty$. The proof of Theorem 4.4 is complete. ■

The comparison theorem of solutions to RSDE (4.1) is a useful tool. For example, by using the comparison Theorem 4.2 we can get the existence of a strong solution for (4.1) with non-Lipschitzian drift coefficient b, where the diffusion coefficient σ can still be degenerate. In this case the Girsanov theorem can not be applied.

4.6 Theorem. Assume that the conditions $1^{\circ\prime} - 5^{\circ\prime}$ for (b, σ, c) in Theorem 4.2 hold, and assume that

$6^{\circ\prime\prime\prime}$ $b(t, x)$ is jointly continuous.

Then (4.1) for (b, σ, c) has a strong solution (x_t, ϕ_t, η_t).

Proof. The proof is almost completely the same as the first part of that of Theorem 3.4. Take jointly Lipschitzian continuous functions $\overline{b}^n(t,x)$ and $\underline{b}^1(t,x)$ such that $\forall n = 1, 2, \cdots$

$$b(t,x) - 1 < \underline{b}^1(t,x) < b(t,x) < b(t,x) + 1/(n+1) < \overline{b}^n(t,x) < b(t,x) + 1/n.$$

Then by Theorem 3.8 of Chapter 3 there exists a unique $\left(\overline{x}_t^n, \overline{\phi}_t^n, \overline{\eta}_t^n\right)$ satisfying (4.1) with coefficients $\left(\overline{b}^n, \sigma, c\right)$ and there exists a unique $\left(\underline{x}_t^1, \underline{\phi}_t^1, \underline{\eta}_t^1\right)$ satisfying (4.1) with coefficients $\left(\underline{b}^1, \sigma, c\right)$. By the comparison Theorem 4.3 $P - a.s.$

$$\underline{x}_t^1 \le \overline{x}_t^{n+1} \le \overline{x}_t^n \le \overline{x}_t^1, \forall t \ge 0, \forall n.$$

Hence, there exists a finite

$$\overline{x}_t = \lim_{n\to\infty} \overline{x}_t^n, \ P - a.s.$$

Set

$$\overline{y}_t^n = x_0 + \int_0^t \overline{b}^n(s, \overline{x}_s^n)ds + \int_0^t \sigma(s, \overline{x}_s^n)dw_s + \int_0^t \int_Z c(s, \overline{x}_{s-}^n, z)q(ds, dz),$$

$$\overline{y}_t = x_0 + \int_0^t b(s, \overline{x}_s)ds + \int_0^t \sigma(s, \overline{x}_s)dw_s + \int_0^t \int_Z c(s, \overline{x}_{s-}, z)q(ds, dz).$$

By assumption one can show that

$$E[\sup_{t\le T} \left|\underline{x}_t^1\right|^2 + \sup_{t\le T} \left|\overline{x}_t^1\right|^2] \le k_T.$$

Hence, by Lebesgue's dominated convergence theorem, one finds that as $n \to \infty$

$$E \sup_{t\le T} \left|\overline{y}_t^n - \overline{y}_t\right|^2 \to 0.$$

Therefore, there exists a subsequence $\{n_k\}$ of $\{n\}$ such that $n_k \uparrow \infty$, as $k \uparrow \infty$, and

$$\lim_{k\to\infty} \sup_{t\le T} \left|\overline{y}_t^{n_k} - \overline{y}_t\right|^2 = 0.$$

Note that by Theorem 2.9 of Chapter 2 there exists a unique solution $(\widetilde{x}_t, \widetilde{\phi}_t)$ satisfying

$$\begin{cases} \widetilde{x}_t = \overline{y}_t + \widetilde{\phi}_t \in [r_1, r_2], \\ \text{and the other statements in (2.1) of Chapter 2 hold for } (\widetilde{x}_t, \widetilde{\phi}_t). \end{cases}$$

Applying Corollary 2.10 of Chapter 2, one has that

$$\lim_{k\to\infty} \sup_{s\le t} \left|\overline{x}_s^{n_k} - \widetilde{x}_s\right| = 0,$$

$$\lim_{k\to\infty} \sup_{s\le t} \left|\overline{\phi}_s^{n_k} - \widetilde{\phi}_s\right| = 0, \text{ for all } t \ge 0.$$

By the uniqueness of a limit

$$\widetilde{x}_t = \overline{x}_t.$$

Hence $(\widetilde{x}_t, \widetilde{\phi}_t)$ satisfies

$$\begin{cases} d\widetilde{x}_t = b(t,\widetilde{x}_t)dt + \sigma(t,\widetilde{x}_t)dw_t + \int_Z c(t,\widetilde{x}_{t-},z)q(dt,dz) + d\widetilde{\phi}_t, \\ \widetilde{x}_0 = x_0 \in [r_1, r_2], \\ \text{and the other statements in (2.1) of Chapter 2 hold for } (\widetilde{x}_t, \widetilde{\phi}_t). \end{cases}$$

Let

$$\phi_t = \int_0^t I_{(\widetilde{x}_s = r_1)} d\left|\widetilde{\phi}\right|_s, \ \eta_t = \int_0^t I_{(\widetilde{x}_s = r_2)} d\left|\widetilde{\phi}\right|_s.$$

Then, obviously, $(\widetilde{x}_t, \phi_t, \eta_t)$ satisfies (4.1), which is a strong solution. ∎

We can also weaken the continuity assumption on $b(t,x)$ to get the following

4.7 Theorem. Assume that all conditions for (b,σ,c) in Theorem 4.6 hold except that $6^{\circ\prime\prime\prime}$ is replaced by

$\overline{6}^\circ$ (i) $\overline{\lim}_{y\to x-} b(t,y) \leq b(t,x)$, as $x \notin G, \forall t \geq 0$,

$\sigma(t,x) = 0$, as $x \in G, \forall t \geq 0$,

where $G \subset [r_1, r_2]$, $m_1 G = 0$, m_1 is the Lebesgue measure in R^1;

(ii) $\underline{\lim}_{s\uparrow t, y\to x} b(s,y) \geq b(t,x)$, as $x \notin G, \forall t \geq 0$.

Then (4.1) for (b,σ,c) has a strong solution.

The significance of condition $\overline{6}^\circ$ is that $b(t,x)$ can be discontinuous. More than that

$\overline{\lim}_{y\to x-} b(t,y) \leq b(t,x)$

may also be not true at $x \in G, \forall t \geq 0$, if at such point (t,x) $\sigma(t,x)$ is not equal to zero, and the Lebesgue measure of this set G of such points equals zero. Let us give an example.

4.8 Example. Let

$$b(t,x) = \begin{cases} x/|x|, & \text{as } x = 0, \\ -1, & \text{as } x = 0. \end{cases}$$

Then $\overline{6}^\circ$ in Theorem 4.7 is satisfied.

To show Theorem 4.7 we need the following

4.9 Lemma. Assume that $|b| \leq k_0$. Let

$b_n(t,x) = \inf_{t\geq s\geq 0, y}(b(s,y) + n|y-x| + n(t-s) - 2^{-n}$.

Then

1) $b_n(t,x) \uparrow\uparrow b(t,x), \forall t \geq 0, x \in G$;

2) $|b_n(t,x) - b_n(t,y)| \leq n|y-x|, \forall t \geq 0, x,y \in R^1$;

3) $|b_n(s,x) - b_n(t,x)| \leq n(t-s), \forall t \geq s \geq 0, x \in R^1$;

4) $x_n \uparrow x \Rightarrow \lim_{n\to\infty} b_n(t,x_n) = b(t,x). \forall t \geq 0, x \in G$.

Proof. By the left continuity of b in $x \in G$, and the jointly lower semi-continuity in $x \in G$ and the lower semi-continuity in t of b one has that $b_n(t,x)$ is also jointly measurable. Obviously, $b_n < b_{n+1}$. By

$|y-z| - |z-x| \leq |y-x| \leq |y-z| + |z-x|$

one easily sees that

$b_n(t,z) - n|z-x| \leq b_n(t,x) \leq b_n(t,z) + n|z-x|$.

Hence, 2) is obtained and 3) is similarly proved. Note that by the lower semi-continuity of b at $x \in G$ for any given $\varepsilon > 0$, $\exists \delta > 0$ such that as $|y-x| < \delta$ and $0 \leq t-s < \delta$, $b(s,y) \geq b(t,x) - \varepsilon$. Hence,

$\inf_{|y-x|<\delta, 0\leq t-s<\delta, s\geq 0}(b(s,y) + n|y-x| + n|t-s|) \geq b(t,x) - \varepsilon$.

On the other hand, as $n \geq \delta^{-1}(k_0 + b(t,x)), |y-x| \geq \delta$ or $t-s \geq \delta$,

$$b(s,y) + n|y-x| + n|t-s| \geq b(t,x).$$

Since $b(t,x) \geq b_n(t,x)$, as $n \geq \delta^{-1}(k_0 + b(t,x))$

$$b(t,x) \geq b_n(t,x) \geq b(t,x) - \varepsilon - 2^{-n}.$$

Therefore, 1) is proved. Finally, let us show 4). Fix $t \geq 0$ and $x \in G$. For any given $\varepsilon > 0$ by the left continuity of b in x there exists a $\delta > 0$ such that as $x - \delta < y \leq x$,

$$b(t,x) - \varepsilon < b(t,y) < b(t,x) + \varepsilon.$$

Since $b_n \uparrow\uparrow b, \exists n_0$ such that

$$b(t,x) - \varepsilon < b_{n_0}(t,x) \leq b(t,x).$$

Note that b_{n_0} is continuous. Hence $\exists \delta > \delta_0 > 0$ such that as $|y-x| \leq \delta_0$

$$b_{n_0}(t,x) - \varepsilon < b_{n_0}(t,y) < b_{n_0}(t,x) + \varepsilon.$$

Now since $x_n \uparrow x$. $\exists N_0 \geq n_0$ such that as $n \geq N_0$

$$x_n \in (x - \delta_0, , x].$$

Therefore, as $n \geq N_0$

$$b(t,x) + \varepsilon > b(t,x_n) \geq b_n(t,x_n) > b_{n_0}(t,x_n) > b_{n_0}(t,x) - \varepsilon > b(t,x) - 2\varepsilon.$$

The proof is complete. ■

By using Lemma 4.9 it is not difficult to derive Theorem 4.7.

Proof. Take $b_n(t,x)$ from Lemma 4.9. Obviously, $b_n(t,x)$ is jointly Lipschitzian continuous. Hence by Theorem 4.6 there exists a solution x_t^n satisfying (4.1) with coefficients (b_n, σ, c) for each $n = 1, 2, \cdots$. Moreover, by properties 1) and 2) of Lemma 4.9 applying the comparison Theorem 4.3 one finds that $x_t^n \leq x_t^{n+1}, \forall n$. Hence there exists a limit $P-a.s.$

$$x_t = \lim_{n\to\infty} x_t^n, \forall t \geq 0.$$

Note that as $0 \leq T < \infty$, $E \sup_{t \leq T} |x_t^n|^2 \leq k_T < \infty$, $\forall n$. By Fatou's lemma $\forall t \in [0,T]$

$$E|x_t|^2 \leq k_T < \infty.$$

Note also that

$$x_t^n = x_0 + A_t^n + \int_0^t \sigma(s, x_s^n) dw_s + \int_0^t \int_Z c(s, x_{s-}^n, z) q(ds, dz),$$

where $A_t^n = \int_0^t b_n(s, x_s^n) ds$. However, by the assumptions on σ and c it is easily shown that as $n \to \infty$ (take a subsequence if necessary)

$$\int_0^t \sigma(s, x_s^n) dw_s \to \int_0^t \sigma(s, x_s) dw_s,$$

$$\int_0^t \int_Z c(s, x_{s-}^n, z) q(ds, dz) \to \int_0^t \int_Z c(s, x_{s-}, z) q(ds, dz).$$

Hence, there exists a limit

$$A_t = \lim_{n\to\infty} A_t^n,$$

and we have

$$x_t = x_0 + A_t + \int_0^t \sigma(s, x_s) dw_s + \int_0^t \int_Z c(s, x_{s-}, z) q(ds, dz).$$

Since by $|b| \leq k_0$ one has that $|A^n|_T \leq k_0 T$. Hence, one sees that $|A|_T \leq k_0 T$, i.e. A_t is a finite variational process. Let us show that

$$(4.5) \quad A_t = \int_0^t b(s, x_s) ds.$$

Indeed, by $m_1(G) = 0$

$$0 = \int_G L_t^a(x) da = \int_0^t I_{(x_s \in G)} \sigma(s, x_s)^2 ds.$$

Since by assumption $\sigma(s, x_s)^2 > 0$, as $x_s \in G$. Hence,

$m_1(s\in[0,t]:x_s\in G)=0.$
Therefore, by conclusion 4) of Lemma 4.9
$\int_0^t |b(s,x_s)-b_n(s,x_s^n)|\,ds \le \int_0^t I_{(x_s\notin G)}\,|b(s,x_s)-b_n(s,x_s^n)|\,ds \to 0.$
(4.5) is derived. . ■

4.5 Uniqueness of Solutions

Let us discuss the uniqueness of solutions to (3.1) with coefficients (b,σ,c). There are two kinds of conditions: one is for the components of coefficients, another one is for the whole coefficients. The first one is an application of Theorem 3.1. The second one is based on the formula (10.14) in Chapter 1. We discuss the first one.

5.1 Theorem. Assume that conditions $1^\circ-3^\circ$ in Theorem 3.1 for coefficients (b,σ,c) hold. Moreover, assume that
4° $\sum_{i=1}^d I_{(x_i>y_i)}(b_i(t,x)-b_i(t,y))\le k_N(t)\rho_N\left(|x-y|\right)$, as $|x|,|y|\le N$,
where $k_N(t)$ and $\rho_N(u)$ satisfy the same conditions as that in 5° of Theorem 4.1. Then the pathwise uniqueness of solutions to (3.1) with coefficients (b,σ,c) holds.
Proof. Assume that $x_t^i, i=1,2,$ are two solutions of (3.1) with coefficients (b,σ,c) on the same probability space with the same BM and the same $q(dt,dz)$. Set
$\tau_N=\inf\left\{t\ge 0:\left|x_t^1\right|+\left|x_t^2\right|>N\right\}.$
By Theorem 3.1
$E\left(x_{i,t\wedge\tau_N}^1-x_{i,t\wedge\tau_N}^2\right)^+\le E\int_0^{t\wedge\tau_N}k_N(s)\rho_N\left(\left|x_s^1-x_s^2\right|\right)ds,$
and
$E\left(x_{i,t\wedge\tau_N}^2-x_{i,t\wedge\tau_N}^1\right)^+\le E\int_0^{t\wedge\tau_N}k_N(s)\rho_N\left(\left|x_s^1-x_s^2\right|\right)ds,$
where we have used the result
$\int_0^t I_{\left(x_{i,s}^1>x_{i,s}^2\right)}d(\phi_{i,s}^1-\phi_{i,s}^2)=-\int_0^t I_{\left(x_{i,s}^1>x_{i,s}^2=0\right)}d\phi_{i,s}^2\le 0.$
Hence
$E\left|x_{t\wedge\tau_N}^1-x_{t\wedge\tau_N}^2\right|\le 2dE\int_0^{t\wedge\tau_N}k_N(s)\rho_N\left(\left|x_s^1-x_s^2\right|\right).$
From this it is not difficult to show that $P-a.s$
$x_t^1=x_t^2,$
$\phi_t^1=\phi_t^2,\forall t\ge 0.$
(cf. the proof of Theorem 6.1 in Chapter 1). ■

5.2 Remark. 1) If
$\left|\widetilde{b}(t,x)-\widetilde{b}(t,y)\right|\le k_N(t)\rho_N\left(|x-y|\right)$, as $|x|,|y|\le N$,
where $k_N(t)$ and $\rho_N(u)$ satisfy the same conditions as that in 5° of Theorem 4.1, then condition 4° is satisfied by $\widetilde{b}$.
2) Let
$b_i(t,x)=-I_{(x_i=0)}\frac{x_i}{|x_i|}+k_0\widetilde{b}(t,x),$

where $\widetilde{b}$ comes from 1). Then condition 4° is satisfied by b.

Indeed,

$(x_i - y_i)(-I_{(x_i=0)}\frac{x_i}{|x_i|} + I_{(y_i=0)}\frac{y_i}{|y_i|}) = -|x_i| - |y_i| + I_{(y_i=0)}\frac{y_i x_i}{|y_i|} + I_{(x_i=0)}\frac{x_i y_i}{|x_i|}$
$\leq 0.$

Hence,

$I_{(x_i>y_i)}(-I_{(x_i=0)}\frac{x_i}{|x_i|} + I_{(y_i=0)}\frac{y_i}{|y_i|}) \leq 0.$

Hence, it is easily shown that condition 4° is satisfied by b.

Next, we have

5.3 Theorem. Assume that

1° $|b(t,x,\omega)|^2 + \|\sigma(t,x,\omega)\|^2 + \int_Z \|c(t,x,z,\omega)\|^2 \pi(dz) \leq g(|x|)$,
where $g(u)$ is a continuous function on $u \geq 0$;

2° $\|\sigma(t,x,\omega) - \sigma(t,y,\omega)\| \leq k_N(t)\rho_N(|x-y|)$, as $|x|,|y| \leq N$;
where $0 \leq k_N(t)$, $\int_0^T k_N^2(t)dt < \infty$, for each $T < \infty; 0 < \rho_N(u)$, as $u > 0; \rho_N(0) = 0$, and

$\int_{0+} du/\rho_N^2(u) = \infty$,

3° there exist real continuous functions G_N; $N = 1,2,\cdots$ such that $N = 1,2,\cdots$

$\|\sigma(t,x,\omega) - \sigma(t,y,\omega)\|^2 \leq G_N(|x-y|)\,|x-y|$, as $|x|,|y| \leq N$;

where $G_N(u) \geq 0$, as $u \geq 0$, it is concave, strictly increasing with $G_N(0) = 0$ such that

$\int_{0+} du/G_N(u) = \infty$;

4° $\sum_{i=1}^d (sgn(x-y))_i \cdot (b(s,x,\omega) - b(s,y,\omega))_i \leq \overline{G}_N(|x-y|)$,
as $|x|,|y| \leq N$,

where $\overline{G}_N(u)$ has the same property as that of $G_N(u)$,

$sgn x = I_{x=0}(\frac{x_1}{|x|},\cdots,\frac{x_d}{|x|})$,

5° $\int_Z |c(t,x,z,\omega)|\,\pi(dz) < \infty$,

$\int_Z |c(t,x,z,\omega) - c(t,x,z,\omega)|\,\pi(dz) \leq G'_N(|x-y|)$, as $|x|,|y| \leq N$;

where $G'_N(u)$ has the same property as that of $G_N(u)$.

Then the pathwise uniqueness of solutions for (3.1) with coefficients (b,σ,c) holds.

Proof. Assume that $x_t^i, i = 1,2$, are two solutions of (3.1) with the same coefficients (b,σ,c) on the same probability space with the same $A_t, M_t, q(dt,dz)$. Then by Theorem 10.7 in Chapter 1

$$\begin{aligned}
|x_t^1 - x_t^2| &\leq |x_0^1 - x_0^2| + \int_0^t sgn(x_{s-}^1 - x_{s-}^2)\cdot d(x_s^1 - x_s^2) \\
&+\frac{1}{2}\sum_{i,j,k=1}^d \int_0^t I_{(x_s^1 = x_s^2)} \frac{|x_s^1 - x_s^2|^2\delta_{ij} - (x_{is}^1 - x_{is}^2)(x_{js}^1 - x_{js}^2)}{|x_s^1 - x_s^2|^3} \\
&\cdot(\sigma_{ik}(s,x_s^1,\omega) - \sigma_{ik}(s,x_s^2,\omega))(\sigma_{jk}(s,x_s^1,\omega) - \sigma_{jk}(s,x_s^2,\omega))ds \\
&+\int_0^t \int_Z [|x_{s-}^1 - x_{s-}^2 + c(s,x_{s-}^1,z,\omega) - c(s,x_{s-}^2,z,\omega)| - |x_{s-}^1 - x_{s-}^2| \\
&-sgn(x_{s-}^1 - x_{s-}^2)\cdot(c(s,x_{s-}^1,z,\omega) - c(s,x_{s-}^2,z,\omega))]p(ds,dz),
\end{aligned}$$

where we have used the result that

$$\int_0^t sgn(x_s^1 - x_s^2)\cdot d(\phi_s^1 - \phi_s^2) = -\sum_{i=1}^d \int_0^t I_{(x_{is}^1 > x_{is}^2 \geq 0)} \frac{(x_{is}^1 - x_{is}^2)}{|x_s^1 - x_s^2|} d\phi_{is}^2$$
$$-\sum_{i=1}^d \int_0^t I_{(x_{is}^2 > x_{is}^1 \geq 0)} \frac{(x_{is}^2 - x_{is}^1)}{|x_s^1 - x_s^2|} d\phi_{is}^1 \leq 0.$$

Now, applying the same technique as in the proof of Theorem 10.2 of Chapter 1, by assumption 3° and 4°, one arrives at the conclusion. ■

Example 10.4 in Chapter 1 can also be an example for this Theorem 5.3.

4.6 Convergence of Solutions in Half space

Finally, we will discuss the convergence of solutions to RSDE in a half space, i.e. to RSDE (1.1) and (1.2)' in Chapter 3. More precisely, let us consider the RSDEs as follows:

$$(6.1)\begin{cases} dx_t^n = b^n(t,x_t^n)dt + \sigma^n(t,x_t^n)dw_t + \int_Z c^n(t,x_{t-}^n,z)q(dt,dz) + d\phi_t^n, \\ x_0^n = x_0^n \in \overline{\Theta}, \\ \text{and } (x_t^n,\phi_t^n),\ t\geq 0 \text{ satisfies the statements in (1.2)' of Chapter 3;} \\ \text{where } \Theta = \left\{x\in R^d : x = (x_1,\cdots,x_d), x_d > 0\right\}; \end{cases}$$

$n = 0,1,2,.....$

We have

6.1 Theorem. Assume that

1° there exists a constant $k_0 \geq 0$, and for each $N = 1,2,...$ there exists a $\delta_N > 0$ such that

$$|b^n(t,x)|^2 + \|\sigma^n(t,x)\|^2 + \int_Z |c^n(t,x,z)|^2 \pi(dz) \leq k_0,$$
$$\langle A^n\lambda,\lambda\rangle \geq \delta_N |\lambda|^2, \text{ for all } |x|\leq N, \lambda\in R^d, n = 0,1,2,...$$

where $N = 1,2,...$, and $A^n = \frac{1}{2}\sigma^n\sigma^{n*}$;

2° $2(x-y)\cdot(b^0(t,x)-b^0(t,y)) \leq k(t)\rho(|x-y|^2)$,

$$\left\|\sigma^0(t,x)-\sigma^0(t,y)\right\|^2 \leq k(t)\rho(|x-y|^2),$$
$$\int_Z \left|c^0(t,x,z)-c^0(t,y,z)\right|^2 \pi(dz) \leq k(t)\rho(|x-y|^2),$$

where $k(t)\geq 0, \rho(u)$ is continuous, strictly increasing, concave, $\rho(0) = 0$, and they are such that for any $0 < T < \infty$

$$\int_0^T k(t)dt < \infty,\ \int_{0+} du/\ \rho(u) = \infty;$$

3° as $n\to\infty$

$$\left\|\left|b^n - b^0\right|^2\right\|_{p,[0,\infty)\times[-N,N]} \to 0,\ \left\|\left\|\sigma^n-\sigma^0\right\|^2\right\|_{p,[0,\infty)\times[-N,N]} \to 0,$$
$$\left\|\int_Z \left|c^n - c^0\right|^2 \pi(dz)\right\|_{p,[0,\infty)\times[-N,N]} \to 0,$$

for some $p\geq d+1$, and for each $N = 1,2,...$;

4° $E\left|x^n - x^0\right|^2 \to 0$, as $n\to\infty$,

Then for all $t\geq 0$, as $n\to\infty$

$$E\sup_{s\leq t}\left|x_s^n - x_s^0\right|^2 \to 0.$$

In addition, if

$$\lim_{y\to x}\sup_{s\leq t}\left|b_d^0(s,x)-b_d^0(s,y)\right|^2 = 0,$$

where $b^0 = (b_1^0,\cdots,b_d^0)$,

then for all $t\geq 0$, as $n\to\infty$

$$E\sup_{s\leq t}|\phi_s^n - \phi_s^n|^2 \to 0.$$

Proof. By Corollary 4.1 in Chapter 3

$$(1.2)\ \left|x_t^n - x_t^0\right|^2 \leq \left|x_0^n - x_0^0\right|^2 + \int_0^t [2(x_s^n - x_s^0) \cdot (b^n(s, x_s^n) - b^0(s, x_s^0))$$
$$+ \left\|\sigma^n(s, x_s^n) - \sigma^0(s, x_s^0)\right\|^2]ds + 2\int_0^t (x_s^n - x_s^0) \cdot (\sigma^n(s, x_s^n) - \sigma^0(s, x_s^0))dw_s$$
$$+2\int_0^t (x_s^n \quad x_s^0) \cdot \int_Z (c^n(s, x_{s-}^n, z) - c^0(s, x_{s-}^0, z))q(ds, dz)$$
$$+ \int_0^t \int_Z \left|c^n(s, x_{s-}^n, z) - c^0(s, x_{s-}^0, z)\right|^2 p(ds, dz).$$

Now, following the proof of Theorem 1.1 we arrive at the conclusion. ■

Similarly, one can also have the versions of Theorem 1.2 - 1.5 for solutions to the RSDEs in (6.1). We omit them.

Chapter 5

Non-linear Filtering of RSDEs

In this chapter we will discuss the non-linear filtering for RSDEs with jumps by using the martingale representation theorem. For a more concrete case, a Zakai equation is also derived. All results here seem to be new.

5.1 Representation of Martingales (Functional Coefficient Case)

5.1.1 Innovation Process and Girsanov Theorem (Functional Coefficient Case)

Let us consider the following partially observed Reflecting Stochastic Differential System:

$$(1.1)\begin{cases} h_t = h_0 + \int_0^t H_s ds + y_t + \eta_t, \\ \xi_t = \xi_0 + \int_0^t A_s(\omega)ds + \int_0^t B_s(\xi,\phi)dw_s + \int_0^t \int_Z c_{s-}(\xi,\phi,z)q(ds,dz) \\ +\phi_t,\ h_0, \xi_0 \in \Theta^{cl}, \\ \text{and } (h_t,\eta_t), (\xi_t,\phi_t),\ t \geq 0 \text{ satisfy the statements in (1.2) of Chapter 3,} \\ \text{respectively, moreover, } \eta_t \text{ and } \phi_t \text{ are continuous;} \end{cases}$$

where y_t is a $d-$dimensional cadlag square integrable martingale, and we call h_t a signal process, ξ_t an observable process, and they are all $d-$dimensional processes. For simplicity we assume that the existence domains for h_t and ξ_t are the same, both being Θ.

In this chapter we will derive the SDE satisfied by the non-linear filtering processes $E(h_t \big| \Im_t^{\xi,\phi})$, where $\Im_t^{\xi,\phi} = \sigma(\xi_s, \phi_s; s \leq t)$. Naturally, in (3.1) we assume that h, H, y, η, ξ, A, c and ϕ are $d-$dimensional vectors, B is a $d \times d$

matrix, and w_t is a d -dimensional standard Brownian Motion process, $q(ds,dz)$ is a Poisson martingale measure with the compensator $\pi(dz)dt$ such that

$$q(ds,dz) = p(ds,dz) - \pi(dz)dt,$$

where $\pi(dz)$ is a non-random $\sigma-$finite measure in $(Z, \Re(Z))$. More precisely, let us denote

$D = D([0,\infty); R^d) =$ The totality of cadlag maps from $[0,\infty)$ to R^d,

$C = C([0,\infty); R^d) =$ The totality cf continuous maps from $[0,\infty)$ to R^d,

$\Im_t = \sigma((x_s \phi_s); s \leq t, x \in D, \phi \in C)$,

and suppose that (1.1) is defined on the probability space $(\Omega, \Im, (\Im_t), P)$, where $\Omega = D \times C$, $\Im = \vee_{t \geq 0} \Im_t$, i.e. w_t and $q(dt,dz)$ are $\Im_t-$BM and $\Im_t-$ Poisson martingale measure, respectively, etc. We always make the following assumption

(1.2) $A_t(\omega) : [0,\infty) \times D \to R^d$,

$B_t(\xi,\phi) = B(t,\xi(r),\phi(r), r \leq t) : [0,\infty) \times D \times C \to R^{d\otimes d}$,

are jointly measurable and $\Im_t-$adapted,

$c_t(\xi,\phi,z) = c(t,\xi(r),\phi(r), r \leq t; z) : [0,\infty) \times D \times C \times Z \to R^d$,

is jointly measurable and $\Im_t-$adapted. (Hence, if we denote

$c_{t-}(\xi,\phi,z) = c(t,\xi(r),\phi(r), r < t; z)$,

then $c_{t-}(\xi,\phi,z)$ is $\Im_t-$predictable).

We have the following

1.1 Theorem. Assume that for all $T > 0$

$P(\int_0^T |A_s(\omega)|\, ds < \infty) = 1$, $P(\int_0^T \|B_s(\xi,\phi)\|^2 ds < \infty) = 1$.

$P(\int_0^T \int_Z |c_s(\xi,\phi,z)|^2 \pi(dz)ds < \infty) = 1$,

and B^{-1} exists such that

$\|B^{-1}\| \leq k_0$.

Then

1) $\overline{M}_t = \int_0^t B_{s-}^{-1}(\xi,\phi)[d\xi_s - E(A_s \,|\Im_s^{\xi,\phi})ds - d\phi_s]$,

is a martingale on probability space $(\Omega, \Im, (\Im_t^{\xi,\phi}), P)$. The 2nd equation of (1.1) can be rewritten on probability space $(\Omega, \Im, (\Im_t^{\xi,\phi}), P)$ as

$$(1.1)_1 \begin{cases} \xi_t = \xi_0 + \int_0^t E(A_s \,|\Im_s^{\xi,\phi})ds + \int_0^t B_{s-}(\xi,\phi)d\overline{M}_s + \phi_t, \\ \qquad \xi_0 \in \Theta^{cl}, \\ \text{and } (\xi_t,\phi_t),\ t \geq 0 \text{ satisfy the statements in (1.2) of} \\ \text{Chapter 3, respectively, moreover, } \phi_t \text{ is continuous.} \end{cases}$$

2) Furthermore, if

$c_s(\xi,\phi,z) = c_s(\xi,\phi) I_\Gamma(z)$,

where $\Gamma \in \Re(Z)$ such that $\pi(\Gamma) < \infty$, and $c_{s-}(\xi,\phi)^{-1}$ exists, then

$\overline{M}_t = \overline{w}_t + \int_0^t B_{s-}^{-1}(\xi,\phi)c_{s-}(\xi,\phi)d\widetilde{N}_s$,

where

$\overline{w}_t = \int_0^t B_s^{-1}(\xi,\phi)(A_s - E(A_s \,|\Im_s^{\xi,\phi}))ds + w_t = \overline{M}_t^c$,

the continuous martingale part of $\overline{M}_t$, is a BM on $(\Omega, \Im, (\Im_t^{\xi,\phi}), P)$; and

$\widetilde{N}_t = q((0,t],\Gamma) = \overline{M}_t^d$,

the pure discontinuous martingale part of $\overline{M}_t$, is a centralized Poisson random process with constant deterministic density function $\pi(\Gamma)$ on $(\Omega, \Im, (\Im_t^{\xi,\phi}), P)$ such that

$\widetilde{N}_t = p((0,t],\Gamma) - \pi(\Gamma)t = N_t - \pi(\Gamma)t,$

i.e. N_t is the Poisson random process such that $EN_t = \int_0^t \pi(\Gamma)ds$. Usually, we call $\overline{w}_t$ and $\widetilde{N}_t$ the innovation processes of w_t and $\widetilde{N}_t$. The 2nd equation of (1.1) can be rewritten on probability space $(\Omega, \Im, (\Im_t^{\xi,\phi}), P)$ as

$$(1.1)_2 \begin{cases} \xi_t = \xi_0 + \int_0^t E(A_s \,|\Im_s^{\xi,\phi})ds + \int_0^t B_s(\xi,\phi)d\overline{w}_s \\ \quad + \int_0^t c_{s-}(\xi,\phi)d\widetilde{N}_s + \phi_t, \\ \xi_0 \in \Theta^{cl}, \\ \text{and } (\xi_t,\phi_t),\ t \geq 0 \text{ satisfy the statements in (1.2) of} \\ \text{Chapter 3, respectively, moreover, } \phi_t \text{ is continuous.} \end{cases}$$

Before showing Theorem 1.1, we should note the following fact:

$$E\int_0^T |B_s(\xi,\phi) - B_{s-}(\xi,\phi)|^2 ds$$
$$= E\int_0^T |B(s,\xi(r),\phi(r), r\leq s) - B(s,\xi(r),\phi(r), r< s)|^2 ds = 0,$$

since ξ and ϕ are cadlag functions. Now let us show Theorem 1.1.

Proof. 1): In fact, for any $G \in \Im_s^{\xi,\phi}$, $s \leq t$

$$\int_\Omega (\overline{M}_t - \overline{M}_s)I_G dP = \int_\Omega \int_s^t I_G B_{r-}^{-1}(\xi,\phi)[d\xi_r - E(A_r\,|\Im_r^{\xi,\phi})dr - d\phi_r]dP$$
$$= \int_\Omega \int_s^t E(I_G B_{r-}^{-1}(\xi,\phi)[d\xi_r - A_r - d\phi_r\,|\Im_r^{\xi,\phi})dr]dP = 0,$$

by Fubini's theorem. Hence $\overline{M}_t$ is a $\Im_t^{\xi,\phi}$–martingale. Now $(1.1)_1$ is obviously true.

2): Note that

$$\Delta\xi_t = c_{t-}(\xi,\phi) \,\Delta\, N_t.$$

Hence,

$$N_t = \sum_{0<s\leq t} c_{s-}(\xi,\phi)^{-1} \,\Delta\, \xi_s$$

is $\Im_t^{\xi,\phi}$–adapted, so is $\widetilde{N}_t$. Let us show that $\widetilde{N}_t$ is still a $\Im_t^{\xi,\phi}$– centralized Poisson process with the same density function $\pi(\Gamma)$. For this we only need to show that N_t is still a $\Im_t^{\xi,\phi}$– Poisson process with the same density function $\pi(\Gamma)$. In fact, for any $s\leq t$,and $\lambda>0$, by Ito's formula

$$f(t) - f(s) = \exp\{-\lambda N_t\} - \exp\{-\lambda N_s\}$$
$$= \exp\{-\lambda p((0,t],\Gamma)\} - \exp\{-\lambda p((0,s],\Gamma)\}$$
$$= \int_s^t \int_Z f(r)(e^{-\lambda I_\Gamma(z)} - 1)\pi(dz)dr + (\Im_t^{\xi,\phi}\text{–martingale}),\ P-a.s.$$

Hence, for any $s\leq t$,and $\lambda>0, G\in \Im_s^{\xi,\phi}$

$$EI_G[f(t)/f(s)] - EI_G = (e^{-\lambda}-1)\pi(\Gamma)\int_s^t EI_G[f(r)/f(s)]dr.$$

Therefore $EI_G[f(t)/f(s)]$ satisfies the following ordinary differential equation

$$dy_t = (e^{-\lambda}-1)\pi(\Gamma)y_t dt,\ t\geq s;\ y_s = EI_G.$$

Solving it, one obtains

$$E_P\exp\{-\lambda(N_t - N_s)\}I_G = P(G)\exp[(t-s)(e^{-\lambda}-1)\pi(\Gamma)].$$

This means that N_t is still a $\Im_t^{\xi,\phi}$–Poisson process with the same density $\pi(\Gamma)$.
Now write

$$\overline{w}_t = \int_0^t B_{s-}^{-1}(\xi,\phi)(d\xi_s - E(A_s\,|\Im_s^{\xi,\phi})\,ds - c_{s-}(\xi,\phi)d\widetilde{N}_s - d\phi_s)$$

$= \int_0^t B_s^{-1}(\xi,\phi)(A_s - E(A_s |\Im_s^{\xi,\phi}))ds + w_t.$

Obviously, $\overline{w}_t$ is $\Im_t^{\xi,\phi}-$ adapted by the first equality. By Ito's formula

$$e^{i\langle\lambda,\overline{w}_t-\overline{w}_s\rangle} - 1 = i\left\langle\lambda, \int_s^t e^{i\langle\lambda,\overline{w}_r-\overline{w}_s\rangle}[B_r^{-1}(\xi,\phi)(A_r - E(A_r |\Im_r^{\xi,\phi}))dr + dw_r],\right\rangle$$

$$-\frac{|\lambda|^2}{2}\int_s^t e^{i\langle\lambda,\overline{w}_r-\overline{w}_s\rangle}dr = i\left\langle\lambda, \int_s^t e^{i\langle\lambda,\overline{w}_r-\overline{w}_s\rangle}[dM_r - c_{r-}(\xi,\phi)d\widetilde{N}_r]\right\rangle$$

$$-\frac{|\lambda|^2}{2}\int_s^t e^{i\langle\lambda,\overline{w}_r-\overline{w}_s\rangle}dr,$$

where M_t and $\int_0^t c_{r-}(\xi,\phi)d\widetilde{N}_r$ are both $\Im_t^{\xi,\phi}-$martingale. Hence for any $G \in \Im_s^{\xi,\phi}$, $s \le t$, $EI_G e^{i\langle\lambda,\overline{w}_t-\overline{w}_s\rangle}$ satisfies the following ordinary differential equation:

$$dy_t = -\frac{|\lambda|^2}{2}y_t dt,\ t \ge s;\ y_s = P(G).$$

Solving it gives

$$EI_G e^{i\langle\lambda,\overline{w}_t-\overline{w}_s\rangle} = P(G)e^{-|\lambda|^2(t-s)/2},$$

i.e. $\overline{w}_t$ is a $\Im_t^{\xi,\phi}-$BM. Obviously,

$$\overline{M}_t = \overline{w}_t + \int_0^t B_s^{-1}(\xi,\phi)c_s(\xi,\phi)d\widetilde{N}_s.$$

Now $(1.1)_2$ is easily derived. ■

To derive a non-linear filtering equation we need a Girsanov theorem for RSDE with functional coefficients as follows: Suppose that (x_t,ϕ_t) solve the following RSDE with jumps and with the functional coefficients on $(\Omega,\Im,(\Im_t),P)$:

$$(1.3)\quad \begin{cases} x_t = x_0 + \int_0^t [b(s,x(.),\phi(.)) - \sigma\widetilde{b}(s,x(.),\phi(.))]ds + \int_0^t \sigma(s,x(.),\phi(.))dw_s \\ \qquad + \int_0^t \int_Z c(s-,x(.),\phi(.),z)q(ds,dz) + \phi_t, \\ \qquad x_0 \in \Theta^{cl}, \\ \text{and } (x_t,\phi_t),\ t \ge 0 \text{ satisfy the statements in (1.2) of Chapter 3,} \\ \qquad \text{moreover, } \phi_t \text{ is continuous.} \end{cases}$$

where we make the following assumption

(1.4) $b(s,x,\phi) : [0,\infty) \times D \times C \to R^d$

is jointly measurable and $\Im_t-$adapted, σ and c satisfy the same conditions in (1.2) as that for B and c.

Thus we have

1.2 Theorem. Assume that

1° $|b(t,x,\phi)|^2 + \left|\widetilde{b}(t,x,\phi)\right|^2 + \|\sigma(t,x,\phi)\|^2$

$$+\int_Z |c(t,x,\phi,z)|^2 \pi(dz) \le k_0^2(1 + \|x\|_t^2 + \|\phi\|_t^2),$$

where $k_0 \ge 0$ is a constant, and write $\|x\|_t = \sup_{s\le t}|x(s)|$, where $\|\phi\|_t$ is similarly defined.

Let

$$d\widetilde{P}_t = z_t(\widetilde{b})dP,$$

where

$$z_t(\widetilde{b}) = \exp[\int_0^t \widetilde{b}(s,x,\phi)\cdot dw_s - \frac{1}{2}\int_0^t \left|\widetilde{b}(s,x,\phi)\right|^2 ds],$$

and (x_t,ϕ_t) is the solution of (1.3).

Then

1) $E\ z_t(\widetilde{b}) = 1$,

2) there exists a probability measure $\widetilde{P}$ defined on $(\Omega,\Im)$ such that

$(d\widetilde{P}/dP)|_{\Im_t} = z_t(\widetilde{b})$

3) let

$\widetilde{w}_t = w_t - \int_0^t \widetilde{b}(s,x,\phi)ds,$

then $\widetilde{w}_t$ is a $\widetilde{P}$ -BM,

4) $q(dt,dz)$ is still a $\widetilde{P}$ -Poisson martingale measure with the same compensator $\pi(dz)dt$,

5) (x_t,ϕ_t) also solve the following RSDE on $(\Omega,\Im,(\Im_t),\widetilde{P})$:

$$(1.5)\quad \begin{cases} x_t = x_0 + \int_0^t b(s,x(.),\phi(.))ds + \int_0^t \sigma(s,x(.),\phi(.))d\widetilde{w}_s \\ \quad + \int_0^t \int_Z c(s-,x(.),\phi(.),z)q(ds,dz) + \phi_t, \\ \quad x_0 \in \Theta^{cl}, \\ \text{and } (x_t,\phi_t),\ t \geq 0 \text{ satisfy the statements in (1.2) of} \\ \quad \text{Chapter 3, moreover, } \phi_t \text{ is continuous.} \end{cases}$$

Proof. Let

$$\widetilde{b}_N(t,x,\phi) = \begin{cases} \widetilde{b}(t,x,\phi), & \text{as } \|x\|_t^2 + \|\phi\|_t^2 \leq N^2, \\ 0, & \text{otherwise.} \end{cases}$$

Then

$$\left|\widetilde{b}_N(t,x,\phi)\right|^2 \leq k_0^2(1+N^2).$$

By Girsanov type Lemma 3.9, Theorem 3.10 and the proof of Theorem 3.12 in Chapter 3 the conclusions of 1)-4) hold when $\widetilde{b}_N$ is replaced by b. Hence (x_t,ϕ_t) also solve the following RSDE on $(\Omega,\Im,(\Im_t),\widetilde{P}_N)$:

$$(1.6)\quad \begin{cases} x_t = x_0 + \int_0^t b(s,x(.),\phi(.))ds + \int_0^t \sigma(s,x(.),\phi(.))d\widetilde{w}_s^N \\ \quad + \int_0^t \int_Z c(s-,x(.),\phi(.),z)q(ds,dz) + \phi_t, \\ \quad x_0 \in \Theta^{cl}, \\ \text{and } (x_t,\phi_t),\ t \geq 0 \text{ satisfy the statements in (1.2) of} \\ \quad \text{Chapter 3, respectively, moreover, } \phi_t \text{ is continuous,} \end{cases}$$

where $(d\widetilde{P}_N/dP)|_{\Im_t} = z_t(\widetilde{b}_N)$, and

$\widetilde{w}_t = w_t - \int_0^t \widetilde{b}_N(s,x(.),\phi(.))ds.$

is a BM under probability measure $\widetilde{P}_N$. By Ito's formula $\widetilde{P}_N - a.s.$

$$(1.7)\ |x_t - x_0|^2 = 2\int_0^t (x_s - x_0)\cdot b(s,x(.),\phi(.))ds$$
$$+2\int_0^t (x_s - x_0)\cdot \sigma(s,x(.),\phi(.))d\widetilde{w}_s^N$$
$$+2\int_0^t \int_Z (x_{s-} - x_0)\cdot c(s-,x(.),\phi(.),z)q(ds,dz) + 2\int_0^t (x_s - x_0)\cdot d\phi_s$$
$$+\int_0^t \|\sigma(s,x(.),\phi(.))\|^2 ds + \int_0^t \int_Z |c(s-,x(.),\phi(.),z)|^2 p(ds,dz), \text{ as } t \geq 0.$$

Note that

$|x_t - x_0|^2 = |x_t|^2 + |x_0|^2 - 2x_t \cdot x_0$

Hence,

$|x_t|^2 \leq \frac{1}{4}|x_t|^2 + 4|x_0|^2 +$ the right hand side of (1.7).

Denote

$\tau_N = \inf\{t \geq 0 : \|x\|_t + \|\phi\|_t > N\},\ N = 1,2,....$

Then by the martingale inequality

$E_{\widetilde{P}_N} \sup_{t \leq T \wedge \tau_N} \left|\int_0^t (x_s - x_0)\cdot \sigma(s,x(.),\phi(.))d\widetilde{w}_s^N\right|$

$\leq \widetilde{k}_0 E_{\widetilde{P}_N}[\int_0^{T\wedge\tau_N} |(x_s - x_0)\cdot\sigma(s,x(.),\phi(.))|^2 ds]^{1/2}$
$\leq k_0 E_{\widetilde{P}_N}[\int_0^{T\wedge\tau_N} |(|x_s|+|x_0|)\cdot(1+\|x\|_s+\|\phi\|_s))|^2 ds]^{1/2}$.
A similar estimate holds for the third term of the right side of (1.7). Now applying Young's inequality for $a, b \geq 0$
$2a\cdot b \leq \varepsilon a^2 + \varepsilon^{-1}b^2$,
one gets that
$E_{\widetilde{P}_N}\|x\|^2_{t\wedge\tau_N} \leq k_T[1 + E_{\widetilde{P}_N}\int_0^{t\wedge\tau_N}(\|x\|_s^2 + \|\phi\|_s^2)ds$
$\leq k_T[1 + \int_0^t E_{\widetilde{P}_N}(\|x\|^2_{s\wedge\tau_N} + \|\phi\|^2_{s\wedge\tau_N})ds.$
By (1.5) and the above inequality one also gets that
$E_{\widetilde{P}_N}\|\phi\|^2_{t\wedge\tau_N} \leq \widetilde{k}_T[1 + \int_0^t E_{\widetilde{P}_N}(\|x\|^2_{s\wedge\tau_N} + \|\phi\|^2_{s\wedge\tau_N})ds.$
Hence,
$E_{\widetilde{P}_N}(\|x\|^2_{t\wedge\tau_N} + \|\phi\|^2_{t\wedge\tau_N}) \leq \widetilde{k}_T[1 + \int_0^t E_{\widetilde{P}_N}(\|x\|^2_{s\wedge\tau_N} + \|\phi\|^2_{s\wedge\tau_N})ds.$
By Gronwall's inequality it easily shows that for any given $0 \leq T < \infty$
$2N^2\widetilde{P}_N(\tau_N \leq T) \leq E_{\widetilde{P}_N}(\|x\|^2_{T\wedge\tau_N} + \|\phi\|^2_{T\wedge\tau_N}) \leq k_T < \infty.$
Hence, for any given $0 \leq T < \infty$
$\lim_{N\to\infty}\widetilde{P}_N(\tau_N \leq T) = 0.$
Therefore,
$Ez_T(\widetilde{b}) \geq \lim_{N\to\infty}\int_{\{\tau_N > T\}} z_T(\widetilde{b})dP = \lim_{N\to\infty}\widetilde{P}_N(\tau_N > T) = 1.$
This shows that for all $0 \leq T < \infty$
$Ez_T(\widetilde{b}) = 1.$
1) is true. 2) is also true, because $\Omega = D$ is a standard measurable space. (See the proof of Theorem 3.10 in Chapter 3). Now 3) and 4) are derived in the proof of Theorem 3.12 in Chapter 3. 5) follows from 1)-4) and RSDE (1.3). ■

5.1.2 Martingale Representation Theorems (Functional Coefficient Case)

To derive the non-linear filtering equation we also need some martingale representation theorems for the RSDE (1.3) in the special case with functional coefficients:

$$(1.8)_0 \begin{cases} x_t = x_0 + \int_0^t [b(s,x(.),\phi(.)) - \sigma\widetilde{b}(s,x(.),\phi(.))]ds \\ + \int_0^t \sigma(s,x(.),\phi(.))dw_s + \int_0^t c(s-,x(.),\phi(.))d\widetilde{N}_s + \phi_t, \\ x_0 \in \Theta^{cl}, \\ \text{and } (x_t,\phi_t),\ t \geq 0 \text{ satisfy the statements in (1.2) of} \\ \text{Chapter 3, moreover, } \phi_t \text{ is continuous,} \end{cases}$$

where $\widetilde{N}_t$ is a centralized Poisson process with constant density $\pi(\Gamma) < \infty, \Gamma \in \Re(Z)$, such that $EN_t = E\widetilde{N}_t + \pi(\Gamma)t = \pi(\Gamma)t$. (In this case we also say that $\widetilde{N}_t$ has a compensator $\pi(\Gamma)t$). We have

1.3 Theorem. Assume that
1° $|b(t,x,\phi)|^2 + \|\sigma(t,x,\phi)\|^2$
$+ |c(t,x,\phi)|^2\pi(\Gamma) \leq k_0^2(1 + \|x\|_t^2 + \|\phi\|_t^2)$,
where $k_0 \geq 0$ is a constant, and c^{-1} exist;

2° σ^{-1}exists such that

$$\|\sigma^{-1}\| \leq k_0,$$

3° $\|\sigma(t,x,\phi) - \sigma(t,y,\phi)\|^2 + |c(t,x,\phi) - c(t,y,\psi)|^2 \pi(\Gamma)$

$$\leq k_0(|x_t - y_t|^2 + |\phi_t - \psi_t|^2), \text{ for all } x,y \in D,\ t \geq 0,\ \phi,\psi \in C,$$

Now assume that (x_t, ϕ_t) satisfies $(1.8)_0$ with $\widetilde{b} = 0$.

If ξ_t is a cadlag $R^{d_1}-$valued square integrable $\Im_t^{x,\phi}-$martingale, where $\Im_t^{x,\phi} = \sigma(x_s, \phi_s, s \leq t)$, then there exist

$$f(t,\omega) : [0,\infty) \times \Omega \to R^{d_1 \otimes d},\ g(t,\omega) : [0,\infty) \times \Omega \to R^{d_1},$$

which is $\Im_t^{x,\phi}-$adapted, $\Im_t^{x,\phi}-$predictable, respectively, satisfying for any $0 \leq T < \infty$

$$E[\int_0^T \|f(t,\omega)\|^2 dt + \int_0^T |g(t,\omega)|^2 \pi(\Gamma)dt] < \infty,$$

such that for all $t \geq 0$

$$\xi_t = \xi_0 + \int_0^t f(s,\omega)dw_s + \int_0^t g(s,\omega)d\widetilde{N}_s.$$

Moreover, the representation is unique.

Proof. Let

$$z_{0,t}^w = z_{0,t}^w(-\sigma^{-1}b) = \exp(-\int_0^t \sigma^{-1}b(s,x,\phi)\cdot dw_s - \frac{1}{2}\int_0^t \left\|\sigma^{-1}b(s,x,\phi)\right\|^2 ds),$$

$$d\widetilde{P}_t = z_{0,t}^w dP,$$

where (x_t, ϕ_t) is assumed to be the solution of $(1.8)_0$ with $\widetilde{b} = 0$ given in the assumption. Then by Theorem 1.2 above there exists a probability measure $\widetilde{P}$ on $(\Omega, \Im)$ such that

$$\widetilde{P}|_{\Im_t} = \widetilde{P}_t,$$

and

$$\widetilde{w}_t = w_t + \int_0^t \sigma^{-1}b(s,x,\phi)ds,\ \ 0 \leq t$$

is a BM under the probability measure $\widetilde{P}$, moreover, $\widetilde{N}_t$ is still a centralized Poisson process under $\widetilde{P}$ with the same compensator $\pi(\Gamma)t$. (In fact, instead of $(1.8)_0$ we only need to consider the following RSDE

$$\begin{cases} x_t = x_0 + \int_0^t \sigma(\sigma^{-1}b)(s,x(.),\phi(.))]ds \\ +\int_0^t \sigma(s,x(.),\phi(.))dw_s + \int_0^t c(s-,x(.),\phi(.))d\widetilde{N}_s + \phi_t, \\ x_0 \in \Theta^{cl}, \\ \text{and } (x_t,\phi_t),\ t \geq 0 \text{ satisfy the statements in (1.2) of} \\ \text{Chapter 3, moreover, } \phi_t \text{ is continuous,} \end{cases}$$

and to apply Theorem 1.2 for this RSDE, then the above facts are derived). Hence (x_t, ϕ_t) also satisfies

$$(1.8)_1 \begin{cases} x_t = x_0 + \int_0^t \sigma(s,x(.),\phi(.))d\widetilde{w}_s \\ +\int_0^t c(s-,x(.),\phi(.))d\widetilde{N}_s + \phi_t, \\ x_0 \in \Theta^{cl}, \\ \text{and } (x_t,\phi_t),\ t \geq 0 \text{ satisfy the statements in (1.2) of} \\ \text{Chapter 3, moreover, } \phi_t \text{ is continuous.} \end{cases}$$

Since by condition 1° and 2° the solution of $(1.8)_1$ is pathwise unique, so it is a strong solution by Theorem 1.3 in Chapter 3, which is still true for this case. Hence,

$\Im_t^{x,\phi} \subset \Im_t^{\widetilde{w},\widetilde{N}}$, for all $t \geq 0$.
On the other hand, as in the proof of 2) in Theorem 1.1

$N_t = \sum_{0<s\leq t} c_{s-}(x,\phi)^{-1} \triangle x_s$

is $\Im_t^{x,\phi}-$adapted, so is $\widetilde{N}_t = N_t - \pi(\Gamma)t$. However, $\widetilde{w}_t$ is obviously $\Im_t^{x,\phi,\widetilde{N}}-$adapted. Hence,

$\Im_t^{x,\phi} = \Im_t^{\widetilde{w},\widetilde{N}}$.

Note that $\xi_t \cdot (z_{0,t}^w)^{-1}$ is a $\Im_t^{\widetilde{w},\widetilde{N}}-$ adapted martingale under probability $\widetilde{P}$. Indeed, for all $A \in \Im_s^{\widetilde{w},\widetilde{N}}, s \leq t$

$\int_A \xi_t \cdot (z_{0,t}^w)^{-1} d\widetilde{P} = \int I_A \xi_t \cdot E(z_{t,T}^w \left| \Im_t^{\widetilde{w},\widetilde{N}} \right.) \, dP = \int_A \xi_s \cdot dP = \int_A \xi_s \cdot (z_{0,s}^w)^{-1} d\widetilde{P},$

and

$\sup_{t\leq T} E_{\widetilde{P}} \left| \xi_t \cdot (z_{0,t}^w)^{-1} \right| = \sup_{t\leq T} E \left| \xi_t \right| = \sup_{t\leq T} E \left| E(\xi_T \left| \Im_t^{\widetilde{w},\widetilde{N}} \right.) \right|$

$\leq E \left| \xi_T \right| < \infty,$

since ξ_t is a $(\Im_t^{\widetilde{w},\widetilde{N}}, P)-$martingale, and we have applied the Jensen inequality, and

$E(z_{t,T}^w \left| \Im_t^{\widetilde{w},\widetilde{N}} \right.) = E[E(z_{t,T}^w | \Im_t) \left| \Im_t^{\widetilde{w},\widetilde{N}} \right.] = 1$, for all $0 \leq t \leq T$.

Moreover, let

$\sigma_N = \inf \left\{ t \in [0,T] : \left| (z_{0,t}^w)^{-1} \right| + \left| \xi_t \right| > N \right\}.$

Then $\sigma_N \uparrow \infty$, as $N \uparrow \infty$, since $(z_{0,t}^w)^{-1}$ is a continuous process, and ξ_t is a cadlag process. We also have

$\sup_{t\leq T} E_{\widetilde{P}} \left| \xi_{t\wedge\sigma_N} \cdot (z_{0,t\wedge\sigma_N}^w)^{-1} \right|^2 \leq N \sup_{t\leq T} E_{\widetilde{P}} [\left| \xi_{t\wedge\sigma_N} \right|^2 \cdot (z_{0,t\wedge\sigma_N}^w)^{-1}]$

$= N \sup_{t\leq T} E \left| \xi_{t\wedge\sigma_N} \right|^2 \leq N \sup_{t\leq T} E \left| \xi_t \right|^2 < \infty,$

since ξ_t is a square integrable $(\Im_t^{\widetilde{w},q}, P)-$martingale.

Therefore $\xi_t \cdot (z_{0,t}^w)^{-1}$ is a local square integrable $(\Im_t^{\widetilde{w},\widetilde{N}}, \widetilde{P})-$martingale. Now by Theorem 1.4 below for each given $0 \leq T < \infty$ there exists a

$f_T(t,\omega) : [0,T] \times \Omega \to R^{d_1 \otimes d},$

which is $\Im_t^{\widetilde{w},\widetilde{N}}-$adapted, satisfying

$\widetilde{P}(\int_0^T \left\| f_T(t,\omega) \right\|^2 dt < \infty) = 1$, and there exists a

$g_T(t,\omega) : [0,T] \times \Omega \to R^{d_1},$

which is $\Im_t^{\widetilde{w},\widetilde{N}}-$predictable, satisfying

$\widetilde{P}(\int_0^T \left| g_T(t,\omega) \right|^2 \pi(\Gamma) dt < \infty) = 1,$

such that $\widetilde{P} - a.s.$

$\overline{\xi}_t = \xi_t \cdot (z_{0,t}^w)^{-1} = \overline{\xi}_0 + \int_0^t f_T(s,\omega) d\widetilde{w}_s + \int_0^t g_T(s,\omega) d\widetilde{N}_s$, for all $t \in [0,T]$.

Since $\widetilde{P} \sim P$. Hence $P - a.s.$

$\overline{\xi}_t = \overline{\xi}_0 + \int_0^t f_T(s,\omega) \cdot (\sigma^{-1} b(s,x,\phi)) ds + \int_0^t f_T(s,\omega) dw_s$

$+ \int_0^t g_T(s,\omega) d\widetilde{N}_s$, for all $t \in [0,T]$.

Note that $\xi_t = \overline{\xi}_t \cdot (z_{0,t}^w)$. Applying Ito's formula, one verifies that $P - a.s$ for all $t \in [0,T]$

$(1.9)_1$ $\xi_t = \xi_0 + \int_0^t z_{0,s}^w f_T(s,\omega) dw_s$

$+\int_0^t z_{0,s}^w f_T(s,\omega)(\sigma^{-1}b(s,x,\phi))ds - \int_0^t z_{0,s}^w f_T(s,\omega)(\sigma^{-1}b(s,x,\phi))ds$
$+\int_0^t z_{0,s}^w g_T(s,\omega)d\widetilde{N}_s - \int_0^t \xi_s(\sigma^{-1}b(s,x,\phi)\cdot dw_s)$
$= \xi_0 + \int_0^t \overline{f}_T(s,\omega)\cdot dw_s + \int_0^t \overline{g}_T(s,\omega)d\widetilde{N}_s,$

where (denote A^* = the transpose of A)

$\overline{f}_T(s,\omega) = z_{0,s}^w f_T(s,\omega) - \xi_s\cdot(\sigma^{-1}b(s,x,\phi))^*,$
$\overline{g}_T(s,\omega) = z_{0,s}^w g_T(s,\omega).$

Moreover, it is evident that $\overline{f}_T(t,\omega)$ is $\Im_t^{x,\phi}$−adapted,

$P(\int_0^T \left\|\overline{f}_T(t,\omega)\right\|^2 dt < \infty) = 1,$

and $\overline{g}_T(t,\omega)$ is $\Im_t^{x,\phi}$−predictable,

$P(\int_0^T \left|\overline{g}_T(t,\omega)\right|^2 \pi(\Gamma)dt < \infty) = 1.$

However, by assumption, ξ_t is R^{d_1}−valued and square integrable. Hence applying the representation $(1.9)_1$, it yields that

$E\int_0^{t\wedge\tau_N}[\left\|\overline{f}_T(t,\omega)\right\|^2 + \left|\overline{g}_T(t,\omega)\right|^2\pi(\Gamma)]dt$
$= E\left|\xi_{t\wedge\tau_N} - \xi_0\right|^2 \leq 2[E\left|\xi_T\right|^2 + E\left|\xi_0\right|^2] < \infty,$

where

$\tau_N = \inf\left\{t\in[0,T] : \int_0^t \left\|\overline{f}_T(s,\omega)\right\|^2 + \left|\overline{g}_T(s,\omega)\right|^2\pi(\Gamma)]ds > N\right\},$
$\tau_N = T$, for $\inf\{\Phi\}$.

Now as $N\to\infty$, $\tau_N\to T$. Hence by Fatou's lemma

$(1.9)_2\ E\int_0^T[\left\|\overline{f}_T(t,\omega)\right\|^2 + \left|\overline{g}_T(t,\omega)\right|^2\pi(\Gamma)]dt$
$\leq 2[E\left|\xi_T\right|^2 + E\left|\xi_0\right|^2] < \infty.$

Recall that by the definition of stochastic integral if $f^i(t,\omega), g(t,z,\omega), i=1,2,$ are such that

$(1.10)_1\ E\int_0^T[\left\|f^1(t,\omega) - f^2(t,\omega)\right\|^2 + \left|g^1(t,\omega) - g^2(t,\omega)\right|^2\pi(\Gamma)]dt = 0,$

then $P-a.s.$ for all $t\in[0,T]$

$$(1.10)_2\left\{\begin{array}{l}\int_0^t f^1(s,\omega)dw_s = \int_0^t f^1(s,\omega)dw_s,\\ \int_0^t g^1(s,\omega)d\widetilde{N}_s = \int_0^t g^2(s,\omega)d\widetilde{N}_s,\end{array}\right.$$

and f^1, g^1 are not distinguished from f^2, g^2 in $(1.10)_2$, and denote

$f^1 \doteq f^2,\ g^1 \doteq g^2.$

Now if there exist $f_T^i, g_T^i, i=1,2,$ for each i both satisfying $(1.9)_1$ and $(1.9)_2$, then $(1.10)_1$ holds. Hence,

$f_T^1 \doteq f_T^2,\ g_T^1 \doteq g_T^2.$

Now define

$$f(t,\omega) = \left\{\begin{array}{c}\overline{f}_1(t,\omega),\ \text{as } t\in[0,1],\\ \overline{f}_n(t,\omega),\ \text{as } t\in(n,n+1], n=1,2,....,\end{array}\right.$$

$$g(t,\omega) = \left\{\begin{array}{c}\overline{g}_1(t,\omega),\ \text{as } t\in[0,1],\\ \overline{g}_n(t,\omega),\ \text{as } t\in(n,n+1], n=1,2,....\end{array}\right.$$

Apparently, $f(t,\omega)$ is $\Im_t^{x,\phi}$−adapted, and $g(t,\omega)$ is $\Im_t^{x,\phi}$−predictable such that $P-a.s$ for all $t\geq 0$

$\xi_t = \xi_0 + \int_0^t f(s,\omega)\cdot dw_s + \int_0^t g(s,\omega)d\widetilde{N}_s.$

Moreover, for any $0\leq T<\infty$

$E[\int_0^T \|f(t,\omega)\|^2 dt + \int_0^T |g(t,\omega)|^2 \pi(\Gamma)dt] < \infty.$ ■

1.4 Theorem. If $\xi_t, t \in [0,T]$, is a R^{d_1}-valued cadlag $\Im_t^{w,\widetilde{N}}$- adapted local square integrable martingale under probability measure P, where $w_t, t \geq 0$, is a $P-$BM, and $\widetilde{N}_t$ is a $P-$ centralized Poisson process with compensator $\pi(\Gamma)t, \pi(\Gamma) < \infty$, then there exists a

$f_T(t,\omega) : [0,T] \times \Omega \to R^{d_1 \otimes d}$,

which is $\Im_t^{w,\widetilde{N}}$-adapted, satisfying

$P(\int_0^T \|f_T(t,\omega)\|^2 dt < \infty) = 1$, and there also exists a

$g_T(t,\omega) : [0,T] \times \Omega \to R^{d_1}$,

which is $\Im_t^{w,\widetilde{N}}$-predictable, satisfying

$P(\int_0^T |g_T(t,\omega)|^2 \pi(\Gamma)dt < \infty) = 1$,

such that $P - a.s.$

(1.11) $\xi_t = \xi_0 + \int_0^t f_T(s,\omega)dw_s + \int_0^t g_T(s,\omega)d\widetilde{N}_s$, for all $t \in [0,T]$.

Moreover, the representation is unique.

Proof. Assume that $\{\sigma_N\}_{N=1}^\infty$ is a sequence of stopping times such that $.\tau_N \uparrow \infty$, as N does, which make $\xi_{t\wedge\sigma_N}, t \geq 0$, a square integrable martingale, for each $N = 1, 2, \cdots$.Then by Theorem 1.7 below there exists a

$f_T^N(t,\omega) : [0,T] \times \Omega \to R^{d_1 \otimes d}$,

which is $\Im_t^{w,\widetilde{N}}$-adapted, satisfying

$E\int_0^{T\wedge\tau_N} \left\|f_T^N(t,\omega)\right\|^2 dt < \infty$, and there exists a

$g_T^N(t,\omega) : [0,T] \times \Omega \to R^{d_1}$,

which is $\Im_t^{w,\widetilde{N}}$-predictable, satisfying

$E\int_0^{T\wedge\tau_N} \left|g_T^N(t,\omega)\right|^2 \pi(\Gamma)dt < \infty$,

such that $P - a.s.$

$\xi_{t\wedge\tau_N} = \xi_0 + \int_0^t f_T^N(s,\omega)dw_s + \int_0^t g_T^N(s,\omega)d\widetilde{N}_s$
$= \xi_0 + \int_0^{t\wedge\tau_N} f_T^N(s,\omega)dw_s + \int_0^{t\wedge\tau_N} g_T^N(s,\omega)d\widetilde{N}_s$
$= \xi_0 + \int_0^t f_T^N(s,\omega)I_{s\leq\tau_N}dw_s + \int_0^t g_T^N(s,\omega)I_{s\leq\tau_N}d\widetilde{N}_s$, for all $t \in [0,T]$.

Moreover, the representation is unique for each $N = 1, 2, ...$ Note that

$\xi_{t\wedge\tau_{N_2}} = \xi_{t\wedge\tau_{N_2}\wedge\tau_{N_1}} = \xi_0 + \int_0^{t\wedge\tau_{N_2}} f_T^{N_1}(s,\omega)I_{s\leq\tau_{N_1}}dw_s$
$+ \int_0^{t\wedge\tau_{N_2}} g_T^{N_1}(s,\omega)I_{s\leq\tau_{N_1}}d\widetilde{N}_s$, for $N_1 \geq N_2$.

Therefore by uniqueness

$f_T^{N_1}(s,\omega)I_{s\leq\tau_{N_2}} = f_T^{N_2}(s,\omega)I_{s\leq\tau_{N_2}}, \; g_T^{N_1}(s,\omega)I_{s\leq\tau_{N_2}} = g_T^{N_2}(s,\omega)I_{s\leq\tau_{N_2}}.$

Now put

$$f_T(t,\omega) = \begin{cases} f_T^1(t,\omega), & \text{as } t \in [0,\tau_1], \\ f_T^2(t,\omega), & \text{as } t \in (\tau_1,\tau_2], \\ \cdots\cdots, & \end{cases}$$

and

$$g_T(t,\omega) = \begin{cases} g_T^1(t,\omega), & \text{as } t \in [0,\tau_1], \\ g_T^2(t,\omega), & \text{as } t \in (\tau_1,\tau_2], \\ \cdots\cdots. & \end{cases}$$

Then, obviously, $f_T(t,\omega)$ is $\Im_t^{w,\widetilde{N}}-$ adapted and $g_T(t,\omega)$ is $\Im_t^{w,\widetilde{N}}-$ predictable. ($I_{[0,\tau_1]}$ and $I_{(\tau_n,\tau_{n+1}]}$ are $\Im_t^{w,\widetilde{N}}-$ predictable, see Theorem 2 in §2, Chapter 1 of Liptser and Shiryayev, 1989). Furthermore, for each $N = 1, 2, ...,$ except a $\omega-$null set

$$\left\{\omega : \int_0^T \|f_T(t,\omega)\|^2 dt = \infty\right\} \subset \left\{\omega : \int_0^T \left\|f_T(t,\omega) - f_T^N(t,\omega)\right\|^2 dt > 0\right\}$$
$$\subset \left\{\omega : \sup_{t\leq T} |\xi_t| \geq N\right\} \cup \{\omega : \sigma_N \leq T\}.$$

However, $\xi_t, t \in [0,T]$, is right continuous with left limit. Hence, as $N \to \infty$

$$P\left\{\omega : \sup_{t\leq T} |\xi_t| \geq N\right\} \to 0.$$

In fact, if not, write $\Lambda = \cap_{N=1}^{\infty} \left\{\omega : \sup_{t\leq T} |\xi_t| \geq N\right\}$. Then $P(\Lambda) > 0$. Hence, as $\omega \in \Lambda$,

$$\sup_{t\leq T} |\xi_t| = \infty.$$

This is in contradiction with the cadlagness of $|\xi_t|$. Now, since $\sigma_N \uparrow \infty$, as N increases, one has that as $N \to \infty$

$$P\{\omega : \sigma_N \leq T\} \to 0.$$

Therefore,

$$P\left\{\omega : \int_0^T \|f_T(t,\omega)\|^2 dt < \infty\right\} = 1.$$

Similarly, one finds that

$$P\left\{\omega : \int_0^T |g_T(t,\omega)|^2 \pi(\Gamma) dt < \infty\right\} = 1. \blacksquare$$

From the proof of Theorem 1.4 one sees that the following corollary is true.

1.5 Corollary. In Theorem 1.4 if $\xi_t, t \in [0,T]$, is a $R^{d_1}-$valued cadlag $\Im_t^{w,\widetilde{N}}-$ adapted martingale such that

(1.12) $|\triangle\xi_t| = |\xi_t - \xi_{t-}| \leq k_0, \forall t \in [0,T]$,

i.e. ξ_t only has bounded jumps; then the conclusion of Theorem 1.4 still holds.

Proof. Let

$$\tau_N = \inf\{t \in [0,T] : |\xi_t| > N\}; \tau_N = T, \text{ for } \inf\{\Phi\}; N = 1, 2, ...$$

Then $\xi_{t\wedge\tau_N}, t \in [0,T]$, is a square integrable martingale. Hence, the proof of Theorem 1.4 still goes through. $\blacksquare$

The proof of Theorem 1.4 can also be applied to generalize Theorem 1.3 by using Theorem 1.3 itself as follows, we do not need to repeat it:

1.6 Theorem. If $\xi_t, t \geq 0$, in Theorem 1.3 is only a cadlag $R^{d_1}-$valued local square integrable $\Im_t^{x,\phi}-$martingale, then there exist

$$f(t,\omega) : [0,\infty) \times \Omega \to R^{d_1 \otimes d}, \; g(t,\omega) : [0,\infty) \times \Omega \to R^{d_1},$$

which is $\Im_t^{x,\phi}-$adapted, $\Im_t^{x,\phi}-$predictable, respectively, satisfying for any $0 \leq T < \infty$

$$P\left(\int_0^T \|f(t,\omega)\|^2 dt + \int_0^T |g(t,\omega)|^2 \pi(\Gamma) dt < \infty\right) = 1,$$

such that for all $t \geq 0$

$$\xi_t = \xi_0 + \int_0^t f(s,\omega) dw_s + \int_0^t g(s,\omega) d\widetilde{N}_s.$$

Moreover, the representation is unique.

Now we give the basic martingale representation theorem as follows:

1.7 Theorem. If $\xi_t, t \in [0,T]$, in Theorem 1.4 is a $R^{d_1}-$valued cadlag $\Im_t^{w,\widetilde{N}}-$adapted square integrable martingale, then the conclusion of Theorem 1.4 is true with the stronger property that

$E\int_0^T \|f_T(t,\omega)\|^2 dt < \infty,$

$E\int_0^T |g_T(t,\omega)|^2 \pi(\Gamma)dt < \infty.$

Proof. Consider the $d+1-$dimensional local martingale $M_t = (w_t, \widetilde{N}_t)$. Suppose that Q is a probability measure such that $Q \sim P$, and $M_t = (w_t, \widetilde{N}_t)$ is still a $d+1-$dimensional local martingale under Q. Then by theorem 12.3 in Yan (1981)

$$[M,M]_t(Q) = [M,M]_t(P) = \begin{pmatrix} tI_{d\times d} & 0 \\ 0 & \pi(\Gamma)t \end{pmatrix},$$

where $I_{d\times d}$ is the unit $d\times d$ matrix. Hence w_t and $\widetilde{N}_t$ is still a BM and a centralized Poisson process with compensator $\pi(\Gamma)t$, respectively. By the uniqueness of finite-dimensional probability distributions of BM and Poisson process

$$P\Big|_{\vee_{t\geq 0}\Im_t^{w,\widetilde{N}}} = Q\Big|_{\vee_{t\geq 0}\Im_t^{w,\widetilde{N}}}.$$

Hence applying theorem 15.6 and 15.4 of Yan (1981), one arrives at the conclusion. ■

A slightly different martingale representation theorem for systems with jumps is presented in Tang, S and Li, X. (1994). However, our proof here is much simpler than theirs, and can also be applied to show their result.

To weaken condition 1° in Theorem 1.3, i.e. we do not assume that b is less that linear growth in x, we have to strengthen the condition that all coefficients do not depend on the reflection $\phi(.)$, and we also need to consider the RSDE without jumps, since in our proof random coefficients are involved. Actually, we have

1.8 Theorem. Assume that all coefficients in (1.3) do not depend on $\phi(.)$, and all conditions in Theorem 1.3 hold except 1°. Instead, we assume

1°′ for arbitrary $0 \leq T < \infty$

$P(\int_0^T |\sigma^{-1}b(s,x(\cdot))|^2 ds < \infty) = 1,$

and

$\|\sigma(t,x)\|^2 \leq k_0^2(1 + \|x\|_t^2).$

Moreover, we assume that

$c(t,x,z) = 0.$

Then the conclusion of Theorem 1.3 still holds. (In this case $g \equiv 0$).

Proof. Given $T < \infty$. Set

$$\tau_n = \inf\left\{t \in [0,T] : \int_0^T |\sigma^{-1}b(s,x(\cdot))|^2 ds \geq n\right\};\ \tau_n = T, \text{ for } \inf\{\phi\}.$$

By Theorem 2.9 in Chapter 3 (actually, by the proof of that theorem) there exists a pathwise unique continuous strong solution (x_t^n, ϕ_t^n) satisfying the following continuous RSDE: $P-a.s$ for all $t \in [0,T]$

$$(1.13)_1 \begin{cases} x_t^n = x_{t\wedge\tau_n} + \int_0^t (1 - I^n(s))\sigma(s,x^n(.))dw_s + \phi_t^n \in \Theta^{cl}, \\ \text{and } (x_t^n, \phi_t^n)\text{satisfy the statements in (1.2) of Chapter 3;} \end{cases}$$

where

$I^n(t) = I_{(\int_0^t |\sigma^{-1}b(s,x(\cdot))|^2 ds < n)}.$
Note that $(1.13)_1$ can be written as

$$(1.13)_2 \begin{cases} x_t^n = x_{t\wedge\tau_n} + \int_{t\wedge\tau_n}^t \sigma(s,x^n(.))dw_s + \phi_t^n \in \Theta^{cl}, \ t \in [0,T], \\ \ldots\ldots \end{cases}$$

Since as $t \le \tau_n$, $(x_t, 0)$ also satisfies $(1.13)_1$. By uniqueness
$(1.14)_1$ $x_t^n = x_t, \phi_t^n = 0$, as $t < \tau_n$.
Therefore, by $(1.13)_1$ and $(1.13)_2$

$$x_t^n = x_0 + \int_0^{t\wedge\tau_n} b(s,x(.))ds + \int_0^{t\wedge\tau_n} \sigma(s,x(.))dw_s + \phi_{t\wedge\tau_n}$$
$$+ \int_{t\wedge\tau_n}^t \sigma(s,x^n(.))dw_s + \phi_t^n$$
$$= x_0 + \int_0^t b^n(s,x^n(.))ds + \int_0^t \sigma(s,x^n(.))dw_s + \overline{\phi}_t^n,$$

where

$$b^n(t,x(.)) = b(t,x(.))I_{(\int_0^t |\sigma^{-1}b(s,x(\cdot)|^2 ds < n)},$$

$$(1.14)_2 \ \overline{\phi}_t^n = \begin{cases} \phi_t, \text{ as } t \le \tau_n, \\ \phi_{\tau_n} + \phi_t^n, \text{ as } t > \tau_n; \end{cases}$$

and we have used the fact that if $t > \tau_n$, then

$$(1.15) \int_{\tau_N}^t b^n(s,x^n(.))ds = \int_{\tau_N}^t b(s,x^n(.))I_{(\int_0^s |\sigma^{-1}b(r,x^n(\cdot))|^2 dr < n)}ds = 0.$$

Note that

$$\int_0^T \left|\sigma^{-1}b^n(s,x^n(\cdot))\right|^2 ds \le n.$$

Hence,

$$E\exp((1/2)\int_0^T \left|\sigma^{-1}b^n(s,x^n(\cdot))\right|^2 ds) < \infty.$$

Now by Novikov's theorem (1973)

$$Ez_{0,T}^w(-\sigma^{-1}b^n) = 1,$$

where

$$z_{0,t}^w(-\sigma^{-1}b^n) = \exp(-\int_0^t \sigma^{-1}b^n(s,x^n)\cdot dw_s - \tfrac{1}{2}\int_0^t \left|\sigma^{-1}b^n(s,x^n)\right|^2 ds).$$

Hence, by Lemma 3.9 and Theorem 3.12 in Chapter 3

$$\widetilde{w}_t = w_t + \int_0^t \sigma^{-1}b^n(s,x^n(\cdot))ds$$

is a BM under probability measure $\widetilde{P}$, where

$$d\widetilde{P} = z_{0,T}^w(-\sigma^{-1}b^n)dP.$$

Denote

$$\tau_n(x^n(\cdot)) = \inf\left\{t \in [0,T] : \int_0^t \left|\sigma^{-1}b(s,x^n(\cdot))\right|^2 ds \ge n\right\};$$
$$\tau_n(x^n(\cdot)) = T, \text{ for } \inf\{\phi\}.$$

Then by $(1.14)_1$

$$\tau_n(x^n(\cdot)) = \tau_n, \ P - a.s.$$

From this it is not difficult to show that

$$\Im_{t\wedge\tau_n(x^n(\cdot))}^{x^n,\overline{\phi}^n} = \Im_{t\wedge\tau_n}^{x^n,\overline{\phi}^n} = \Im_{t\wedge\tau_n}^{x,\phi}.$$

Now, by virtue of the proof of Theorem 1.3 there exists a $\Im_t^{x^n,\overline{\phi}^n}$ –adapted process $f_T^n(t,\omega)$, which satisfies,

$$E\int_0^T |f_T^n(t,\omega)|^2 dt < \infty,$$

such that

$$E(\xi_t \left| \Im_t^{x^n,\overline{\phi}^n} \right.) = \xi_0 + \int_0^t f_T^n(s,\omega)dw_s, \text{ for all } t \in [0,T].$$

In particular,

(1.16) $\xi_{t\wedge\tau_n} = E(\xi_{t\wedge\tau_n}\left|\Im^{x^n,\overline{\phi}^n}_{t\wedge\tau_n}\right.) = \xi_0 + \int_0^{t\wedge\tau_n} f^n_T(s,\omega)dw_s.$
Note that by $1^{o\prime}$ as $n \uparrow \infty$,

$\tau_n \uparrow \infty;$

and by the above expression as $n \le m$

(1.17) $0 = E\left|\xi_{t\wedge\tau_n} - \xi_{t\wedge\tau_n\wedge\tau_m}\right|^2$

$= E\left|\int_0^{t\wedge\tau_n}(f^n_T(s,\omega) - f^m_T(s,\omega))dw_s\right|^2$

$= E\int_0^{t\wedge\tau_n}\left|(f^n_T(s,\omega) - f^m_T(s,\omega))\right|^2 ds.$

Now define

$$f_T(t,\omega) = \begin{cases} f^1_T(t,\omega), \text{ as } 0 \le t \le \tau_1; \\ f^n_T(t,\omega), \text{ as } \tau_n < t \le \tau_{n+1}; \text{ for } n = 1,2,... \end{cases}$$

It is not difficult to show that $f_T(t,\omega)$ is $\Im^{x,\phi}_t$–adapted. Moreover, by (1.16) and (1.17)

$\xi_{t\wedge\tau_n} = \xi_0 + \int_0^{t\wedge\tau_n} f_T(s,\omega)dw_s$, as $t \in [0,T]$;

$E\int_0^{t\wedge\tau_n}\|f_T(s,\omega)\|^2 ds \le E\left|\xi_T - \xi_0\right|^2 < \infty$, as $t \in [0,T]$.

By Fatou's lemma the conclusion is true for $t \in [0,T]$. Now the proof can be completed as in Theorem 1.3. ■

5.2 Non-linear Filtering Equation

In this section we will derive the non-linear filtering equation for the partially observed RSDE system (1.1). More precisely, we will show the following

2.1 Theorem. Assume that $A_t(\omega)$ satisfies conditions in Theorem 1.1, $B_t(\xi,\phi)$ and $c_t(\xi,\phi,z)$ satisfy the conditions stated in Theorem 1.3 for $\sigma(t,x,\phi)$ and $c(t,x,z,\phi)$, furthermore, for simplicity let us assume that

$(2.1)_0$ $E\left|h_0\right|^2 < \infty$, $E\int_0^T \left|H_t\right|^2 dt < \infty$,

$c_s(\xi,\phi,z) = c_s(\xi,\phi)I_\Gamma(z)$, where $\Gamma \in \Re(Z)$ such that $0 < \pi(\Gamma) < \infty$, and $c_s(\xi,\phi)^{-1}$ exists. Moreover, assume that as $0 \le t \le T$, $|f_t(\omega)| \le k_T$, where $f_t(\omega) = h_0, H_t, A_t$, and $0 \le k_T$ is a constant depending on T only.

Write

$\Pi_t(h) = E(h_t\left|\Im^{\xi,\phi}_t\right.)$, etc.

Then the following non-linear filtering equation holds:

(2.1) $\Pi_t(\widetilde{h}) = \Pi_0(\widetilde{h}) + \int_0^t \Pi_s(H)ds + \int_0^t(\Pi_s(D)+$

$\cdot(\Pi_s(\widetilde{h}A^*) - \Pi_s(\widetilde{h})\Pi_s(A^*))B_s^{-1}(\xi,\phi)^* d\overline{w}_s + \int_0^t E(y_s - y_{s-}\left|\Im^{\xi,\phi}_{s-}\right.)) d\widetilde{N}_s,$

where A^* means the transpose of A, $\overline{w}_t$ is the innovation process defined in Theorem 1.1,

$\widetilde{h}_t = h_t - \eta_t,$

$$D_t dt = d\left\langle y^c, w\right\rangle_t = d\begin{bmatrix} \langle y^c_1, w_1\rangle_t & \cdots & \langle y^c_1, w_d\rangle_t \\ \cdots & \cdots & \cdots \\ \langle y^c_d, w_1\rangle_t & \cdots & \langle y^c_d, w_d\rangle_t \end{bmatrix},$$

$\langle y^c, w\rangle_t$ is the quadratic variational process corresponding to the martingale y^c and w, and such D_t exists, since y_t is a square integrable martingale.(See Theorem 5.3 in Chapter 5, Liptser, R.S. and A.N. Shiryayev, 1977)..

Proof. First of all, by (1.1) it is easily seen that

$$\Pi_t(\widetilde{h}) = E(h_0 \Big| \Im_t^{\xi,\phi}) + E(\int_0^t H_s ds \Big| \Im_t^{\xi,\phi}) + E(y_t \Big| \Im_t^{\xi,\phi}),$$

where $\widetilde{h}_t = h_t - \eta_t$, and $\widetilde{h}_0 = h_0$.

In the following we will use the martingale representation theorem Theorem 1.4 to get the expression of each term on the above right side. Indeed, we have the following lemmas.

2.2 Lemma. $E(h_0 \Big| \Im_t^{\xi,\phi}), t \in [0,T]$, is a square integrable martingale having the following representation

$$E(h_0 \Big| \Im_t^{\xi,\phi}) = \Pi_0(\widetilde{h}) + \int_0^t f_s^{\widetilde{h}}(\xi,\phi) d\overline{w}_s + \int_0^t g_s^{\widetilde{h}}(\xi,\phi) d\widetilde{N}_s,$$

with

$$E\int_0^T (\left\| f_s^{\widetilde{h}}(\xi,\phi)\right\|^2 + \left| g_s^{\widetilde{h}}(\xi,\phi)\right|^2 \pi(\Gamma)) ds < \infty,$$

where $f_t^{\widetilde{h}}(\xi,\phi)$ is $\Im_t^{\xi,\phi}-$adapted, and $g_t^{\widetilde{h}}(\xi,\phi)$ is $\Im_t^{\xi,\phi}-$predictable.

Proof. By Theorem 1.1 (ξ_t,ϕ_t) satisfies $(1.1)_3$. Now the conclusion can be derived immediately by Theorem 1.3. ■

2.3 Lemma. $E(y_t \Big| \Im_t^{\xi,\phi}), t \in [0,T]$, is a square integrable martingale having the following representation

$$E(y_t \Big| \Im_t^{\xi,\phi}) = \int_0^t f_s^y(\xi,\phi) d\overline{w}_s + \int_0^t g_s^y(\xi,\phi) d\widetilde{N}_s,$$

with

$$E\int_0^T (\left\| f_s^y(\xi,\phi)\right\|^2 + \int_Z \left| g_s^y(\xi,\phi)\right|^2 \pi(\Gamma)) ds < \infty,$$

where $f_t^y(\xi,\phi)$ is $\Im_t^{\xi,\phi}-$adapted, and $g_t^y(\xi,\phi)$ is $\Im_t^{\xi,\phi}-$predictable.

Proof. Write $z_t = E(y_t \Big| \Im_t^{\xi,\phi})$. Then as $s \leq t$

$$E(z_t \left| \Im_s^{\xi,\phi}\right.) = E[E(y_t \Big| \Im_t^{\xi,\phi}) \left| \Im_s^{\xi,\phi}\right.)] = E(y_t \left| \Im_s^{\xi,\phi}\right.)$$

$$= E(E(y_t \left| \Im_s\right.) \left| \Im_s^{\xi,\phi}\right.) = E(y_s \left| \Im_s^{\xi,\phi}\right.) = z_s.$$

Hence z_t is a $\Im_t^{\xi,\phi}-$martingale, and by $(1.1)_3$ Theorem 1.3 applies. ■

2.4 Lemma. Let $\alpha = (\alpha_t, \Im_t), t \in [0,T]$, be a random process with $\int_0^T E\left|\alpha_t\right| dt < \infty$, and let $\widehat{\Im} \subset \Im$ be some sub-$\sigma-$algebra. Then

$$E(\int_0^t \alpha_s ds \Big| \widehat{\Im}) = \int_0^t E(\alpha_s \Big| \widehat{\Im})\, ds, \ P-a.s. \text{ for all } t \in [0,T].$$

Proof. The proof of Lemma 2.4 is easily derived from the definition of the conditional expectation and Fubini's theorem. Indeed, $\forall A \in \widehat{\Im}$

$$E[I_A E(\int_0^t \alpha_s ds \Big| \widehat{\Im})] = E(I_A \int_0^t \alpha_s ds) = \int_0^t E(I_A \alpha_s) ds$$

$$= \int_0^t E(I_A E(\alpha_s \Big| \widehat{\Im})) ds = E[I_A \int_0^t E(\alpha_s \Big| \widehat{\Im})) ds]. \ ■$$

2.5 Lemma. The process

$E(\int_0^t H_s ds \Big| \Im_t^{\xi,\phi}) - \int_0^t \pi_s(H)ds$

is $\Im_t^{\xi,\phi}$−adapted, with the representation

$E(\int_0^t H_s ds \Big| \Im_t^{\xi,\phi}) - \int_0^t \pi_s(H)ds$

$= \int_0^t f_s^H(\xi,\phi)d\overline{w}_s + \int_0^t g_s^H(\xi,\phi)d\widetilde{N}_s,$

such that

$E\int_0^T (\left\| f_s^H(\xi,\phi)\right\|^2 + \left| g_s^H(\xi,\phi)\right|^2 \pi(\Gamma))ds < \infty,$

where $f_t^H(\xi,\phi)$ is $\Im_t^{\xi,\phi}$−adapted, and $g_t^H(\xi,\phi)$ is $\Im_t^{\xi,\phi}$−predictable.

Proof. As $u \leq t$, by Lemma 2.4

$E\left\{ E(\int_0^t H_s ds \Big| \Im_t^{\xi,\phi}) - \int_0^t \pi_s(H)ds \left| \Im_u^{\xi,\phi} \right.\right\}$

$= E(\int_0^u H_s ds \left| \Im_u^{\xi,\phi}\right.) - \int_0^u \pi_s(H)ds$

$+E(\int_u^t H_s ds \left| \Im_u^{\xi,\phi}\right.) - \int_u^t E(\pi_s(H) \left| \Im_u^{\xi,\phi}\right.)\, ds.$

$= E(\int_0^u H_s ds \left| \Im_u^{\xi,\phi}\right.) - \int_0^u \pi_s(H)ds.$

Hence $E(\int_0^t H_s ds \Big| \Im_t^{\xi,\phi}) - \int_0^t \pi_s(H)ds$ is a $\Im_t^{\xi,\phi}$−martingale, and Theorem 1.3 is applied. ■

Now we are in a position to show Theorem 1.1.

Proof. By Lemma 2.2-2.5 one sees that

(2.2) $\Pi_t(\widetilde{h}) = \Pi_0(\widetilde{h}) + \int_0^t \Pi_s(H)ds + \int_0^t f_s(\xi,\phi)d\overline{w}_s + \int_0^t g_s(\xi,\phi)d\widetilde{N}_s,$

where

$f_s(\xi,\phi) = f_s^{\widetilde{h}}(\xi,\phi) + f_s^y(\xi,\phi) + f_s^H(\xi,\phi),$

$g_s(\xi,\phi) = g_s^{\widetilde{h}}(\xi,\phi) + g_s^y(\xi,\phi) + g_s^H(\xi,\phi),$

with

$E\int_0^T (\left\| f_s(\xi,\phi)\right\|^2 + \left| g_s(\xi,\phi)\right|^2 \pi(\Gamma)ds < \infty,$

where $f_t(\xi,\phi) \in R^{d\otimes d}$ is $\Im_t^{\xi,\phi}$−adapted, and $g_t(\xi,\phi) \in R^d$ is $\Im_t^{\xi,\phi}$−predictable.

Let

$\overline{y}_t = \int_0^t f_s(\xi,\phi)d\overline{w}_s + \int_0^t g_s(\xi,\phi)d\widetilde{N}_s,$

and

$\overline{z}_t = \int_0^t \lambda_1(s,\xi,\phi)d\overline{w}_s + \int_0^t \lambda_2(s,\xi,\phi)d\widetilde{N}_s,$

where $\lambda_1(t,\xi,\phi) \in R^{d\otimes d}$ is a bounded $\Im_t^{\xi,\phi}$−adapted process with $|\lambda_1(t,\xi,\phi)| \leq k_0$, and $\lambda_2(t,\xi,\phi) \in R^d$ is a bounded $\Im_t^{\xi,\phi}$− predictable process with $|\lambda_2(t,\xi,\phi)| \leq k_0$. Then

(2.3) $E\overline{y}_t \cdot \overline{z}_t = E\int_0^t tr.(\lambda_1(s,\xi,\phi)f_s^*(\xi,\phi))ds$

$+E\int_0^t \lambda_2(s,\xi,\phi,z) \cdot g_s(\xi,\phi,z)\pi(\Gamma)ds.$

On the other hand, by (2.2)

$\overline{y}_t = \Pi_t(\widetilde{h}) - \Pi_0(\widetilde{h}) - \int_0^t \Pi_s(H)ds.$

Note that

$E\overline{z}_t \cdot \Pi_0(\widetilde{h}) = E[\Pi_0(\widetilde{h}) \cdot E(\overline{z}_t \Big| \Im_0^{\xi,\phi})] = 0,$

and

$E\overline{z}_t \cdot \int_0^t \Pi_s(H)ds = \int_0^t E\overline{z}_t \cdot \Pi_s(H)ds = \int_0^t E[\Pi_s(H) \cdot E(\overline{z}_t \left| \Im_s^{\xi,\phi}\right.)]ds$

$= \int_0^t E[\overline{z}_s \cdot \Pi_s(H)]ds.$

Hence,

$$E\overline{y}_t \cdot \overline{z}_t = E[\overline{z}_t \cdot \Pi_t(\widetilde{h})] - \int_0^t E[\overline{z}_s \cdot \Pi_s(H)]ds = E[\overline{z}_t \cdot E(\widetilde{h}_t \,|\Im_t^{\xi,\phi})]$$

$$- \int_0^t E[\overline{z}_s \cdot E(H_s \,|\Im_s^{\xi,\phi})]ds = E[\overline{z}_t \cdot \widetilde{h}_t - \int_0^t \overline{z}_s \cdot H_s ds].$$

Now by Theorem 1.1

$$\overline{w}_t = \int_0^t B_s^{-1}(\xi,\phi)(A_s - E(A_s \,|\Im_s^{\xi,\phi}))ds + w_t.$$

Hence,

$$\overline{z}_t = \widetilde{z}_t + \int_0^t \lambda_1(s,\xi,\phi)B_s^{-1}(\xi,\phi)(A_s - E(A_s \,|\Im_s^{\xi,\phi}))ds \, ,$$

where

$$\widetilde{z}_t = \int_0^t \lambda_1(s,\xi,\phi)dw_s + \int_0^t \lambda_2(s,\xi,\phi)d\widetilde{N}_s.$$

Now one finds that

$$(2.4)\quad E\overline{y}_t \cdot \overline{z}_t = E[\widetilde{z}_t \cdot \widetilde{h}_t - \int_0^t \widetilde{z}_s \cdot H_s ds]$$
$$+E\widetilde{h}_t \cdot \int_0^t \lambda_1(s,\xi,\phi)B_s^{-1}(\xi,\phi)(A_s - E(A_s \,|\Im_s^{\xi,\phi}))ds$$
$$-E\int_0^t H_u \cdot \int_0^u \lambda_1(s,\xi,\phi)B_s^{-1}(\xi,\phi)(A_s - E(A_s \,|\Im_s^{\xi,\phi}))ds\, du.$$

However, $\widetilde{z}_t$ is a $\Im_t$−adapted square integrable martingale with respect to probability measure P. Hence

$$E\widetilde{z}_t \cdot h_0 = E[h_0 \cdot E(\widetilde{z}_t \,|\Im_0)] = E[h_0 \cdot \widetilde{z}_0] = 0.,$$

and

$$E\int_0^t \widetilde{z}_s \cdot H_s ds = E\int_0^t [E(\widetilde{z}_t \,|\Im_s) \cdot H_s]ds = E(\widetilde{z}_t \cdot \int_0^t H_s ds).$$

Therefore,

$$E[\widetilde{z}_t \cdot \widetilde{h}_t - \int_0^t \widetilde{z}_s \cdot H_s ds] = E[\widetilde{z}_t \cdot (\widetilde{h}_t - h_0 - \int_0^t H_s ds)]$$
$$= E[\widetilde{z}_t \cdot y_t] = E\,\langle \widetilde{z}^c, y^c \rangle_t + E\sum_{0<s\le t} \triangle\widetilde{z}_s \cdot \triangle y_s.$$

Hence,

$$(2.5)_1\quad E[\widetilde{z}_t \cdot \widetilde{h}_t - \int_0^t \widetilde{z}_s \cdot H_s ds] = E\int_0^t tr.(\lambda_1(s,\xi,\phi)D(s)^*)ds$$
$$+E\sum_{0<s\le t} \triangle y_s \cdot \lambda_2(s,\xi,\phi) \triangle N_s,$$

where

$$D_t dt = d\,\langle y^c, w \rangle_t = d \begin{bmatrix} \langle y_1^c, w_1 \rangle_t & \cdots & \langle y_1^c, w_d \rangle_t \\ \cdots & \cdots & \cdots \\ \langle y_d^c, w_1 \rangle_t & \cdots & \langle y_d^c, w_d \rangle_t \end{bmatrix},$$

exists. Computing now the second item on the right side of (2.4), one obtains

$$E\widetilde{h}_t \cdot \int_0^t \lambda_1(s,\xi,\phi)B_s^{-1}(\xi,\phi)(A_s - \Pi_s(A))ds$$
$$= E\int_0^t \lambda_1(s,\xi,\phi)B_s^{-1}(\xi,\phi)(A_s - \Pi_s(A)) \cdot \widetilde{h}_s ds$$
$$+E\int_0^t \lambda_1(s,\xi,\phi) \cdot B_s^{-1}(\xi,\phi)(A_s - \Pi_s(A))) \cdot (\widetilde{h}_t - \widetilde{h}_s)ds$$
$$= E\int_0^t tr.(\lambda_1(s,\xi,\phi)(\Pi_s[B_s^{-1}(\xi,\phi)(A_s\widetilde{h}_s^*)] - [B_s^{-1}(\xi,\phi)(\Pi_s(A)\Pi_s[\widetilde{h}_s]^*))ds$$
$$+E\int_0^t \lambda_1(s,\xi,\phi) \cdot B_s^{-1}(\xi,\phi)(A_s - \Pi_s(A)) \cdot (\widetilde{h}_t - \widetilde{h}_s)ds.$$

Note that

$$\widetilde{h}_t - \widetilde{h}_s = \int_s^t H_u du + (y_t - y_s),$$

and $E(y_t - y_s \,|\Im_s) = 0$. Hence

$$E\int_0^t \lambda_1(s,\xi,\phi) \cdot B_s^{-1}(\xi,\phi)(A_s - \Pi_s(A)) \cdot (\widetilde{h}_t - \widetilde{h}_s)ds$$
$$= E\int_0^t \lambda_1(s,\xi,\phi) \cdot B_s^{-1}(\xi,\phi)(A_s - \Pi_s(A)) \cdot (y_t - y_s)ds$$
$$+E\int_0^t \lambda_1(s,\xi,\phi) \cdot B_s^{-1}(\xi,\phi)(A_s - \Pi_s(A)) \cdot \int_s^t H_u du ds$$
$$= E\int_0^t \int_0^u \lambda_1(s,\xi,\phi) \cdot B_s^{-1}(\xi,\phi)(A_s - \Pi_s(A))ds \cdot H_u du.$$

From this it follows that

$(2.5)_2$ $E\widetilde{h}_t \cdot \int_0^t \lambda_1(s,\xi,\phi)B_s^{-1}(\xi,\phi)(A_s - \Pi_s(A))ds$
$-E\int_0^t H_u \cdot \int_0^u \lambda_1(s,\xi,\phi)B_s^{-1}(\xi,\phi)(A_s - E(A_s \left| \Im_s^{\xi,\phi}\right.))ds\, du$
$= E\int_0^t tr.\lambda_1(s,\xi,\phi)[B_s^{-1}(\xi,\phi)(\Pi_s(A\widetilde{h}^*) - (\Pi_s(A)\Pi_s(\widetilde{h})^*))]ds.$

Hence, by (2.4), $(2.5)_1$ and $(2.5)_2$ we get

(2.6) $E\overline{y}_t \cdot \overline{z}_t = E\sum_{0<s\leq t} \triangle y_s \cdot \lambda_2(s,\xi,\phi) \triangle N_s$
$+E\int_0^t tr.\lambda_1(s,\xi,\phi)[D(s)^* + B_s^{-1}(\xi,\phi)(\Pi_s(A\widetilde{h}^*) - (\Pi_s(A)\Pi_s(\widetilde{h})^*))]ds$
$= E\sum_{0<s\leq t} \lambda_2(s,\xi,\phi) \cdot E[\triangle y_s \left| \Im_{s-}^{\xi,\phi}\right.] \triangle N_s$
$+E\int_0^t tr.\lambda_1(s,\xi,\phi)[D(s)^* + B_s^{-1}(\xi,\phi)(\Pi_s(A\widetilde{h}^*) - (\Pi_s(A)\Pi_s(\widetilde{h})^*))]ds$
$= E\int_0^t \lambda_2(s,\xi,\phi) \cdot E[\triangle y_s \left| \Im_{s-}^{\xi,\phi}\right.]\, \pi(\Gamma)ds$
$+E\int_0^t tr.\lambda_1(s,\xi,\phi)[D(s)^* + B_s^{-1}(\xi,\phi)(\Pi_s(A\widetilde{h}^*) - (\Pi_s(A)\Pi_s(\widetilde{h})^*))]ds$

Comparing (2.3) and (2.6), one finds that

$f_s(\xi,\phi)^* = D(s)^* + B_s^{-1}(\xi,\phi)(\Pi_s(A\widetilde{h}^*) - (\Pi_s(A)\Pi_s(\widetilde{h})^*)),$
$g_s(\xi,\phi)\pi(\Gamma) = E[\triangle y_s \left| \Im_{s-}^{\xi,\phi}\right.]\, \pi(\Gamma).$

The conclusion is derived. ■

2.6 Corollary. 1) If $\triangle y_t \equiv 0$, i.e. for y_t without jumps, then the non-linear filtering equation becomes

(2.7) $\Pi_t(\widetilde{h}) = \Pi_0(\widetilde{h}) + \int_0^t \Pi_s(H)ds + \int_0^t(\Pi_s(D)+$
$\cdot(\Pi_s(\widetilde{h}A^*) - \Pi_s(\widetilde{h})\Pi_s(A^*))B_s^{-1}(\xi,\phi)^* d\overline{w}_s.$

2) If $c_t(\xi,\phi,z) \equiv 0$, i.e. the observation process ξ_t has no jumps, one also finds that (2.7) is the filtering equation. In this case, the assumption $(2.1)_0$ in Theorem 2.1 can be cancelled by the proof. Moreover, for the martingales in Lemma 2.2-2.5 we can apply the representation Theorem 1.3. Then we see that all of them have continuous versions.

3) If $\eta_t \equiv 0$, i.e., the signal process h_t has no reflection, or say, the first SDE in (1.1) is a usual SDE without reflection such that $h_t \in R^d$, then the non-linear filtering equation becomes

(2.8) $\Pi_t(h) = \Pi_0(h) + \int_0^t \Pi_s(H)ds + \int_0^t(\Pi_s(D)+$
$\cdot(\Pi_s(hA^*) - \Pi_s(h)\Pi_s(A^*))B_s^{-1}(\xi,\phi)^* d\overline{w}_s + \int_0^t E(y_s - y_{s-} \left| \Im_{s-}^{\xi,\phi}\right.)\, d\widetilde{N}_s.$

In a special case, one can solve the non-linear filtering equation.

5.3 Zakai Equation

In this section, we will derive the Zakai equation for a more concrete partial observed system. Suppose we are given a signal process x_t satisfying the following RSDE with a reflection η_t :

$$
(3.1)_1 \begin{cases} x_t = x_0 + \int_0^t b(s, x(.), \eta(.))ds + \int_0^t \sigma(s, x(.), \eta(.))dw'_s \\ \quad + \int_0^t \int_Z c_{s-}(x(.), \eta(.), z)q'(ds, dz) + \eta_t, \\ \quad x_0 \in \Theta^{cl}, \\ \text{and } (x_t, \eta_t),\ t \geq 0 \text{ satisfy the statements in (1.2) of} \\ \quad \text{Chapter 3, moreover, } \eta_t \text{ is continuous,} \end{cases}
$$

where $w'_t, t \geq 0$, is a $d-$dimensional BM, $q'(ds, dz)$ is a Poisson martingale measure with the compensator $\pi'(dz)ds$ such that

$q'(ds, dz) = p'(ds, dz) - \pi'(dz)ds,$

where $\pi'(dz)$ is a $\sigma-$finite measure on the measurable space $(Z, \Re(Z))$,and x_0 is supposed to be independent of $\Im_0^{w',q'}$, and suppose that the observation (ξ_t, ϕ_t) is given as follows:

$$
(3.1)_2 \begin{cases} \xi_t = \xi_0 + \int_0^t A_s(x(.), \eta(.))ds + \int_0^t B_s(\xi(.), \phi(.))dw_s \\ \quad + \int_0^t C_{s-}(\xi(.), \phi(.))d\widetilde{N}_s + \phi_t, \\ \text{and } (\xi_t, \phi_t),\ t \geq 0 \text{ satisfy the statements in (1.2) of} \\ \quad \text{Chapter 3, moreover, } \phi_t \text{ is continuous,} \end{cases}
$$

where A_s is assumed to be a bounded function, B, C satisfy the same conditions stated in Theorem 2.1, and $\widetilde{N}_t$ is a centralized Poisson process as that given in 2) of Theorem 1.1. Write

$\langle w'^i, w^j \rangle = \int_0^t \rho_s^{ij} ds,\ 1 \leq i, j \leq d.$

For the random process $A_t(x(.), \eta(.))$ write $\Pi_t(A) = E(A_t(x(.), \eta(.)) \big| \Im_t^{\xi,\phi})$, etc.

We have

3.1 Theorem. Suppose $f \in C_b^2([0, \infty); R^1)$ such that

$f(x_t) = f(x_0) + \int_0^t Lf(x, \eta, s)ds + \int_0^t \nabla f(x_s) \cdot \sigma(s, x(.), \eta(.))dw'_s$
$+ \int_0^t \int_Z (f(x_{s-} + c(s-, x(.), \eta(.))) - f(x_{s-}))q'(ds, dz) + \overline{\eta}_t,$

where $\overline{\eta}_t = \int_0^t \nabla f(x_s) \cdot d\eta_s$, and for each $x(.) \in D([0, \infty), R^d), \eta(.) \in D([0, \infty), R^d)$,

$Lf(x, \eta, s) = b(s, x(.), \eta(.)) \cdot \nabla f(x_s)$
$+ \frac{1}{2} tr.(\sigma(s, x(.), \eta(.))^* (\partial^2 f(x_s)/\partial x^2)\sigma(s, x(.), \eta(.)))$
$+ \int_Z (f(x_{s-} + c(s-, x(.), \eta(.), z)) - f(x_{s-})$
$- \nabla f(x_s) \cdot c(s-, x(.), \eta(.), z))\pi'(dz).$

Suppose

$\langle w'_i, w_j \rangle_t = \int_0^t \rho_s^{ij} ds.$

Then

(3.2) $\Pi_t(f) - \Pi_t(\overline{\eta}) = \Pi_0(f) - \Pi_0(\overline{\eta}) + \int_0^t \Pi_s(Lf)ds + \int_0^t [\Pi_s(\nabla f \cdot \sigma \cdot \rho) +$
$\cdot(\Pi_s(fA^*) - \Pi_s(f)\Pi_s(A^*))B_s^{-1}(\xi, \phi)^* - (\Pi_s(\overline{\eta}A^*) - \Pi_s(\overline{\eta})\Pi_s(A^*))B_s^{-1}(\xi, \phi)^*]d\overline{w}_s$
$+ \int_0^t E(\int_Z (f(x_{s-} + c(s-, x(.), \eta(.))) - f(x_{s-}))p'(\{s\}, dz) \big| \Im_{s-}^{\xi,\phi}) d\widetilde{N}_s.$

Theorem 3.1 is a direct corollary of Theorem 2.1. One only needs to see $f(x_t)$ as the signal process h_t in (1.1). Then applying Theorem 2.1, the conclusion is immediately derived.

3.2 Corollary. Assume that in (3.1) the signal and observation noise are independent, the jump noise in the signal and observation have no common

jump time, i.e. $\rho \equiv 0, \triangle y_t \cdot \triangle \widetilde{N}_t \equiv 0$; and the signal process h_t has no reflection, i.e. $\eta_t \equiv 0$, and the conditional distribution of h_t given $\Im_t^{\xi,\phi}$ $P[h_t \leq x \Big| \Im_t^{\xi,\phi}]$ has a density $\widehat{p}(t,x) = dP[h_t \leq x \Big| \Im_t^{\xi,\phi}]/dx$,which satisfies suitable differential hypothesis, and $A(t,\omega) = \widetilde{A}(h_t)$, a bounded function.
Then one has the following Zakai equation satisfied by the conditional density $\widehat{p}(t,x)$

(3.3) $d\widehat{p} = L^*\widehat{p}dt + \widehat{p}(\widetilde{A}^* - \Pi_t(\widetilde{A}^*))B_t^{-1*}d\overline{w}_t,$

where L^* is the adjoint operator of L.

Proof. To show Corollary 3.2 one only needs to note that

$$\Pi_t(\widetilde{A}^*) = \int_{R^d} \widetilde{A}^*(x)\widehat{p}(t,x)dx = (\widetilde{A}^*\widehat{p}(t,\cdot)) = (\widetilde{A}^*, \widehat{p}),$$

and to apply the integration by parts. Indeed, by assumption $\rho \equiv 0, \triangle y_t \cdot \triangle \widetilde{N}_t \equiv 0, \eta_t = 0$; hence (3.2) becomes

$$\Pi_t(f) = \Pi_0(f) + \int_0^t \Pi_s(Lf)ds$$
$$+ \int_0^t (\Pi_s(fA^*) - \Pi_s(f)\Pi_s(A^*))B_s^{-1}(\xi,\phi)^* d\overline{w}_s, \ \forall f \in C_b^2([0,\infty); R^1) \ .$$

Or, $\forall f \in C_b^2([0,\infty); R^1)$

$$(f, d_t\widehat{p}(t,\cdot)) = d_t(f, \widehat{p}(t,\cdot)) = (f, L^*\widehat{p}(t,\cdot))dt + (f, \widehat{p}(t,\cdot)(A^* - \Pi_t(A^*))B_t^{-1*}d\overline{w}_t.$$

Hence (2.3) is obtained. ■

Chapter 6

Stochastic Control

In this chapter we will discuss the stochastic optimal control problem for RSDEs with jumps by using the Girsanov theorem and martingale method. The necessary and the sufficient conditions for the optimal stochastic control are also derived.

6.1 Girsanov Theorem with Weaker conditions

In this section we discuss the Girsanov theorem under some weaker conditions and in a more suitable version for our stochastic control problem.

Suppose that (x_t, ϕ_t) solve the following RSDE with jumps and with the functional coefficients on $(\Omega, \Im, (\Im_t), P)$:

$$(1.1)\quad \begin{cases} x_t = x_0 + \int_0^t \sigma(s, x(.), \phi(.))dw_s + \int_0^t \int_Z c(s-, x(.), \phi(.), z)q(ds, dz) \\ \qquad +\phi_t, \\ x_0 \in \Theta^{cl}, \\ \text{and } (x_t, \phi_t),\ t \geq 0 \text{ satisfy the statements in (1.2) of Chapter 3,} \\ \text{moreover, } \phi_t \text{ is continuous.} \end{cases}$$

where we denote

$D = D([0, \infty); R^d)$ = The totality of cadlag maps from $[0, \infty)$ to R^d,

$C = C([0, \infty); R^d)$ = The totality of continuous maps from $[0, \infty)$ to R^d,

$\Im_t = \sigma((x_s, \phi_s); s \leq t, x \in D, \phi \in C)$,

$\Omega = D \times C$, $\Im = \vee_{t \geq 0} \Im_t$, i.e. w_t and $q(dt, dz)$ are $\Im_t-$BM and $\Im_t-$ Poisson martingale measure as that in (1.1) of Chapter 3, respectively, and we always make the following assumption

(1.2) $\sigma(s, x, \phi) : [0, \infty) \times D \times C \to R^d$,

$c(s, x, \phi, z) : [0, \infty) \times D \times C \times Z \to R^d$

are jointly measurable and $\Im_t-$adapted.

We have the following Girsanov type theorem.

1.1 Theorem. Assume that (1.2) holds, and

1° $b(s,x,\phi):[0,\infty)\times D\times C\to R^d$,
is also jointly measurable and $\Im_t$−adapted such that

$$\langle x_t, b(t,x,\phi)\rangle \le \widetilde{k}_0(t)(1+|x_t|^2\,\Pi_{i=1}^{n_0} g_i(|x_t|)),$$
$$\|\sigma(t,x,\phi)\|^2 \le \widetilde{k}_0(t)(1+|x_t|^2\,\Pi_{i=1}^{n_0} g_i(|x_t|)),$$
$$\int_Z |c(t,x,\phi,z)|^2\,\pi(dz) \le \widetilde{k}_0(t)$$

for all $x\in D,\phi\in C,t\ge 0$, where $\widetilde{k}_0(t)\ge 0$ such that for any $T<\infty$

$$\int_0^T \widetilde{k}_0(t)dt < \infty,$$

and for $i=1,2,...,n_0$, for any $u\in R^1$,

$$g_i(u) = 1+\log g_{i-1}(u),\ g_0(u) = 1+|u|^{2n_1},$$

and $n_1\ge 1$ is a constant; n_0 is a given natural number; and

$$|b(t,x,\phi)| + \|\sigma(t,x,\phi)\| + (\int_Z |c(t,x,\phi,z)|^2\,\pi(dz))^{1/2} \le k_N,$$

as $\|x\|_t\le N, \|\phi\|_t\le N,\ (\|x\|_t \,\hat{=}\, \sup_{s\le t}|x_s|)$,
where k_N is a constant depending on N only;
2° σ^{-1} exists and is bounded: $\|\sigma^{-1}\|\le k_0$.
Let

$$d\widetilde{P}_t = z_t^w(\sigma^{-1}b)dP = \exp(\int_0^t(\sigma^{-1}b)(s,x(\cdot)\phi(\cdot))dw_s - \tfrac{1}{2}\int_0^t |(\sigma^{-1}b)|^2 ds)dP.$$

Then
1) $\widetilde{P}_t$ is a probability measure for each $t\ge 0$; moreover, there exists a probability measure $\widetilde{P}$ defined on $(\Omega,\Im)$ such that $\widetilde{P}|_{\Im_t} = \widetilde{P}_t$, for each $t\ge 0$;
2) $\widetilde{w}_t = w_t - \int_0^t \sigma^{-1}b(s,x,\phi)ds$
is a BM under probability $\widetilde{P}$;

$$q(dt,dz) = p(dt,dz) - \pi(dz)dt$$

is still a Poisson random martingale measure with the same compensator $\pi(dz)dt$ under probability $\widetilde{P}$;
3) (x_t,ϕ_t) solve the following RSDE on $(\Omega,\Im,(\Im_t),\widetilde{P})$:

$$(1.2)\quad\left\{\begin{array}{l} x_t = x_0 + \int_0^t b(s,x(.),\phi(.))ds + \int_0^t \sigma(s,x(.),\phi(.))d\widetilde{w}_s \\ \qquad + \int_0^t\int_Z c(s-,x(.),\phi(.),z)q(ds,dz) + \phi_t, \\ \qquad\qquad x_0\in\Theta^{cl}, x_0 \text{ is non-random;} \\ \text{and } (x_t,\phi_t),\ t\ge 0 \text{ satisfy the statements in (1.2) of} \\ \qquad\qquad \text{Chapter 3, moreover, } \phi_t \text{ is continuous.} \end{array}\right.$$

Proof. Let

$$b^N(t,x,\phi) = \left\{\begin{array}{ll} b(t,x,\phi), & \text{as } \|x\|_t\le N, \|\phi\|_t\le N, \\ & 0, \text{ otherwise.} \end{array}\right.$$

Then by our Girsanov theorem (Lemma 3.9 and Theorem 3.12 in Chapter 3)

$$\widetilde{w}_t^N = w_t - \int_0^t \sigma^{-1}b^N(s,x,\phi)ds$$

is a BM under probability $\widetilde{P}^N$;

$$q(dt,dz) = p(dt,dz) - \pi(dz)dt$$

is still a Poisson random martingale measure with the same compensator $\pi(dz)dt$ under probability $\widetilde{P}^N$; where $d\widetilde{P}_t^N = z_t^w(\sigma^{-1}b^N)dP$. Hence, (x_t,ϕ_t) solve the following RSDE on $(\Omega,\Im,(\Im_t),\widetilde{P}^N)$:

$$(1.2)' \begin{cases} x_t = x_0 + \int_0^t [b^N(s, x(.), \phi(.)) + \int_0^t \sigma(s, x(.), \phi(.)) d\widetilde{w}_s^N \\ \quad + \int_0^t \int_Z c(s-, x(.), \phi(.), z) q(ds, dz) + \phi_t, \\ \quad x_0 \in \Theta^{cl}, \ x_0 \text{ is non-random;} \\ \text{and } (x_t, \phi_t), t \geq 0 \text{ satisfy the statements in (1.2) of} \\ \quad \text{Chapter 3, moreover, } \phi_t \text{ is continuous.} \end{cases}$$

Note that for $g_{no+1}(|x|)$ we have as $x \in R^d$

$\frac{\partial}{\partial x_j} g_{no+1}(|x|) = \Pi_{i=1}^{no} g_i(|x|)^{-1} 2n_1 x_j |x|^{2n_1-2} (1 + |x|^{2n_1})^{-1}$,

$\frac{\partial^2}{\partial x_j \partial x_m} g_{no+1}(|x|) = \Pi_{i=1}^{no} g_i(|x|)^{-1} (1 + |x|^{2n_1})^{-1} [(4n_1^2 - 4n_1) |x|^{2n_1-4} x_j x_m$

$+2n_1 \delta_{jm} |x|^{2n_1-2} - (1 + |x|^{2n_1})^{-1} 4n_1 |x|^{4n_1-4} x_j x_m$

$-4n_1^2 |x|^{4n_1-4} x_j x_m (1 + |x|^{2n_1})^{-1} \sum_{k=1}^{no} \Pi_{i=1}^{k} g_i(|x|)^{-1}]$.

Therefore, as $\|x\|_t \leq N$, $\|\phi\|_t \leq N$, $(x, \phi) \in D \times C$,

$\nabla_x g_{no+1}(|x_t|) b^N(t, x, \phi) + \frac{1}{2} tr.(\sigma(t, x, \phi)^* g''_{no+1}(|x_t|) \sigma(t, x, \phi))$

$+ \int_Z [g_{no+1}(|x_{t-} + c(t-, x, \phi, z)|) - g_{no+1}(|x_{t-}|)$

$- \nabla_x g_{no+1}(|x_{t-}|) \cdot c(t-, x, \phi, z)] \pi(dz)$

$\leq \nabla_x g_{no+1}(|x_t|) b(t, x, \phi) + \frac{1}{2} tr.(\sigma(t, x, \phi)^* g''_{no+1}(|x_t|) \sigma(t, x, \phi))$

$+2^{-1} \int_Z g''_{no+1}(|x_{t-} + \theta c(t-, x, \phi, z)|) \cdot |c(t-, x, \phi, z)|^2 \pi(dz)$

$= \Pi_{i=1}^{no} g_i(|x_t|)^{-1} 2n_1 |x_t|^{2n_1-2} (1 + |x_t|^{2n_1})^{-1} x_t \cdot b(t, x, \phi)$

$+\frac{1}{2} tr.(\sigma(t, x, \phi)^* g''_{no+1}(|x_t|) \sigma(t, x, \phi))$

$+2^{-1} \int_Z g''_{no+1}(|x_{t-} + \theta c(t-, x, \phi, z)|) \cdot |c(t-, x, \phi, z)|^2 \pi(dz)$

$\leq k'_0(1 + \widetilde{k}_0(t))$,

where k'_0 is a constant depending on n_1 only, and we have applied the Taylor formula. Now let

$\tau_N^T = \inf \{t \in [0, T] : \|x\|_t + \|\phi\|_t > N\}$.

Applying Ito's formula to $g_{no+1}(|x_t - x_0|)$,where (x_t, ϕ_t) solve (1.2)', one finds that

$E_{\widetilde{P}^N} g_{no+1}(|x(\tau_N^T) - x_0|) \leq k'_0 \int_0^T (1 + \widetilde{k}_0(t)) dt < \infty$,

where we have used Lemma 2.2 in Chapter 2

$\int_0^t \nabla_x g_{no+1}(|x_s - x_0|) \cdot d\phi_s$

$= \int_0^t \Pi_{i=1}^{no} g_i(|x_s - x_0|)^{-1} 2n_1 |x_s - x_0|^{2n_1-2} (1 + |x_s - x_0|^{2n_1})^{-1} (x_s - x_0) \cdot d\phi_s \leq 0$

However, by assumption, $x_0 \in R^d$ is a given fixed non-random vector,

$E_{\widetilde{P}^N} g_{no+1}(|x(\tau_N^T) - x_0|) \geq g_{no+1}(N - |x_0|) \widetilde{P}^N(\tau_N^T < T)$.

Thus,

$\lim_{N \to \infty} \widetilde{P}^N(\tau_N^T < T) = 0$.

Hence, as $N \to \infty$

$E z_T^w(\sigma^{-1} b) \geq \int_{\{\tau_N^T = T\}} z_T^w(\sigma^{-1} b^N) dP = \widetilde{P}^N \{\tau_N^T = T\} \to 1$.

Therefore,

$E z_T^w(\sigma^{-1} b) = 1$.

This shows that $\widetilde{P}_T$ is a probability measure, where

$d\widetilde{P}_T = z_T^w(\sigma^{-1} b) dP$.

Now since $\Omega = D \times C$ is a standard measurable space, the proof of 1) and 2) can be completed as in Theorem 3.12 in Chapter 3. 3) is obviously derived by 1) and 2). ■

We can also have another kind of Girsanov type theorem.

1.2 Theorem. If we only change the assumption 1° of Theorem 1.1 into the following

$1^{\circ\prime}$ $b(s,x,\phi) : [0,\infty) \times D \times C \to R^d$,

is also jointly measurable and $\Im_t$-adapted such that

$$|b(t,x,\phi)|^2 + \|\sigma(t,x,\phi)\|^2 + \int_Z |c(t,x,\phi,z)|^2 \pi(dz) \leq \widetilde{k}_0(t)(1 + \|x\|_t^2 + \|\phi\|_t^2),$$

for all $x \in D, \phi \in C, t \geq 0$, where $\widetilde{k}_0(t) \geq 0$ such that for any $T < \infty$

$$\int_0^T \widetilde{k}_0(t)dt < \infty;$$

and keep the other assumption unchanged, then the conclusion of Theorem 1.1 still holds.

To show Theorem 1.2 we only need to apply Ito's formula to $|x_t - x_0|^2$ in the above proof of Theorem 1.1. Then the proof of Theorem 1.2 can be completed almost the same as that of Theorem 1.2 in Chapter 5.

Applying Theorem 1.1 we can also get the following theorem on the existence of solution to RDSE (1.2) with greater than linear growth condition on coefficients b and σ, moreover, b can be discontinuous.

1.3 Theorem. Under assumption of Theorem 1.1 if Assumption 2.7 in Chapter 2 holds, and for all $x, y \in D$; $t, s \geq 0$, $\phi, \psi \in C$

$$\|\sigma(t,x,\phi) - \sigma(t,y,\psi)\|^2 + \int_Z |c(t,x,\phi) - c(t,y,\psi)|^2 \pi(dz)$$
$$\leq k_0(|x_t - y_t|^2 + |\phi_t - \psi_t|^2),$$
$$\|\sigma(t,x,\phi) - \sigma(s,x,\phi)\|^2 + \int_Z |c(t,x,\phi) - c(s,x,\phi)|^2 \pi(dz)$$
$$\leq k_0 |t-s|^2,$$

then (1.2) has a weak solution. Furthermore, if

$$2(x-y) \cdot (b(t,x(.),\phi(.)) - b(t,y(.),\psi(.)))$$
$$\leq k_N(t)\rho_N(\sup_{s \leq t} |x_s - y_s|^2 + \sup_{s \leq t} |\phi_s - \psi_s|^2),$$
as $\sup_{s \leq t} |x_s|, \sup_{s \leq t} |y_s| \leq N, \sup_{s \leq t} |\phi_s|, \sup_{s \leq t} |\psi_s| \leq N$,

where $0 \leq k_N(t)$, $\int_0^t k_N(t)ds < \infty$, and $\rho_N(u)$ is concave and strictly increasing in $u \geq 0$ such that $\rho_N(0) = 0$, and $\int_{0+} du/\ \rho_N(u) = \infty$, for each $N = 1, 2, ...$, then (1.2) has a pathwise unique strong solution.

Proof. First of all, by Theorem 3.20 in Chapter 3 (1.1) has a pathwise unique strong solution. Secondly, by Theorem 1.1 (1.2) has a weak solution. Finally, applying Theorem 3.22 and Theorem 1.3 both in Chapter 3, one finds that (1.2) has a pathwise unique strong solution. ■

6.2 Martingale Method, Necessary and Sufficient Conditions for Optimal Control

6.2.1 Maximum Principle

Let us consider the optimal stochastic control problem with respect to the following RSDE with jumps:

$$
(2.1)\begin{cases} x_t = x_0 + \int_0^t [b(s, x(.), \phi(.), u(.))ds + \int_0^t \sigma(s, x(.), \phi(.))dw_s \\ \qquad + \int_0^t c(s-, x(.), \phi(.))d\widetilde{N}_s + \phi_t, \\ x_0 \in \Theta^{cl}, \\ \text{and } (x_t, \phi_t),\ t \geq 0 \text{ satisfy the statements in (1.2) of Chapter 3,} \\ \text{moreover, } \phi_t \text{ is continuous;} \end{cases}
$$

where we use the same notation as in $(1.8)_1$ of Chapter 5, and $u(.)$ is a control, which will be explained below. From now on let us always make the following

2.1 Assumption. 1° $\|\sigma(t, x, \phi)\|^2 \leq k_0(1 + \|x\|_t^2 + \|\phi\|_t^2)$,

$|c(t, x, \phi)|^2 \pi(\Gamma) \leq k_0$, and c^{-1} exists,

where $k_0 \geq 0$ is a constant, $\pi(\Gamma) < \infty$ is the density of the centralized Poisson process $\widetilde{N}_t$ (see $(1.8)_1$ in Chapter 5 there), and we denote $\|x\|_t = \sup_{s \leq t} |x(s)|$; moreover, assume that Assumption 2.7 in Chapter 2 holds;

2° $\langle A\lambda, \lambda\rangle \geq \delta |\lambda|^2$, for all $\lambda \in R^d$;

where $A = \frac{1}{2}\sigma\sigma^*$, and $\delta > 0$ is a constant;

3° $\|\sigma(t, x, \phi) - \sigma(s, x, \phi)\|^2 + |c(t, x, \phi) - c(s, x, \phi)|^2 \pi(\Gamma)$

$\leq k_0 |t - s|^2$,

$\|\sigma(t, x, \phi) - \sigma(t, y, \psi)\|^2 + |c(t, x, \phi) - c(t, y, \psi)|^2 \pi(\Gamma)$

$\leq k_0(|x_t - y_t|^2 + |\phi_t - \psi_t|^2)$, for all $x, y \in D$; $t, s \geq 0$; $\phi, \psi \in C$.

Under Assumption 2.1 by Theorem 3.23 in Chapter 3 RSDE

$$
(2.2)\begin{cases} x_t = x_0 + \int_0^t \sigma(s, x(.), \phi(.))dw_s + \int_0^t c(s-, x(.), \phi(.))d\widetilde{N}_s + \phi_t, \\ x_0 \in \Theta^{cl}, \\ \text{and } (x_t, \phi_t),\ t \geq 0 \text{ satisfy the statements in (1.2) of Chapter 3,} \\ \text{moreover, } \phi_t \text{ is continuous.} \end{cases}
$$

has a pathwise unique strong solution (x_t, ϕ_t), $t \geq 0$, on the probability space $(\Omega, \Im, (\Im_t), P)$ with the BM w_t and Poisson martingale measure $q(dt, dz)$ and with jump coefficient

$$c(s-, x(.), \phi(.), z) = c(s-, x(.), \phi(.))I_\Gamma(z)$$

where $d\widetilde{N}_t = q(dt, \Gamma)$. Hence

$$\Im_t^{w, \widetilde{N}_t} \supset \Im_t^{x, \phi}, \text{ for all } t \geq 0.$$

On the other hand, by the existence of c^{-1} and σ^{-1} and by

$$N_t = \widetilde{N}_t + \pi(\Gamma)t = \textstyle\sum_{0 < s \leq t} c^{-1} \triangle x_s$$

one also has that

$$\Im_t^{w, \widetilde{N}_t} \subset \Im_t^{x, \phi}, \text{ for all } t \geq 0.$$

Hence,

$$\Im_t^{w, \widetilde{N}_t} = \Im_t^{x, \phi}, \text{ for all } t \geq 0.$$

Now assume that $R^d \supset U$ - a compact set. Write

(2.3) $\widetilde{U} = \left\{ u : u = u(t,\omega) \in U, \text{ it is } \Im_t^{w,\widetilde{N}_t} - \text{adapted} \right\}.$

Obviously, if $u \in \widetilde{U}$, then it can be expressed as $u(t, x_s, \phi_s, s \leq t)$. $\widetilde{U}$ is called the admissible control set. Let us make

2.2 Assumption. $|b(t,x,\phi,u)|^2 \leq k_0(1 + \|x\|_t^2 + \|\phi\|_t^2)$,for all $x \in D$, $t \geq 0$, $\phi \in C, u \in U$.

For simplicity sometimes we just write

$b^u = b(t,x,\phi,u)$.

Then under Assumption 2.1 and 2.2 by Theorem 1.2

(2.4) $(dP^u/dP)|\Im_t = z_t(\sigma^{-1}b^u) = z_t^w(\sigma^{-1}b^u)$

$= \exp(\int_0^t \sigma^{-1}b(s,x(.),\phi(.),u_s)dw_s - \frac{1}{2}\int_0^t |\sigma^{-1}b(s,x(.),\phi(.),u_s)|^2 ds)$

defined a probability measure P^u on $(\Omega, \Im)$ and (x_t, ϕ_t), $t \geq 0$, satisfies: $P^u - a.s.$

$$
(2.5) \begin{cases} x_t = x_0 + \int_0^t b(s,x(.),\phi(.),u_s)ds + \int_0^t \sigma(s,x(.),\phi(.))dw_s^u \\ \qquad + \int_0^t c(s-,x(.),\phi(.))d\widetilde{N}_s + \phi_t, \\ \qquad x_0 \in \Theta^{cl}, \\ \text{and } (x_t,\phi_t),\ t \geq 0 \text{ satisfy the statements in (1.2) of} \\ \qquad \text{Chapter 3, moreover, } \phi_t \text{ is continuous.} \end{cases}
$$

where

$w_t^u = w_t - \int_0^t \sigma^{-1}b^u ds$, $t \geq 0$,

$\widetilde{N}_t = N_t - \pi(\Gamma)t$,

is a BM and a centralized Poisson process under probability P^u, respectively. Now the optimization problem here is to minimize

(2.6) $J(u) = E_{P^u}(\int_0^T e(t,x(.),\phi(.),u_t)dt + h(x_T,\phi_T)$,

among all $u \in \widetilde{U}$, where (x_t, ϕ_t) is subject to (2.2), $t \in [0,T]$, for the finite horizon case; and to minimize

$J(u) = E_{P^u}\int_0^\infty e(t,x(.),\phi(.),u_t)dt$

among all $u \in \widetilde{U}$, where (x_t,ϕ_t) is subject to (2.2), $t \geq 0$, for the infinite horizon case, where we assume that $e(t,x(.),\phi(.),u)$ is jointly measurable with respect to $(t,x(.),\phi(.),u) \in [0,T] \times D \times C \times R^d$, and $h(x,\phi)$ is jointly measurable with respect to $(x,\phi) \in R^d \times R^d$.

Let us consider the finite horizon and the simpler case in this section only, i.e. consider (2.6), hence (2.2) and (2.5) are all considered on $t \in [0,T]$ only, and assume that in the cost functional $J(u)$

$e(t,x(.),\phi(.),u) = e(t,x(.),\phi(.))$

does not depend on u.

For the existence of an ε - optimal control and the maximum principle of optimal control problem (2.2)-(2.6) we have the following

2.3 Theorem. Under assumption 2.1 and 2.2 and assume that

1° $\|\sigma^{-1}b^u\| \leq k_0$, for all $u \in \widetilde{U}$,

2° $|e(t,x(.),\phi(.))| + |h(x_1,y_1)| \leq k_0$, for all $x_1 \in \overline{\Theta}, y_1 \in R^d$, $(x(.),\phi(.)) \in D \times C$.

Then **1)** for any $\varepsilon > 0$ there exists a control $u^\varepsilon \in \widetilde{U}$ such that

$J(u^\varepsilon) \leq \inf_{v \in \widetilde{U}} J(v) + \varepsilon,$

and for all $v \in \widetilde{U}$

$J(v) \geq J(u^\varepsilon) - \varepsilon d(v, u^\varepsilon),$

where for $u, v \in \widetilde{U}$ (denote $m = \frac{\lambda}{T}$, λ - Lebesgue measure on $[0, T]$)

(2.7) $d(u, v) = (m \times P)\{(t, \omega) \in [0, T] \times \Omega : u_t(\omega) = v_t(\omega)\}$;

2) there exist a R^d-valued $\Im_t^{w, \widetilde{N}}$ - adapted process $\gamma_t^\varepsilon(\omega)$ and a R^d-valued $\Im_t^{w, \widetilde{N}}$ - predictable process $\delta_t^\varepsilon(\omega)$, which satisfy

$E \int_0^T |\gamma_t^\varepsilon(\omega)|^2 dt < \infty, E \int_0^T |\delta_t^\varepsilon(\omega)|^2 \pi(\Gamma) dt < \infty,$

and $P - a.s.$ for all $t \in [0, T]$

(2.8) $E_{P^{u^\varepsilon}}(g(x(.), \phi(.)) \Big| \Im_t^{w, \widetilde{N}}) = J(u^\varepsilon) + \int_0^t \gamma_s^\varepsilon(\omega) dw_s^{u^\varepsilon} + \int_0^t \delta_s^\varepsilon(\omega) d\widetilde{N}_s,$

where

$g(x(.), \phi(.)) = \int_0^T e(t, x_t, \phi_t, u_t) dt + h(x_T, \phi_T),$

such that for all $v \in \widetilde{U}$

(2.9) $z_t(\sigma^{-1} b^{u^\varepsilon}) \gamma_t^\varepsilon \sigma^{-1} b^v \geq z_t(\sigma^{-1} b^{u^\varepsilon}) \gamma_t^\varepsilon \sigma^{-1} b^{u^\varepsilon} - \varepsilon$, $m \times P - a.e.$,

3) if u^ε with γ_t^ε and δ_t^ε verifies (2.8) and (2.9), then for all $v \in \widetilde{U}$

$J(v) \geq J(u^\varepsilon) - \exp(2k_0^2 T) \varepsilon T,$

where $0 \leq k_0$ is the constant from 1°;

4) if u^0 is an optimal control, then 2) holds for $\varepsilon = 0$; moreover, for all $v \in \widetilde{U}$

$\gamma_t^0(\sigma^{-1} b^v) \geq \gamma_t^0(\sigma^{-1} b^{u^0})$, $m \times P - a.e.$

5) if $u^0 \in \widetilde{U}$, $\gamma_t^0(\omega)$ is a $\Im_t^{w, \widetilde{N}}$ - adapted process, and $\delta_t^0(\omega)$ is a $\Im_t^{w, \widetilde{N}}$ - predictable process such that (denote $E_{u^0}(.) = E_{P^{u^0}}(.)$)

$E_{u^0}(g(x(.), \phi(.)) \Big| \Im_t^{w, \widetilde{N}}) = J(u^0) + \int_0^t \gamma_s^0(\omega) dw_s^0 + \int_0^t \delta_s^0(\omega) d\widetilde{N}_s,$

i.e. γ_t^0 and δ_t^0 are the integrands in the martingale representation for the $E_{u^0}(g(x(.), \phi(.)) \Big| \Im_t^{w, \widetilde{N}})$, where (x_t, ϕ_t), $t \in [0, T]$, satisfies (2.5) with respect to u^0, $P^{u^0} - a.s.$; and if for all $v \in \widetilde{U}$

$\gamma_t^0(\sigma^{-1} b^v) \geq \gamma_t^0(\sigma^{-1} b^{u^0})$, $m \times P - a.e.$,

then u^0 is an optimal control.

To prove Theorem 2.3 we need some preparations. First of all, it is easily seen that $d(u, v)$ defined by (2.7) is a complete distance on $\widetilde{U}$. Hence $(\widetilde{U}, d)$ is a complete metric space. (See Elliott & Kohlmann, 1980). Indeed, we have

2.4 Lemma. (Elliott & Kohlmann, 1980). $(\widetilde{U}, d)$ is a complete metric space.

Secondly, we have

2.5 Lemma. $J(u) : (\widetilde{U}, d) \to R$ is a continuous map.

2.6 Remark. In Elliott & Kohlmann, 1980, for a continuous SDE system without reflection they also consider the optimization of functional

$J(u) = E_{P^u}(\int_0^T e(t, x_t, u_t) dt + h(x_T)).$

They have proved a Lemma 3 as follows: Under condition that

$|\sigma^{-1}(t, x) b(t, x, u)| \leq k_0(1 + |x|)$

$J(u):(\widetilde{U},d)\to R$ is a continuous map. However, their proof is incomplete. Since the following incorrect fact was used: d-convergence implies convergence $a.e.$ with respect to $m\times P$. In fact, we have a counter example as follows (see Situ Rong, 1983):

Let

$$v_k^{i,k}(t,\omega)=\begin{cases}1, & \text{as } t\in[\frac{i-1}{2^k},\frac{i}{2^k}],\\ 0, & \text{otherwise},\end{cases}$$

as $t\in[0,1]$, for all $i=1,2,...2^k;k=0,1,2,.....;$
and by order let $u_1=v_0^{1,0},u_2=v_1^{1,1},u_3=v_1^{2,1},\cdots\cdots$; then it is evident that

$$d(u_n,0)\to 0, \text{ as } n\to\infty.$$

However, for all $(t,\omega)\in[0,1]\times\Omega$

$$\lim_{n\to\infty}u_n=0$$

is not true. Actually, we can only get the following fact: $d(u_n,0)\to 0$, as $n\to\infty$, implies that there exists a subsequence u_{n_k} of u_n such that as $k\to\infty$

$$u_{n_k}\to u, \ a.e. \text{ w.r.t. } m\times P.$$

Hence it is doubtful that their Lemma 3 is true under the condition

$$\left|\sigma^{-1}(t,x)b(t,x,u)\right|\le k_0(1+|x|).$$

Now let us show Lemma 2.5.

Proof. Assume that $u^n,u\in\widetilde{U}$

$$d(u^n,u)\to 0, \text{ as } n\to\infty.$$

Denote

$$g_T(x(.),\phi(.))=\int_0^T e(t,x(\cdot),\phi(\cdot))dt+h(x_T,\phi_T),$$

$$A_n=\int_0^T(\sigma^{-1}b^{u^n})dw_s-\tfrac{1}{2}\int_0^T\left|\sigma^{-1}b^{u^n}\right|^2ds,$$

and A is similarly defined for u. Note that by Girsanov's theorem and the condition 1°, it is easily seen that

$$E\exp(2(1-\alpha)A_n+2\alpha A)\le k_1, \text{ for any } \alpha\in(0,1),$$

where $0\le k_1$ is a constant. Note that

$$E\int_0^T\left|\sigma^{-1}(b^{u^n}-b^u)\right|^2ds\le 4k_0^2d(u^n,u),$$

$$\left|E\int_0^T(\left|\sigma^{-1}b^{u^n}\right|^2-\left|\sigma^{-1}b^u\right|^2)ds\right|\le 2k_0^2d(u^n,u).$$

Hence as $n\to\infty$

$$|J(u^n)-J(u)|\le E\left|g_T(x(.),\phi(.))\right|\left|z_T(\sigma^{-1}b^{u^n})-z_T(\sigma^{-1}b^u)\right|$$

$$\le k_0(T+1)E([\exp(1-\alpha_n)A_n+\alpha_nA)](\int_0^T(\sigma^{-1}(b^{u^n}-b^u)dw_s$$

$$-\tfrac{1}{2}\int_0^T(\left|\sigma^{-1}b^{u^n}\right|^2-\left|\sigma^{-1}b^u\right|^2)ds)\to 0,$$

where we have used

$$e^a-e^b=e^{\alpha b+(1-\alpha)a}(a-b),$$

where $0<\alpha<1$, and used the Schwarz's inequality ■

Let us quote Ekeland's lemma as follows:

2.7 Lemma. (Ekeland, 1979). Suppose that (V,d) is a complete metric space, where d is the metric, and $F:V\to[0,\infty]$ a lower semi-continuous function, which is not identically $+\infty$, and bounded from below. Then, for any $\varepsilon>0$, there exists a point $u^\varepsilon\in V$ such that

$$F(u^\varepsilon)\le\inf_V F+\varepsilon,$$

and for all $u \in V$

$F(u) \geq F(u^{\varepsilon}) - \varepsilon d(u^{\varepsilon}, u)$.

Now we are in a position to prove Theorem 2.3.

Proof. By Lemma 2.5 and Lemma 2.7 one verifies the conclusion 1). Now for any $v \in \widetilde{U}$ define

$$v^h(s,\omega) = \begin{cases} v(s,\omega), & \text{as } (s,\omega) \in [t,t+h] \times A, \\ u^{\varepsilon}(s,\omega), & \text{otherwise,} \end{cases}$$

where $u^{\varepsilon}(s,\omega)$ is the ε - optimal control from 1), and $t+h \leq T$, $h > 0$, $t \in [0,T)$, $A \in \Im_t^{w,q}$ are arbitrary given. Then by (2.5) (for simplicity just write $w_t^h = w_t^{v^h}, w_t^{\varepsilon} = w_t^{u^{\varepsilon}}$,etc.)

$$dw_s^h = \begin{cases} \sigma^{-1}(s,x(.),\phi(.))(dx_s - b(s,x(.),\phi(.),v_s^h)ds \\ \quad - \int_Z c(s,x(.),\phi(.),z)q(ds,dz) - d\phi_s = dw_s^{\varepsilon}, \\ \quad \text{as } (s,\omega) \notin [t,t+h] \times A, \\ dw_s^h, \text{ otherwise.} \end{cases}$$

Since (x_t, ϕ_t) satisfies (2.5) under probability $P^{\varepsilon} = P^{u^{\varepsilon}}$ with control u^{ε} and BM $w_t^{u^{\varepsilon}}$ and centralized Poisson process $\widetilde{N}_t$, and $\Im_t^{x,\phi} = \Im_t^{w,\widetilde{N}}$. Hence by Theorem 1.3 in Chapter 5 there exist a R^d-valued $\Im_t^{x,\phi}$-adapted process $\gamma_t^{\varepsilon}(\omega)$ and a R^d-valued $\Im_t^{x,\phi}$-predictable process $\delta_t^{\varepsilon}(\omega)$,which satisfy

$$E\int_0^T |\gamma_t^{\varepsilon}(\omega)|^2 dt < \infty, E\int_0^T |\delta_t^{\varepsilon}(\omega)|^2 \pi(\Gamma)dt < \infty,$$

such that $P^{\varepsilon} - a.s.$

$$g(x(.),\phi(.)) = E_{P^{\varepsilon}}(g(x(.),\phi(.)) \Big| \Im_T^{x,\phi}) = J(u^{\varepsilon}) + \int_0^T \gamma_s^{\varepsilon}(\omega)dw_s^{\varepsilon}$$
$$+ \int_0^T \delta_s^{\varepsilon}(\omega)d\widetilde{N}_s.$$

Let us also denote $P^h = P^{v^h}$. Note that $P^{\varepsilon} \sim P \sim P^h$, and

$$\int_t^{t+h} I_A \gamma_s^{\varepsilon} \sigma^{-1}(dx_s - b^v ds - \int_Z cq(ds,dz) - d\phi_s) = \int_t^{t+h} I_A \gamma_s^{\varepsilon} dw_s^h.$$

Hence, $P^h - a.s.$

$$g(x(.),\phi(.)) = J(u^{\varepsilon}) + \int_0^T \gamma_s^{\varepsilon}(\omega)dw_s^h + \int_t^{t+h} I_A \gamma_s^{\varepsilon} \sigma^{-1}(b^v - b^{u^{\varepsilon}})ds$$
$$+ \int_0^T \delta_s^{\varepsilon}(\omega)d\widetilde{N}_s.$$

From this and by 1)

$$J(v^h) = E_{v^h}(g(x(.),\phi(.))) = J(u^{\varepsilon}) + E_{v^h}\int_t^{t+h} I_A \gamma_s^{\varepsilon} \sigma^{-1}(b^v - b^{u^{\varepsilon}})ds$$
$$\geq J(u^{\varepsilon}) - \varepsilon d(u^{\varepsilon}, v^h),$$

where we denote $E_{v^h}(.) = E_{P^{v^h}}$, etc. Therefore

$$\int_t^{t+h}\int_A z_t(\sigma^{-1}b^{u^{\varepsilon}}) z_{t,t+h}(\sigma^{-1}b^v)\gamma_s^{\varepsilon}\sigma^{-1}(b^v - b^{u^{\varepsilon}})dPds \geq -\varepsilon h P(A),$$

where we denote

$$z_{t,t+h}(\sigma^{-1}b^v) = \exp(\int_t^{t+h} \sigma^{-1}b^v \cdot dw_s - \tfrac{1}{2}\int_t^{t+h} |\sigma^{-1}b^v|^2 ds),$$

and we have used

$$E(z_{t+h,T}(\sigma^{-1}b^v)\,|\Im_{t+h}) = 1.$$

Hence, for a.e. $t \in [0,T]$

$$\int_A z_t(\sigma^{-1}b^{u^{\varepsilon}})\gamma_t^{\varepsilon}\sigma_t^{-1}(b_t^v - b_t^{u^{\varepsilon}})dP \geq -\varepsilon P(A), \text{ for all } A \in \Im_t^{w,\widetilde{N}}.$$

Since the integrand above is $\Im_t^{w,\widetilde{N}}$-adapted. Hence

$$z_t(\sigma^{-1}b^{u^{\varepsilon}})\gamma_t^{\varepsilon}\sigma_t^{-1}(b_t^v - b_t^{u^{\varepsilon}}) \geq -\varepsilon, \; m \times P - a.e.,$$

where for the null set we have applied Fubini's theorem. Conclusion 2) then follows. Now note that

$\int_0^T \gamma_s^\varepsilon dw_s^v = \int_0^T \gamma_s^\varepsilon \sigma_s^{-1}(dx_s - b^v ds - cd\widetilde{N}_s - d\phi_s),$

$Ez_t(\sigma_t^{-1}(b_t^v - b_t^{u^\varepsilon}))\exp(\int_0^t(\left|\sigma_s^{-1}b_s^{u^\varepsilon}\right|^2 - (\sigma_s^{-1}b_s^v)(\sigma_s^{-1}b_s^{u^\varepsilon}))ds)$

$\leq e^{2k_0^2 T}.$

Hence, by 2) for all $v \in \widetilde{U}$

$$
\begin{aligned}
&E_v(J(u^\varepsilon) + \int_0^T \gamma_s^\varepsilon(\omega)dw_s^\varepsilon + \int_0^T \delta_s^\varepsilon(\omega)d\widetilde{N}_s \\
&= J(u^\varepsilon) + E_v(\int_0^T \gamma_s^\varepsilon(\omega)dw_s^v + \int_0^T \delta_s^\varepsilon(\omega)d\widetilde{N}_s \\
&+ \int_0^T \gamma_s^\varepsilon \sigma^{-1}(b^v - b^{u^\varepsilon})ds) \geq J(u^\varepsilon) - \varepsilon \int_0^T E[z_t(\sigma^{-1}b^v)(z_t(\sigma^{-1}b^{u^\varepsilon}))^{-1}]dt \\
&\geq J(u^\varepsilon) - \varepsilon T e^{2k_0^2 T}.
\end{aligned}
$$

Applying the martingale representation part of 2), one verifies that for all $v \in \widetilde{U}$

$$J(v) = E_v(g(x(.), \phi(.))) = E_v(E_{u^\varepsilon}(g(x(.), \phi(.)) \left| \Im_T^{w,\widetilde{N}} \right.)) \geq J(u^\varepsilon) - \varepsilon T e^{2k_0^2 T}.$$

Hence, 3) is proved. Now if u^0 is an optimal control, then by 2)

$$z_t(\sigma^{-1}b^{u^0})\gamma_t^0 \sigma_t^{-1}(b_t^v - b_t^{u^0}) \geq 0.$$

Multiply both sides by $(z_t(\sigma^{-1}b^{u^0}))^{-1}$. Then 4) is obtained. Let us show 5). By the assumption in 5), for all $v \in \widetilde{U}$

$$E_v \int_0^T \gamma_s^0 \sigma^{-1}(b^v - b^{u^0})ds \geq 0.$$

Hence, for all $v \in \widetilde{U}$ (denote $w_t^0 = w_t^{u^0}$)

$$
\begin{aligned}
&J(u^0) \leq J(u^0) + E_v(\int_0^T \gamma_s^0(\omega)dw_s^v + \int_0^T \delta_s^0(\omega)d\widetilde{N}_s \\
&+ \int_0^T \gamma_s^0 \sigma^{-1}(b^v - b^{u^0})ds) = E_v(J(u^0) \\
&+ \int_0^T \gamma_s^0(\omega)\sigma^{-1}(dx_s - b^v ds - cd\widetilde{N}_s - d\phi_s) \\
&+ \int_0^T \delta_s^0(\omega)d\widetilde{N}_s + \int_0^T \gamma_s^0 \sigma^{-1}(b^v - b^{u^0})ds) \\
&= E_v(J(u^0) + \int_0^T \gamma_s^0(\omega)dw_s^0 + \int_0^T \delta_s^0(\omega)d\widetilde{N}_s) \\
&= E_v(E_{u^0}(g(x(.), \phi(.)) \left| \Im_T^{w,\widetilde{N}} \right.)) = E_v g(x(.), \phi(.)) = J(v),
\end{aligned}
$$

where we have applied that $g(x(.), \phi(.))$ is $\Im_T^{w,\widetilde{N}}$-measurable. ■

6.2.2 Necessary and Sufficient Conditions for Optimal Control. Martingale Method.

In this section we will consider the existence of an optimal control even in a more general stochastic optimal control problem with respect to RSDE (2.1). More precisely, we want to minimize the cost functional given by

(3.1) $J(u) = E_{P^u}(\int_0^T g(t, x(.), \phi(.), u_t)dt + h(x_T, \phi_T)),$

among all $u \in \widetilde{U}$, where (x_t, ϕ_t) is subject to (2.2), $t \in [0, T]$, and we assume that $g(t, x(.), \phi(.), u)$ is jointly Borel measurable with respect to $(t, x(.), \phi(.), u) \in [0, T] \times D \times C \times R^d$, and $h(x, \phi)$ is jointly Borel measurable with respect to $(x, \phi) \in R^d \times R^d$.

For the necessary, and the sufficient conditions for the existence of an optimal control and the maximum principle of the optimal control problem (2.2)-(2.6) we have the following

3.1 Theorem Under Assumptions 2.1 and 2.2 assume that
1° $|g(t,x(.),\phi(.),y_1)| + |h(y_2,y_3)| \leq k_0(1 + \|x\|_t + \|\phi\|_t + |y_1| + |y_2| + |y_3|)$,
for all $(t,x(.),\phi(.),y_1) \in [0,T] \times D \times C \times R^d$, and $y_2, y_3 \in R^d$.
Then 1) if u^0 is an optimal control, then there exist random processes $\gamma_t(\omega) = \gamma_t(x.(\omega),\phi.(\omega))$ and $\delta_t(\omega) = \delta_t(x.(\omega),\phi.(\omega))$, where
$(3.2)_1$ $\gamma_t(x,\phi) : [0,T]\times D\times C \to R^d$ is jointly Borel measurable, and $\Im_t^{x,\phi}$−adapted,
$(3.2)_2$ $\delta_t(x,\phi) : [0,T]\times D\times C \to R^d$ is jointly Borel measurable, and $\Im_t^{x,\phi}$−predictable,
such that

$$P(\int_0^T (|\gamma_t(\omega)|^2 + |\delta_t(\omega)|^2 \pi(\Gamma))dt < \infty) = 1,$$

and for some $v^0 \in \widetilde{U}$, P^{v^0} - a.s. for all $t \in [0,T]$ (denote $g^v = g(t,x(.),\phi(.),v_t)$ as before)

$$(3.3)\ \int_0^t g^{v^0} ds + \inf_{v\in\widetilde{U}} E_v(\int_t^T g^v ds + h(x_T,\phi_T)\,|\Im_t^{w,q}) = J(v^0) + A_t + \int_0^t \gamma_s dw_s^{v^0} + \int_0^t \int_Z \delta_s q(ds,dz),$$

where A_t is increasing, integrable, $\Im_t^{x,\phi}$−predictable, $A_0 = 0$, and for all $v \in U$,

$$(3.4)\ \gamma_t(x.(\omega),\phi.(\omega))\sigma^{-1}(t,x.(\omega),\phi.(\omega))b(t,x.(\omega),\phi.(\omega),v)) +g(t,x.(\omega),\phi.(\omega),v) \geq \gamma_t(x.(\omega),\phi.(\omega))\sigma^{-1}(t,x.(\omega),\phi.(\omega))b(t,x.(\omega),\phi.(\omega),u_t^0(\omega)) +g(t,x.(\omega),\phi.(\omega),u_t^0(\omega)),\ m\times P - a.e.,$$

(Note that since $\Im_t^{w,\widetilde{N}} = \Im_t^{x,\phi}$, for all $t \in [0,T]$, hence a $\Im_t^{w,\widetilde{N}}$−adapted process $\gamma_t(\omega)$ can always be rewritten as $\gamma_t(x.(\omega),\phi.(\omega))$, more precisely,

$$\gamma_t(x.(\omega),\phi.(\omega)) = \gamma_t(x_s(\omega),\phi_s(\omega); s \leq t),$$

which satisfies $(3.2)_1$. Similarly, a $\Im_t^{w,\widetilde{N}}$−predictable process $\delta_t(\omega)$ can always be rewritten as $\delta_t(\omega) = \delta_t(x_s(\omega),\phi_s(\omega); s < t)$).
2) if $u^0 \in \widetilde{U}$ with $\gamma_t(\omega)$ and $\delta_t(\omega)$ satisfies (3.2), (3.3) (for some $v^0 \in \widetilde{U}$), and (3.4), then u^0 is optimal.

To prove Theorem 3.1 we need two Lemmas.

3.2 Lemma. Under the assumption of Theorem 2.1 denote

$$V_t = \inf_{v\in\widetilde{U}} E_v(\int_t^T g^v ds + h(x_T,\phi_T)\Big|\Im_t^{w,\widetilde{N}}),\ M_t^v = \int_0^t g^v ds + V_t, \text{ for } v \in \widetilde{U},$$

then
1) M_t^v is a submartingale under P_v, for $v \in \widetilde{U}$,
2) $M_t^{u^0}$ is a P_{u^0}−martingale, iff $u^0 \in \widetilde{U}$ is optimal.

Proof. Recall that for all $t \in [0,T]$

$$\Im_t^{w,\widetilde{N}} = \Im_t^{x,\phi}.$$

Denote

$$(3.5)\ \overline{f}^v(t,T) = \int_t^T g^v ds + h(x_T,\phi_T).$$

Note that as $v \in \widetilde{U}$ for all $A \in \Im_t^{x,\phi}$

$$\int_A E_v(\overline{f}^v(t,T)\Big|\Im_t^{x,\phi})\, dP_v = \int_A \overline{f}^v(t,T) z_T(\sigma^{-1}b^v) dP$$

$$= \int_A E(\overline{f}^v(t,T) z_{t,T}(\sigma^{-1}b^v) \Big| \Im_t^{x,\phi}) \, z_{0,t}(\sigma^{-1}b^v) dP$$
$$= \int_A E(\overline{f}^v(t,T) z_{t,T}(\sigma^{-1}b^v) \Big| \Im_t^{x,\phi}) \, dP_v,$$
where we have used the result
$$E(z_{t,T}(\sigma^{-1}b^v) \Big| \Im_t^{x,\phi}) = 1.$$
Hence,

(3.6) $E(\overline{f}^v(t,T) z_{t,T}(\sigma^{-1}b^v) \Big| \Im_t^{w,\widetilde{N}}) = E_v(\overline{f}^v(t,T) \Big| \Im_t^{w,\widetilde{N}})$.

(3.5) and (3.6) show that $E_v(\overline{f}^v(t,T) \Big| \Im_t^{w,\widetilde{N}})$ depends on $v(s), t \leq s \leq T$, only. Assume that 1) holds. Let us prove 2): If $u^0 \in \widetilde{U}$ is optimal, then
$$E_{u^0}(\int_0^T g^{u^0} dt + h(x_T, \phi_T) \Big| \Im_t^{w,\widetilde{N}})$$
$$= \int_0^t g^{u^0} dt + E_{u^0}(\int_t^T g^{u^0} dt + h(x_T, \phi_T) \Big| \Im_t^{w,\widetilde{N}})$$
$$= \int_0^t g^{u^0} dt + V_t = M_t^{u^0},$$
i.e., $\forall 0 \leq t \leq T$
$$E_{u^0}(M_T^{u^0} \Big| \Im_t^{w,\widetilde{N}}) = M_t^{u^0}.$$
However, by 1) $\forall 0 \leq t \leq t_1 \leq T$
$$E_{u^0}(M_T^{u^0} \Big| \Im_t^{w,\widetilde{N}}) \geq E_{u^0}(M_{t_1}^{u^0} \Big| \Im_t^{w,\widetilde{N}}) \geq M_t^{u^0}.$$
Hence, $M_t^{u^0}$ is a $P_{u^0}-$martingale. Conversely, if $M_t^{u^0}$ is a $P_{u^0}-$martingale, where $u^0 \in \widetilde{U}$, then
$$\inf_{v \in \widetilde{U}} J(v) = M_0^{u^0} = E_{u^0}(M_T^{u^0} \Big| \Im_0^{w,\widetilde{N}}) = E_{u^0}(\int_0^T g^{u^0} dt + h(x_T, \phi_T)).$$
Hence, u^0 is optimal.

Next we prove 1): Note that
$$E_v(\int_s^t g^v dt + \int_t^T g^v dt + h(x_T, \phi_T) \Big| \Im_s^{w,\widetilde{N}})$$
$$= E_v(\int_s^T g^v dt + h(x_T, \phi_T) \Big| \Im_s^{w,\widetilde{N}}) \geq V_s.$$
Hence, by interchanging $E_v(\cdot \Big| \Im_s^{w,\widetilde{N}})$ and infimum here (Lemma 16.A.5 of Elliot (1982), pp. 250) one verifies that for all $v \in \widetilde{U}$
$$E_v(\int_s^t g^v dt + V_t \Big| \Im_s^{w,\widetilde{N}}) \geq V_s.$$
Hence,
$$E_v(M_t^v \Big| \Im_s^{w,\widetilde{N}}) = E_v(\int_0^s g^v dt + \int_s^t g^v dt + V_t \Big| \Im_s^{w,\widetilde{N}}) \geq \int_0^s g^v dt + V_s = M_s^v,$$
i.e. $M_t^v$is a P_v- submartingale. ■

Now by the decomposition of submartingale (Doob-Meyer theorem) and Theorem 1.4 in Chapter 5 we have that for any $v \in \widetilde{U}$, $P^v - a.s.$

(3.7) $M_t^v = V_0 + A_t^v + N_t^v = V_0 + A_t^v + \int_0^t \gamma_s^v dw_s^v + \int_0^t \delta_s^v d\widetilde{N}_s$,for all $t \in [0,T]$,

where γ_t^v and $\delta_t^v$are $\Im_t^{w,\widetilde{N}} = \Im_t^{x,\phi}-$ adapted and $\Im_t^{w,\widetilde{N}}-$ predictable, respectively, such that
$$P(\int_0^t (|\gamma_t^v(\omega)|^2 + |\delta_t^v(\omega)|^2 \pi(\Gamma)) dt < \infty) = 1,$$

and A_t^v is a $\Im_t^{w,\widetilde{N}}-$ predictable, integrable and increasing process with $A_0^v = 0$. Now let us show that γ_t^v and δ_t^v in (3.7) do not depend on $v \in \widetilde{U}$. Actually, we have the following

3.3 Lemma. For any $u, v \in \widetilde{U}$,

$v_t^u = v_t^v$, $m \times P - a.e.$

$\delta_t^u = \delta_t^v$, $m \times P \times \pi - a.e.$.

Proof. By

(3.8) $M_t^u = V_0 + A_t^u + \int_0^t \gamma_s^u dw_s^u + \int_0^t \delta_s^u d\widetilde{N}_s$
$= V_0 + A_t^u + \int_0^t \gamma_s^u \sigma^{-1}(dx_s - b^u ds - c_s d\widetilde{N}_s - d\phi_s)$
$+ \int_0^t \delta_s^u d\widetilde{N}_s = V_0 + A_t^u + \int_0^t \gamma_s^u \sigma^{-1}(b^v - b^u)ds$
$+ \int_0^t \gamma_s^u dw_s^v + \int_0^t \delta_s^u d\widetilde{N}_s = \int_0^t g^u ds + V_t.$

Hence,

$V_t = V_0 + A_t^u + \int_0^t \gamma_s^u \sigma^{-1}(b^v - b^u)ds - \int_0^t g^u ds$
$+ \int_0^t \gamma_s^u dw_s^v + \int_0^t \delta_s^u d\widetilde{N}_s,$

and

$M_t^v = \int_0^t g^v ds + V_t = V_0 + A_t^u + \int_0^t (g^v - g^u)ds + \int_0^t \gamma_s^u \sigma^{-1}(b^v - b^u)ds$
$+ \int_0^t \gamma_s^u dw_s^v + \int_0^t \delta_s^u d\widetilde{N}_s.$

On the other hand, by (3.7)

(3.9) $M_t^v = V_0 + A_t^v + \int_0^t \gamma_s^v dw_s^v + \int_0^t \delta_s^v d\widetilde{N}_s.$

Hence, by the uniqueness of decomposition of submartingales and special semimartingales it is seen that $P^v - a.s.$ for all $t \in [0, T]$

(3.10) $A_t^v = A_t^u + \int_0^t (g^v - g^u + \gamma_s^u \sigma^{-1}(b^v - b^u))ds;$

and

$v_t^u = v_t^v$, $m \times P - a.e.$

$\delta_t^u = \delta_t^v$, $m \times P \times \pi - a.e.$ ■

Now we are in a position to prove Theorem 3.1.

Proof. 1): If $u^0 \in \widetilde{U}$ is optimal, then by Lemma 3.2 $M_t^{u^0}$ is a $P^{u^0}-$martingale. Hence if one takes $u = u^0$ in (3.8) then

(3.11) $A_t^{u^0} = 0$, for all $t \in [0, T]$.

Since for any $v \in \widetilde{U}$, A_t^v in (3.10) is increasing. Hence by (3.10) and (3.11) $P^v - a.s.$

$g^v - g^{u^0} + \gamma_s^{u^0} \sigma^{-1}(b^v - b^{u^0}) \geq 0$, $a.e.\ t.$

Applying that $P^v \sim P$ and Lemma 3.3, 1) follows.

2): Set $u = u^0$ in (3.9), then for any $v \in \widetilde{U}$ by (3.9), (3.10) and the assumption one has that

$E_v M_t^v = V_0 + E_v A_t^v \geq V_0 + E_v A_t^{u^0},$

or

$J(v) = E_v M_T^v \geq V_0 + E_v A_T^{u^0}.$

Now take $v_n \in \widetilde{U}$ such that as $n \to \infty$

$J(v_n) \to V_0 = \inf_{v \in \widetilde{U}} J(v).$

Then

$(3.11)_1$ $Ez_{T\wedge\tau_N}(\sigma^{-1}b^{v_n})(A^{u^0}_{T\wedge\tau_N}\wedge N) = Ez_T(\sigma^{-1}b^{v_n})(A^{u^0}_{T\wedge\tau_N}\wedge N)$
$\leq E_{v_n}A^{u^0}_T \to 0.$

Denote
$$\tau_N = \inf\{t\geq 0 : \|x\|_t + \|\phi\|_t > N\}.$$
Then
$$E\left|z_{T\wedge\tau_N}(\sigma^{-1}b^{v_n})\right|^2 \leq k_N < \infty.$$
Hence, there exists a subsequence of $\{n\}$, for simplicity denote it by $\{n\}$ again, such that as $n\to\infty$
$$z_{T\wedge\tau_N}(\sigma^{-1}b^{v_n}) \to \lambda_{T,N}, \text{ weakly in } L^2(\Omega,\Im,P).$$
By Lemma 3.4 below
$$\lambda_{T,N} > 0,\ P-a.s.$$
However, by $(3.11)_1$ as $n\to\infty$ (denote $\overline{A}^{u^0}_t = A^{u^0}_t\wedge N$)
$$E_{v_n}\overline{A}^{u^0}_{T\wedge\tau_N} = Ez_{T\wedge\tau_N}(\sigma^{-1}b^{v_n})\overline{A}^{u^0}_{T\wedge\tau_N} \to E\lambda_{T,N}\overline{A}^{u^0}_{T\wedge\tau_N} = 0,$$
for all $N = 1,2,\cdots$.

Hence, from this for all $N = 1,2,\cdots$
$$\overline{A}^{u^0}_{T\wedge\tau_N} = 0,\ P-a.s.$$
Therefore,
$$A^{u^0}_T = 0,\ P^{u^0}-a.s.$$
It shows that $M^{u^0}_t$ is a $P^{u^0}-$ martingale. By Lemma 3.2 u^0 is optimal. ■

3.4 Lemma. If $|\varphi_n|\leq k_N$, for all n, and as $n\to\infty$
$$z_{T\wedge\tau_N}(\varphi_n) \to A_{T,N}, \text{ weakly in } L^2(\Omega,\Im,P), (*)$$
then
$$A_{T,N} > 0,\ P-a.s.$$
Proof. By (*) there exists a subsequence of $\{n\}$, denote it by $\{n\}$ again, such that as $n\to\infty$
$$g^N_n(T) \to A_{T,N}, \text{ in } L^2(\Omega,\Im,P),$$
where
$$g^N_n(t) = n^{-1}\sum_{i=1}^n z_{t\wedge\tau_N}(\varphi_i).$$
By Ito's formula

(3.12) $dg^N_n(t) = g^N_n(t)\left\langle\overline{\varphi}^N_n(t), dw_t\right\rangle,$

where
$$\overline{\varphi}^N_n(t) = \sum_{i=1}^n z_{t\wedge\tau_N}(\varphi_i)\varphi_i/\sum_{i=1}^n z_{t\wedge\tau_N}(\varphi_i).$$
Obviously, one has that
$$\left|\overline{\varphi}^N_n(t)\right| \leq k_N.$$
Set now
$$\overline{g}^N_n(t) = z_t(\overline{\varphi}^N_n).$$
Then by Ito's formula it satisfies a SDE as (3.12). By the pathwise uniqueness of solution to (3.12)
$$g^N_n(t) = \overline{g}^N_n(t) = z_t(\overline{\varphi}^N_n).$$
Hence, as $n\to\infty$
$$z_t(\overline{\varphi}^N_n) \to A_{T,N}, \text{ in } L^2(\Omega,\Im,P),$$

and there exists a subsequence of $\{n\}$, denote it by $\{n\}$ again, such that as $n \to \infty$

(3.13) $z_t(\overline{\varphi}_n^N) \to A_{T,N}$, $P-a.s.$

Now set

$A = \{\omega : A_{T,N}(\omega) = 0\}$.

Then by (3.13) as $n \to \infty$

$\int_0^T \overline{\varphi}_n^N dw_s - \frac{1}{2}\int_0^T \left|\overline{\varphi}_n^N\right|^2 ds \to -\infty$, as $\omega \in A$.

Since $\left|\overline{\varphi}_n^N(t)\right| \leq k_N$. Hence as $n \to \infty$

$\int_0^T \overline{\varphi}_n^N dw_s \to -\infty$, as $\omega \in A$.

Note that

$E\left|\int_0^T \overline{\varphi}_n^N dw_s\right|^2 = E\int_0^T \left|\overline{\varphi}_n^N\right|^2 ds \leq k_N T < \infty.$

By Fatou's lemma

$E(\lim_{n\to\infty}\left|\int_0^T \overline{\varphi}_n^N dw_s\right| \cdot I_A) \leq (k_N T)^{\frac{1}{2}} < \infty.$

Hence, $P(A) = 0$. ■

Finally, in this section we can derive the following useful corollary. Recall that at the beginning of this Chapter we have already written

$\Omega = D \times C$, $\Im = \vee_{t\geq 0}\Im_t$, $\Im_t = \sigma((x_s, \phi_s); s \leq t, x \in D, \phi \in C)$.

Now let us also introduce the $\Im_t$-predictable σ-algebra as follows: Let $\sum_p$ be the σ-algebra generated by all left continuous $\Im_t$-adapted R^d- valued left continuous processes. In the following corollary we will restrict all processes on $t \in [0,T]$, i.e. we denote by D (C) the totality of all R^d- valued cadlag (continuous) functions defined on $[0,T]$, and $\Im = \vee_{t\in[0,T]}\Im_t$, etc.

3.5 Corollary. Assume that

$b = b(t,x,\phi,u) : [0,T] \times D \times C \times U \to R^d,$

$\sigma = \sigma(t,x,\phi) : [0,T] \times D \times C \to R^{d\otimes d},$

$c = c(t,x,\phi) : [0,T] \times D \times C \to R^d,$

$g = g(t,x,\phi,u) : [0,T] \times D \times C \times U \to R^1,$

$h = h(x,\phi) : R^d \times R^d \to R^1$

are Borel measurable such that for fixed u they are $\Im_t$-predictable (i.e., $\sum_p$-measurable), where $U \subset R^r$ is a compact set. If b and g are continuous in $u \in \widetilde{U}$, then an optimal control exists.

Proof. By assumption there exists a $\Im_t$-predictable map (Stroock & Varadhan, 1979, or R.J. Elliot, 1982)

$u^0(t,x,\phi,\gamma) : [0,T] \times D \times C \times R^d \to U,$

such that for all $(t,x,\phi,\gamma) \in [0,T] \times D \times C \times R^d$

$\gamma\sigma^{-1}(t,x,\phi)b(t,x,\phi,u^0(t,x,\phi,\gamma))$

$+g(t,x,\phi,u^0(t,x,\phi,\gamma))$

$= \inf_{u\in\widetilde{U}}[\gamma\sigma^{-1}(t,x,\phi)b(t,x,\phi,u) + g(t,x,\phi,u)].$

Now set

$u^0 = u^0(t,x.(\omega),\phi.(\omega),\gamma_t(x.(\omega),\phi.(\omega))),$

where γ_t comes from (3.7) with some δ_t, which exist by Lemma 3.2 here and Theorem 1.4 in Chapter 5. Thus $u^0 \in \widetilde{U}$. Moreover, by 2) of Theorem 3.1 u^0 is optimal.. ■

Chapter 7

Stochastic Population Control

In this chapter we will discuss the applications of results of the previous chapters to the stochastic population control and Neurophysiological control problems. The first one is for RSDEs with jumps and the second one is without jumps.

7.1 Stochastic Population Control Model and Maximum Principle

The theory developed in the previous chapter will now be applied to the problem of stochastic population control. Consider the following population control system

$$(1.1)\begin{cases} dx_t = (A(t)x_t + B(t)x_t\beta_t)dt + \sigma(t, x_t, \phi_t)dw_t \\ + \int_Z c(t, x_{t-}, \phi_t, z)q(dt, dz) \; + d\phi_t, \\ x_0 = x, \; t \in [0, T], \\ x_t \in \overline{R}^d_+, \text{for all } t \in [0, T], \\ \text{ where } R^d_+ = \left\{x = (x^1, \cdots, x^d) \in R^d : x^i > 0, 1 \leq i \leq d\right\}, \\ \phi_t \text{ is a } R^d - \text{ valued } \Im_t - \text{adapted continuous process with bounded} \\ \text{variation on any finite interval such that } \phi_0 = 0, \text{ and} \\ \phi_t = \int_0^t n(s)d\left|\phi\right|_s, \\ \left|\phi\right|_t = \int_0^t I_{\partial R^d_+}(x_s)d\left|\phi\right|_s, \\ n(t) \in \aleph_{x_t}, \text{ as } x_t \in \partial R^d_+, \end{cases}$$

where x^i_t can be explained as the size of the population with age between $[i, i+1)$, and $x_t = (x^1_t, \cdots, x^d_t)$; β_t is the specific fertility rate of females.

1.1 Remark. The first equation without the last three terms in (1.1) for the deterministic case (i.e. no random perturbation and no reflection) was studied by Yu, Guo and Zhu (1987), where they consider that

$$A(t)=\begin{pmatrix} -(1+\eta_1(t)) & 0 & \cdots & \cdots & 0 \\ 1 & -(1+\eta_2(t)) & 0 & \cdots & 0 \\ \cdots & \cdots & \cdots & \cdots & \cdots \\ 0 & \cdots & 0 & 1 & -(1+\eta_{r_m}(t)) \end{pmatrix},$$

$$B(t)=\begin{pmatrix} 0 & \cdots & 0 & b_{r_1}(t) & \cdots & b_{r_2}(t) & 0 & \cdots & 0 \\ 0 & \cdots & \cdots & \cdots & \cdots & \cdots & \cdots & \cdots & 0 \end{pmatrix},$$

$$b_i(t)=(1-\mu_{00}(t))k_i(t)h_i(t)=0,\ i=r_1,\cdots,r_2; 1<r_1<r_2<r_m,$$

$\mu_{00}(t)$ is the death rate of babies, $\eta_i(t)$ is the forward death rate by ages, $k_i(t)$ and $h_i(t)$ are the corresponding sex rate and fertility model, respectively, x_t^i is the size of the population with ages between $[i,i+1)$,

$$x_t=(x_t^1,\cdots,x_t^{r_m}),$$

r_m is the largest age of the people, and β_t is the specific fertility rate of females. Here our system (1.1) is disturbed by a stochastic white noise (the usual random fluctuation) and by a centralized random point process (e.g. by a war, earthquake or serious disease etc.). The size of population should be non-negative, i.e. $x_t^i\geq 0, 1\leq i\leq r_m$. Hence we use the system (1.1) as a stochastic population evolution model. We use the following denotation

(1.2) the RSDE=RSDE (1.1) with $A(t)$ and $B(t)$ defined in Remark 1.1.

Write

$$\widetilde{U}=\{\beta:\beta=\beta_t \text{ is } \Im_t-\text{adapted and } \beta_t\in[\beta_0,\beta_1]\},$$

where β_0 and β_1 are constants. The cost functional here is, for all $\beta\in\widetilde{U}$

$$J(\beta)=E_{P^\beta}(\int_0^T g(t,x_t,\phi_t,\beta_t)dt+h(x_T,\phi_T)),$$

where (x_t,ϕ_t) is the pathwise unique strong solution of the following RSDE: $P-a.s.$

$$(1.3)\begin{cases} dx_t=\sigma(t,x_t,\phi_t)dw_t+\int_Z c(t,x_{t-},\phi_t,z)q(dt,dz)+d\phi_t, t\in[0,T], \\ x_0=x\in\overline{R}_+^d \\ x_t\in\overline{R}_+^d,\ \text{for all } t\in[0,T], \\ \text{and other statements in (1.1) holds;} \end{cases}$$

and $dP^\beta=z_T^w(\sigma^{-1}b^\beta)dP$, $b^\beta=A(t)x_t+B(t)x_t\beta_t$.

The population optimal control problem here is to minimize $J(\beta)$ among all $\beta\in\widetilde{U}$. By the above theorem we have

1.2 Theorem. Assume that

1° $A(t)$ and $B(t)$ are bounded,

2° $\sigma(t,x)$ and $c(t,x,\phi,z)$ satisfies Assumption 2.1 of Chapter 6 for all $(t,x,\phi,z)\in[0,T]\times R^d\times R^d\times Z$.

Then 1) (1.3) has a pathwise unique strong solution (x_t,ϕ_t), which is also a weak solution of (1.1) for each $\beta\in\widetilde{U}$ on the probability space $(\Omega,\Im,(\Im_t),P^\beta)$.

Furthermore, assume that

3° condition 1° in Theorem 3.1 of Chapter 6 holds, i.e.

$$|g(t,x,\phi,y_1)|+|h(y_2,y_3)|\leq k_0(1+|x|+|\phi|+|y_1|+|y_2|+|y_3|),$$

for all $(t,x,\phi,y_1,y_2,y_3)\in[0,T]\times R^d\times R^d\times R^d\times R^d\times R^d$,

4° $g(t,x,\phi,\beta)$ is continuous in $\beta \in [\beta_0,\beta_1]$.

Then 2) an optimal control $\beta^0 \in \widetilde{U}$ exists, which satisfies

$$\gamma_t(x_t(\omega),\phi_t(\omega))\sigma^{-1}(t,x_t(\omega),\phi_t(\omega))B(t)x_t(\omega)\beta$$
$$+g(t,x_t(\omega),\phi_t(\omega),\beta)$$
$$\geq \gamma_t(x_t(\omega),\phi_t(\omega))\sigma^{-1}(t,x_t(\omega),\phi_t(\omega))B(t)x_t(\omega)\beta_t^0(\omega)$$
$$+g(t,x_t(\omega),\phi_t(\omega),\beta_t^0(\omega)), \text{ for all } \beta \in\in [\beta_0,\beta_1], m\times P-a.e.,$$

where γ_t comes from (3.7) in chapter 6 with some δ_t, and (x_t,ϕ_t) is the pathwise unique strong solution of (1.1).

Moreover, if

5° $g(t,x,\phi)$ does not depend on the control directly;

then 3) for $(t,x,\phi) \in [0,T]\times \overline{R}_+^d \times \overline{R}_+^d$ set

$$\beta^0(t,x_t(\omega),\phi_t(\omega)) = \beta_0 I_{\gamma_t(x_t(\omega),\phi_t(\omega))\sigma^{-1}(t,x_t(\omega),\phi_t(\omega))B(t)x_t(\omega)\geq 0}$$
$$+\beta_1 I_{\gamma_t(x_t(\omega),\phi_t(\omega))\sigma^{-1}(t,x_t(\omega),\phi_t(\omega))B(t)x_t(\omega)<0},$$

we have that β^0 is an optimal control.

Proof. 1) is derived by Theorem 3.23 in Chapter 3 (i.e. (1.3) has a pathwise unique strong solution) and the Girsanov theorem -Theorem 1.2 in Chapter 6 (i.e. (1.1) has a weak solution for each $\beta \in \widetilde{U}$) Conclusion 2) and 3) are derived from Corollary 3.5 in Chapter 6. ■

1.3 Remark. Under assumption 2.1 in Chapter 6 the jump term in RSDE (1.1) actually can be rewritten as

$$\int_Z c(t,x_{t-},\phi_t,z)q(dt,dz) = c(t,x_{t-},\phi_t)d\widetilde{N}_t,$$

where $d\widetilde{N}_t = q(dt,\Gamma)$.

7.2 Pathwise Stochastic Population Control and Stability of Population

In this section we will discuss the pathwise stochastic population control. For this purpose we consider the admissible control set as

$$\widetilde{U} = \left\{ \begin{array}{c} u = u(t,x(.),\phi(.)) \in U : u(t,x(.),\phi(.)) \text{ is } \Im_t-\text{adapted} \\ \text{such that it makes the following RSDE (2.1)} \\ \text{have a pathwise unique strong solution,} \end{array} \right\}$$

where U is a compact set in R^r :

$$(2.1)\left\{ \begin{array}{l} dx_t = b(t,x(.),\phi(.),u(t,x(.),\phi(.)))dt + \sigma(t,x(.),\phi(.))dw_t \\ \qquad + \int_Z c(t-,x(.),\phi(.),z)q(dt,dz) \ + d\phi_t, \\ x_0 = x_0 \in \overline{R}_+^d \\ \text{the other statements in (1.1) remains true} \end{array} \right.$$

We have the following convergence theorem of solutions to (2.1).

2.1 Theorem Assume that

1° there exists a constant $k_0 \geq 0$ such that

$$|b(t,x(.),\phi(.),u)|^2 + \|\sigma(t,x(.),\phi(.))\|^2$$
$$+\int_Z |c(t,x(.),\phi(.),z)|^2 \pi(dz) \leq k_0(1+\|x\|_t^2+\|\phi\|_t^2),$$

for all $u \in U, (x(.), \phi(.)) \in D \times C$,
where $\|x\|_t^2 = \sup_{s \leq t} |x_s|^2$,
2° $\|\sigma(t, x(.), \phi(.)) - \sigma(t, y(.), \psi(.))\|^2$
$+ \int_Z |c(t, x(.), \phi(.), z) - c(t, y(.), \psi(.), z)|^2 \pi(dz) \leq k(t)\rho(\|x - y\|_t^2)$,
where $k(t) \geq 0, \rho(u)$ is continuous, strictly increasing, concave, $\rho(0) = 0$, and they are such that for any $0 < T < \infty$
$\int_0^T k(t)dt < \infty, \int_{0+} du/\ \rho(u) = \infty$;
3° $b(t, x(.), \phi(.), u)$ is continuous in $(x(.), \phi(.), u) \in D \times C \times U$;
4° for $u^n \in \widetilde{U}, n = 0, 1, 2, \cdots$ one of the following conditions is satisfied:
(i) $2(x_t - y_t) \cdot (b(t, x(.), \phi(.), u^n(t, x(.), \phi(.))) - b(t, y(.), \psi(.), u^n(t, y(.), \psi(.))))$
$\leq k(t)\rho(\|x - y\|_t^2), n = 1, 2, \cdots$
where $k(t) \geq 0, \rho(u)$ satisfy the same condition as that in 2°, and
$\lim_{n \to \infty} u^n(t, x(.), \phi(.)) = u^0(t, x(.), \phi(.))$, for all $(x(.), \phi(.)) \in D \times C$;
(ii) $2(x_t - y_t) \cdot (b(t, x(.), \phi(.), u^0(t, x(.), \phi(.))) - b(t, y(.), \psi(.), u^0(t, y(.), \psi(.))))$
$\leq k(t)\rho(\|x - y\|_t^2)$,
where $k(t) \geq 0, \rho(u)$ satisfy the same condition as that in 2°, and for any $T < \infty$
$\lim_{n \to \infty} \sup_{(t, x(.), \phi(.)) \in [0,T] \times D \times C} |b(t, x(.), \phi(.), u^n(t, x(.), \phi(.)))$
$- b(t, x(.), \phi(.), u^0(t, x(.), \phi(.)))| = 0$;
5° $E|x^n(0) - x^0(0)|^2 \to 0$, as $n \to \infty$,
where we denote
$x^n(t) = x^{u^n}(t),\ \phi^n(t) = \phi^{u^n}(t)$,
and $(x^n(t), \phi^n(t))$ is the pathwise unique strong solution of (2.1) corresponding to $u^n \in \widetilde{U}$.
Then for all $t \geq 0$, as $n \to \infty$
$E \sup_{s \leq t} |x_s^n - x_s^0|^2 \to 0$.
In case that condition 4° is changed into
$4^{\circ\prime}$ for $u^n \in \widetilde{U}, n = 0, 1, 2, \cdots$ one of the following conditions is satisfied:
(i) $|b(t, x(.), \phi(.), u^n(t, x(.), \phi(.))) - b(t, y(.), \psi(.), u^n(t, y(.), \psi(.)))|^2$
$\leq k(t)\rho(\|x - y\|_t^2 + \|\phi - \psi\|_t^2), n = 1, 2, \ldots$,
where $k(t) \geq 0, \rho(u)$ is continuous, strictly increasing, concave, $\rho(0) = 0$, and they are such that for any $0 < T < \infty$
$\int_0^T k(t)dt < \infty, \int_{0+} du/\ \rho(u) = \infty$;
and
$\lim_{n \to \infty} u^n(t, x(.), \phi(.)) = u^0(t, x(.), \phi(.))$, for all $(x(.), \phi(.)) \in D \times C$;
(ii) $|b(t, x(.), \phi(.), u^0(t, x(.), \phi(.))) - b(t, y(.), \psi(.), u^0(t, y(.), \psi(.)))|^2$
$\leq k(t)\rho(\|x - y\|_t^2 + \|\phi - \psi\|_t^2)$,
where $k(t) \geq 0, \rho(u)$ satisfy the same conditions as that in 4°, and for any $T < \infty$
$\lim_{n \to \infty} \sup_{(t, x(.), \phi(.)) \in [0,T] \times D \times C} |b(t, x(.), \phi(.), u^n(t, x(.), \phi(.)))$
$- b(t, x(.), \phi(.), u^0(t, x(.), \phi(.)))| = 0$;
then we will have for all $t \geq 0$, as $n \to \infty$

$E[\sup_{s\leq t}\left|x_s^n - x_s^0\right|^2 + \sup_{s\leq t}\left|\phi_s^n - \phi_s^0\right|^2] \to 0.$

Proof. By Ito's formula

(2.2) $\left|x_t^n - x_t^0\right|^2 = \left|x_0^n - x_0^0\right|^2 + \int_0^t [2(x_s^n - x_s^0)\cdot(b(s,x^n(.),\phi^n(.),u^n(t,x^n(.),\phi^n(.)))$
$-b(s,x^0(.),\phi^0(.),u^0(t,x^0(.),\phi^0(.)))$
$+\left\|\sigma(s,x^n(.),\phi^n(.)) - \sigma(s,x^0(.),\phi^0(.))\right\|^2]ds + 2\int_0^t (x_s^n - x_s^0)\cdot d(\phi_s^n - \phi_s^0)$
$+2\int_0^t (x_s^n - x_s^0)\cdot(\sigma(s,x^n(.),\phi^n(.)) - \sigma(s,x^0(.),\phi^0(.)))dw_s$
$+2\int_0^t (x_s^n - x_s^0)\cdot\int_Z (c(s-,x^n(.),\phi^n(.),z) - c(s-,x^0(.),\phi^0(.),z))q(ds,dz)$
$+\int_0^t\int_Z \left|c(s-,x^n(.),\phi^n(.),z) - c(s-,x^0(.),\phi^0(.),z)\right|^2 p(ds,dz).$

Under conditions $1^\circ - 3^\circ$ assume that condition 4° holds. Then in case (i) by assumption as $t\leq T$

(2.3) $E\sup_{s\leq t}\left|x_s^n - x_s^0\right|^2 \leq E\left|x_0^n - x_0^0\right|^2$
$+k_T\int_0^t k(r)\rho(E\sup_{s\leq r}\left|x_s^n - x_s^0\right|^2)dr$
$+E\int_0^t E\sup_{s\leq r}\left|x_s^n - x_s^0\right|^2 ds$
$+E\int_0^t |b(s,x^0(.),\phi^0(.),u^n(t,x^0(.),\phi^0(.)))$
$-\, b(s,x^0(.),\phi^0(.),u^0(t,x^0(.),\phi^0(.)))|^2 ds$

where we have applied Remark 2.2 in Chapter 3

$\int_0^t (x_s^n - x_s^0)\cdot d(\phi_s^n - \phi_s^0) \leq 0.$

By the condition $1^\circ, 3^\circ$ and Lebesgue's dominated convergence theorem one finds that as $n\to\infty$

the last term of (2.3) $\to 0.$

Hence,

$\overline{lim}_{n\to\infty}E\sup_{s\leq t}\left|x_s^n - x_s^0\right|^2 \leq \int_0^t \widetilde{k}(s)\widetilde{\rho}(\overline{lim}_{n\to\infty}E\sup_{s\leq r}\left|x_s^n - x_s^0\right|^2)dr,$

where $\widetilde{k}(s)$ and $\widetilde{\rho}(u)$ still satisfy the same condition in 2° for $k(s)$ and $\rho(u)$. Therefore,

$\overline{lim}_{n\to\infty}E\sup_{s\leq t}\left|x_s^n - x_s^0\right|^2 = 0.$

Now for case (ii) of 4° as $t\leq T$

E(the 2nd term of (2.2)) $\leq \int_0^t k(s)\rho(E\sup_{s\leq r}\left|x_s^n - x_s^0\right|^2)ds$
$+T\sup_{(t,x(.),\phi(.))\in[0,T]\times D\times C}|b(t,x(.),\phi(.),u^n(t,x(.),\phi(.)))$
$-\, b(t,x(.),\phi(.),u^0(t,x(.),\phi(.)))|.$

Hence, the conclusion still follows.

Now assume that condition $4^{\circ\prime}$ holds. Then for case (i) by (2.2) one gets that

(2.4) $E\sup_{s\leq t}\left|x_s^n - x_s^0\right|^2 \leq E\left|x_0^n - x_0^0\right|^2$
$+k_T\int_0^t k(s)\rho(E\sup_{s\leq r}\left|x_s^n - x_s^0\right|^2 + E\sup_{s\leq r}\left|\phi_s^n - \phi_s^0\right|^2)dr$
$+\int_0^t E\sup_{s\leq r}\left|x_s^n - x_s^0\right|^2 ds + E\int_0^t |b(s,x^0(.),\phi^0(.),u^n(t,x^0(.),\phi^0(.)))$
$-\, b(s,x^0(.),\phi^0(.),u^0(t,x^0(.),\phi^0(.)))|^2 ds$

On the other hand,

(2.5) $|\phi_t^n - \phi_t^n|^2 \leq 5(\left|x_t^n - x_t^0\right|^2 + \left|x_0^n - x_0^0\right|^2$
$+\left|\int_0^t (b(s,x^n(.),\phi^n(.),u^n(t,x^n(.),\phi^n(.)))\right.$

$- b(s,x^0(.),\phi^0(.),u^0(t,x^0(.),\phi^0(.)))ds|^2$
$+\left|\int_0^t(\sigma(s,x^n(.),\phi^n(.)) - \sigma(s,x^0(.),\phi^0(.)))dw_s\right|^2$
$+\left|\int_0^t\int_Z(c(s-,x^n(.),\phi^n(.),z) - c(s-,x^0(.),\phi^0(.),z))q(ds,dz)\right|^2).$

Thus by (2.4) and (2.5)

$$E[\sup_{s\leq t}|x_s^n - x_s^0|^2 + \sup_{s\leq t}|\phi_s^n - \phi_s^0|^2] \leq 6E|x_0^n - x_0^0|^2$$
$$+\tilde{k}_T\int_0^t k(r)\rho(E\sup_{s\leq r}|x_s^n - x_s^0|^2 + E\sup_{s\leq r}|\phi_s^n - \phi_s^0|^2)dr$$
$$+E\int_0^t E\sup_{s\leq r}|x_s^n - x_s^0|^2 ds + E\int_0^t |b(s,x^0(.),\phi^0(.),u^n(t,x^0(.),\phi^0(.)))$$
$$- b(s,x^0(.),\phi^0(.),u^0(t,x^0(.),\phi^0(.)))|^2 ds.$$

Hence,

$$\overline{lim}_{n\to\infty}E[\sup_{s\leq t}|x_s^n - x_s^0|^2 + \sup_{s\leq t}|\phi_s^n - \phi_s^0|^2]$$
$$\leq \int_0^t \tilde{k}(s)\tilde{\rho}(\overline{lim}_{n\to\infty}E[\sup_{s\leq r}|x_s^n - x_s^0|^2 + \sup_{s\leq r}|\phi_s^n - \phi_s^0|^2])dr,$$

where $\tilde{k}(s)$ and $\tilde{\rho}(u)$ still satisfy the same condition in 2° for $k(s)$ and $\rho(u)$. Therefore,

$$\overline{lim}_{n\to\infty}E[\sup_{s\leq t}|x_s^n - x_s^0|^2 + \sup_{s\leq t}|\phi_s^n - \phi_s^0|^2] = 0.$$

In case (ii) of condition 4°′ the proof is also similar. ■

In the case that condition 4° in Theorem 2.1 is changed into some local condition, the conclusion will be weaker. Actually, we have the following

2.2 Corollary. Assume that condition 4° in Theorem 2.1 is changed into 4°″ for $u^n \in \tilde{U}, n = 0,1,2,\cdots$ one of the following conditions is satisfied:

(i) $2(x_t - y_t)\cdot(b(t,x(.),\phi(.),u^n(t,x(.),\phi(.))) - b(t,x(.),\phi(.),u^n(t,y(.),\psi(.))))$
$\leq k_N(t)\rho_N(\|x-y\|_t^2)$, as $\|x\|_t \leq N$, $n = 1,2,\cdots$

where $k_N(t) \geq 0, \rho_N(u)$ satisfy the same condition as that in 4°, which depend on N only and are independent of $n = 1,2,...$; and

$2(x_t - y_t)\cdot(b(t,x(.),\phi(.),u^n(t,x(.),\phi(.))) - b(t,y(.),\psi(.),u^n(t,x(.),\phi(.))))$
$\leq k(t)\rho(\|x-y\|_t^2), n = 1,2,\cdots;$

where $k(t)$ and $\rho(v)$ satisfy the same condition as that in 2° of Theorem 2.1; moreover,

$\lim_{n\to\infty}u^n(t,x(.),\phi(.)) = u^0(t,x(.),\phi(.))$, for all $(x(.),\phi(.)) \in D\times C$;

(ii) $2(x_t - y_t)\cdot(b(t,x(.),\phi(.),u^0(t,x(.),\phi(.))) - b(t,x(.),\phi(.),u^0(t,y(.),\psi(.))))$
$\leq k_N(t)\rho_N(\|x-y\|_t^2)$, as $\|x\|_t \leq N$,

where $k_N(t) \geq 0, \rho_N(u)$ satisfy the same condition as that in 2°, which depend on N only, and

$2(x_t - y_t)\cdot(b(t,x(.),\phi(.),u^0(t,x(.),\phi(.))) - b(t,y(.),\psi(.),u^0(t,x(.),\phi(.))))$
$\leq k(t)\rho(\|x-y\|_t^2)$,

where $k(t)$ and $\rho(v)$ satisfy the same condition as that in 4° of Theorem 2.1; moreover, for any $T < \infty$

$\lim_{n\to\infty}\sup_{(t,x(.),\phi(.))\in[0,T]\times D\times C}|b(t,x(.),\phi(.),u^n(t,x(.),\phi(.)))$
$- b(t,x(.),\phi(.),u^0(t,x(.),\phi(.)))| = 0;$

and assume that the other conditions in Theorem 2.1 are unchanged. Then for all $t \geq 0$, as $n \to \infty$

$E \sup_{s \leq t \wedge \tau_N} \left| x_s^n - x_s^0 \right|^2 \to 0$, for each $N = 1, 2, ...,$

where $\tau_N = \inf \left\{ t \geq 0 : \left\| x^0 \right\|_t > N \right\}$.

Furthermore, if condition $4^{\circ\prime\prime}$ above is changed into

$4^{\circ\prime\prime\prime}$ for $u^n \in \widetilde{U}, n = 0, 1, 2, \cdots$ one of the following conditions is satisfied:

(i) $|b(t, x(.), \phi(.), u^n(t, x(.), \phi(.))) - b(t, x(.), \phi(.), u^n(t, y(.), \psi(.)))|^2$
$\leq k_N(t)\rho_N(\|x - y\|_t^2 + \|\phi - \psi\|_t^2)$,as $\|x\|_t \leq N, \|\phi\|_t \leq N$, for all $n = 1, 2, ...;$

and

$|b(t, x(.), \phi(.), u^n(t, x(.), \phi(.))) - b(t, y(.), \psi(.), u^n(t, x(.), \phi(.)))|^2$
$\leq k(t)\rho(\|x - y\|_t^2 + \|\phi - \psi\|_t^2)$, for all $n = 1, 2, ...;$

where $k(t)$ and $\rho(v)$ satisfy the same condition as that in 4° of Theorem 2.1, so do $k_N(t)$ and $\rho_N(v)$, which depend on N only and are independent of $n = 1, 2, ...;$ moreover,

$\lim_{n \to \infty} u^n(t, x(.), \phi(.)) = u^0(t, x(.), \phi(.))$, for all $(x(.), \phi(.)) \in D \times C;$

(ii) $\left|b(t, x(.), \phi(.), u^0(t, x(.), \phi(.))) - b(t, x(.), \phi(.), u^0(t, y(.), \psi(.)))\right|^2$
$\leq k_N(t)\rho_N(\|x - y\|_t^2 + \|\phi - \psi\|_t^2)$,as $\|x\|_t \leq N, \|\phi\|_t \leq N,$

and

$\left|b(t, x(.), \phi(.), u^0(t, x(.), \phi(.))) - b(t, y(.), \psi(.), u^0(t, x(.), \phi(.)))\right|^2$
$\leq k(t)\rho(\|x - y\|_t^2) + \|\phi - \psi\|_t^2),$

where $k(t)$ and $\rho(v)$ satisfy the same condition as that in 4° of Theorem 2.1, so do $k_N(t)$ and $\rho_N(v)$, which depend on N only and are independent of $n = 1, 2, ...;$ moreover, for any $T < \infty$

$\lim_{n \to \infty} \sup_{(t, x(.), \phi(.)) \in [0,T] \times D \times C} |b(t, x(.), \phi(.), u^n(t, x(.), \phi(.)))$
$- b(t, x(.), \phi(.), u^0(t, x(.), \phi(.)))\Big| = 0;$

and keep the other conditions unchanged, then for all $t \geq 0$, as $n \to \infty$

$E[\sup_{s \leq t \wedge \tau_N} \left| x_s^n - x_s^0 \right|^2 + \sup_{s \leq t \wedge \tau_N} \left| \phi_s^n - \phi_s^0 \right|^2] \to 0$, for each $N = 1, 2, ...,$

where $\tau_N = \inf \left\{ t \geq 0 : \left\| x^0 \right\|_t + \left\| \phi^0 \right\|_t > N \right\}$.

Besides, one has that under condition $4^{\circ\prime\prime}$ for all $t \geq 0$, as $n \to \infty$

$P(\sup_{s \leq t} \left| x_s^n - x_s^0 \right| > \varepsilon) \to 0$, for any given $\varepsilon > 0;$

and under condition $4^{\circ\prime\prime\prime}$ for all $t \geq 0$, as $n \to \infty$

$P(\sup_{s \leq t} \left| x_s^n - x_s^0 \right| > \varepsilon) + P(\sup_{s \leq t} \left| \phi_s^n - \phi_s^0 \right| > \varepsilon) \to 0$, for any given $\varepsilon > 0$.

Proof. The proof of Corollary 2.2 is similar. One only needs to split the 2nd term in the right side of (2.2) into three terms:

$$
\begin{aligned}
(2.6)\ & \int_0^t [2(x_s^n - x_s^0) \cdot (b(s, x^n(.), \phi^n(.), u^n(t, x^n(.), \phi^n(.))) \\
& -b(s, x^0(.), \phi^0(.), u^0(t, x^0(.), \phi^0(.))) \\
& = \int_0^t [2(x_s^n - x_s^0) \cdot (b(s, x^n(.), \phi^n(.), u^n(t, x^n(.), \phi^n(.))) \\
& -b(s, x^0(.), \phi^0(.), u^n(t, x^n(.), \phi^n(.))) \\
& + \int_0^t [2(x_s^n - x_s^0) \cdot (b(s, x^0(.), \phi^0(.), u^n(t, x^n(.), \phi^n(.))) \\
& -b(s, x^0(.), \phi^0(.), u^n(t, x^0(.), \phi^0(.))) \\
& + \int_0^t [2(x_s^n - x_s^0) \cdot (b(s, x^0(.), \phi^0(.), u^n(t, x^0(.), \phi^0(.))) \\
& -b(s, x^0(.), \phi^0(.), u^0(t, x^0(.), \phi^0(.))).
\end{aligned}
$$

Then the proof of

$\lim_{n\to\infty} E\sup_{s\leq t\wedge\tau_N} \left|x_s^n - x_s^0\right|^2 \to 0$, for each $N = 1, 2, ...$

under condition (i) of $4^{\circ\prime\prime}$ and the proof of

$\lim_{n\to\infty} E[\sup_{s\leq t\wedge\tau_N} \left|x_s^n - x_s^0\right|^2 + \sup_{s\leq t\wedge\tau_N} \left|\phi_s^n - \phi_s^0\right|^2] \to 0$, for each $N = 1, 2, ...$

under condition (i) of $4^{\circ\prime\prime\prime}$ can be similarly established.

On the other hand, if one splits the 2nd term in the right side of (2.2) into other three terms:

(2.6) $\int_0^t [2(x_s^n - x_s^0) \cdot (b(s, x^n(.), \phi^n(.), u^n(t, x^n(.), \phi^n(.)))$
$-b(s, x^0(.), \phi^0(.), u^0(t, x^0(.), \phi^0(.)))$
$= \int_0^t [2(x_s^n - x_s^0) \cdot (b(s, x^n(.), \phi^n(.), u^n(t, x^n(.), \phi^n(.)))$
$-b(s, x^n(.), \phi^n(.), u^0(t, x^n(.), \phi^n(.)))$
$+ \int_0^t [2(x_s^n - x_s^0) \cdot (b(s, x^n(.), \phi^n(.), u^0(t, x^n(.), \phi^n(.)))$
$-b(s, x^0(.), \phi^0(.), u^0(t, x^n(.), \phi^n(.)))$
$+ \int_0^t [2(x_s^n - x_s^0) \cdot (b(s, x^0(.), \phi^0(.), u^0(t, x^n(.), \phi^n(.)))$
$-b(s, x^0(.), \phi^0(.), u^0(t, x^0(.), \phi^0(.)))$.

Then the proof of

$\lim_{n\to\infty} E\sup_{s\leq t\wedge\tau_N} \left|x_s^n - x_s^0\right|^2 \to 0$, for each $N = 1, 2, ...$

under condition (ii) of $4^{\circ\prime\prime}$ and the proof of

$\lim_{n\to\infty} E[\sup_{s\leq t\wedge\tau_N} \left|x_s^n - x_s^0\right|^2 + \sup_{s\leq t\wedge\tau_N} \left|\phi_s^n - \phi_s^0\right|^2] \to 0$, for each $N = 1, 2, ...$

under condition (ii) of $4^{\circ\prime\prime\prime}$ can be established in similar fashions.

Finally, since

(2.7) $P(\sup_{s\leq t} \left|x_s^n - x_s^0\right| > \varepsilon) \leq P(\sup_{s\leq t} \left|x_s^0\right|) > N)$
$+P(\sup_{s\leq t\wedge\tau_N} \left|x_s^n - x_s^0\right| > \varepsilon)$,

where $\tau_N = \inf\left\{t \geq 0 : \left\|x^0\right\|_t > N\right\}$, and by assumption one easily derives that

$E\left\|x^0\right\|_t^2 \leq k_T$, as $t \leq T$,

where $k_T \geq 0$ is a constant depending on T only. Hence

$\lim_{N\to\infty} P(\sup_{s\leq t} \left|x_s^0\right|) > N) = 0.$

However, by the above we already have

$\lim_{n\to\infty} E\sup_{s\leq t\wedge\tau_N} \left|x_s^n - x_s^0\right|^2 = 0$, for each $N = 1, 2, ...$

under condition $4^{\circ\prime\prime}$. Hence by (2.7) one easily sees that

$\lim_{n\to\infty} P(\sup_{s\leq t} \left|x_s^n - x_s^0\right| > \varepsilon) = 0$, for any given $\varepsilon > 0$,

under condition $4^{\circ\prime\prime}$. The conclusion for condition $4^{\circ\prime\prime\prime}$ can be established in a similar fashion. ■

Now let us discuss results for the system (1.1). Then we have the following

2.3 Theorem. Consider RSDE (1.1). Assume that

1° $|A(t)| + |B(t)| \leq k_0, |\beta(t, x(.), \phi(.))| \leq k_0$,

where $A(t)$ and $B(t)$ are non-random;

2° $\|\sigma(t, x(.), \phi(.)) - \sigma(t, y(.), \psi(.))\|^2$
$+ \int_Z |c(t, x(.), \phi(.), z) - c(t, y(.), \psi(.), z)|^2 \pi(dz) \leq k(t)\rho(\|x - y\|_t^2)$,

where $k(t) \geq 0, \rho(u)$ satisfy the same condition as that in 2°,
3° $b(t, x(.), \phi(.), u)$ is continuous in $(x(.), \phi(.), u) \in D \times C \times U$;
4° one of the following conditions is satisfied:
(i) for $\beta^n \in \widetilde{U}$, $n = 0, 1, 2, ...$,
$|\beta^n(t, x(.), \phi(.)) - \beta^n(t, y(.), \psi(.))| \leq k_N(t)\rho_N(\|x - y\|_t^2)$,
as $\|x\|_t \leq N$, for all $n = 1, 2, ...$
where $k_N(t)$ and $\rho_N(u)$ satisfy the same conditions as those in 4°'' in Corollary 2.2;
moreover,
$\lim_{n\to\infty}\beta^n(t, x(.), \phi(.)) = \beta^0(t, x(.), \phi(.))$, for all $(x(.), \phi(.)) \in D \times C$;
(ii) for $\beta^n \in \widetilde{U}$, $n = 0, 1, 2, ...$,
$|\beta^0(t, x(.), \phi(.)) - \beta^0(t, y(.), \psi(.))| \leq k_N(t)\rho_N(\|x - y\|_t^2)$, as $\|x\|_t \leq N$,
where $k_N(t)$ and $\rho_N(u)$ satisfy the same conditions as those in 4°'' in Corollary 2.2; moreover, for any $T < \infty$
$\lim_{n\to\infty} \sup_{(t,x(.),\phi(.))\in[0,T]\times D\times C} |\beta^n(t, x(.), \phi(.)) - \beta^0(t, x(.), \phi(.))| = 0$,
for all $(x(.), \phi(.)) \in D \times C$;
5° $E|x^n(0) - x^0(0)|^2 \to 0$, as $n \to \infty$,
where we denote
$x^n(t) = x^{\beta^n}(t)$, $\phi^n(t) = \phi^{\beta^n}(t)$,
and $(x^n(t), \phi^n(t))$ is the pathwise unique strong solution of (2.1) corresponding to $\beta^n \in \widetilde{U}$.
Then,
1) for all $t \geq 0$,
$\lim_{n\to\infty} P(\sup_{s\leq t} |x_s^n - x_s^0| > \varepsilon) = 0$, for any given $\varepsilon > 0$.
Now assume that the condition 4° is changed into
4°' one of the following conditions is satisfied:
(i) for $\beta^n \in \widetilde{U}$, $n = 0, 1, 2, ...$,
$|\beta^n(t, x(.), \phi(.)) - \beta^n(t, y(.), \psi(.))| \leq k_N(t)\rho_N(\|x - y\|_t^2 + \|\phi - \psi\|_t^2)$,
as $\|x\|_t \leq N$, for all $n = 1, 2, ...$
where $k_N(t)$ and $\rho_N(u)$ satisfy the same condition as that in 4°''' in Corollary 2.2; moreover,
$\lim_{n\to\infty}\beta^n(t, x(.), \phi(.)) = \beta^0(t, x(.), \phi(.))$, for all $(x(.), \phi(.)) \in D \times C$;
(ii) for $\beta^n \in \widetilde{U}$, $n = 0, 1, 2, ...$,
$|\beta^0(t, x(.), \phi(.)) - \beta^0(t, y(.), \psi(.))| \leq k_N(t)\rho_N(\|x - y\|_t^2 + \|\phi - \psi\|_t^2)$,
as $\|x\|_t \leq N$,
where $k_N(t)$ and $\rho_N(u)$ satisfy the same condition as that in 4°''' in Corollary 2.2; moreover, for any $T < \infty$
$\lim_{n\to\infty} \sup_{(t,x(.),\phi(.))\in[0,T]\times D\times C} |\beta^n(t, x(.), \phi(.)) - \beta^0(t, x(.), \phi(.))| = 0$,
for all $(x(.), \phi(.)) \in D \times C$;
Then
2) for all $t \geq 0$,
$\lim_{n\to\infty}[P(\sup_{s\leq t} |x_s^n - x_s^0| > \varepsilon) + P(\sup_{s\leq t} |\phi_s^n - \phi_s^0| > \varepsilon)] = 0$,
for any given $\varepsilon > 0$.

Proof. Denote

$$b(t,x(.),\phi(.),\beta(t,x(.),\phi(.))) = A(t)x_t + B(t)x_t\beta_t(x(.),\phi(.)).$$

Then the second term in the right side of (2.2) becomes

$$\int_0^t 2(x_s^n - x_s^0)\cdot(b(s,x^n(.),\phi^n(.),u^n(t,x^n(.),\phi^n(.)))$$
$$-b(s,x^0(.),\phi^0(.),u^0(t,x^0(.),\phi^0(.)))ds$$
$$= \int_0^t 2(x_s^n - x_s^0)\cdot A(s)(x_s^n - x_s^0)ds + \int_0^t 2(x_s^n - x_s^0)\cdot B(s)(x_s^n\beta_s^n(x^n(.),\phi^n(.))$$
$$-x_s^0\beta_s^0(x^0(.),\phi^0(.)))ds = I_{1n} + I_{2n}.$$

In case (i) of 4° we add terms

$$\mp\int_0^t 2(x_s^n - x_s^0)\cdot B(s)x_s^0\beta_s^n(x^n(.),\phi^n(.))ds$$
$$\mp\int_0^t 2(x_s^n - x_s^0)\cdot B(s)x_s^0\beta_s^n(x^0(.),\phi^0(.))ds$$

into I_{2n}, and take the expectation in (2.2). Then after using the assumption and estimating the right side of (2.2), the conclusion 1) in this case is obtained. Similarly, in case (ii) of 4° we add terms

$$\mp\int_0^t 2(x_s^n - x_s^0)\cdot B(s)x_s^0\beta_s^n(x^n(.),\phi^n(.))ds$$
$$\mp\int_0^t 2(x_s^n - x_s^0)\cdot B(s)x_s^0\beta_s^0(x^n(.),\phi^n(.))ds$$

into I_{2n}. Then after similar discussion we still obtain the conclusion 1) for this case. For condition 4°′ the discussion is similar. ■

2.4 Remark. In the case that $\beta^n = \beta^n(t) \in \widetilde{U}$, $n = 0,1,2,...$, do not depend on $(x(.),\phi(.)) \in D\times C$, and

$$\lim_{n\to\infty}\beta^n(t) = \beta^0(t), \text{ for all } t\geq 0,$$

then condition (i) of 4° is satisfied.

For the exponential stability in mean square of solution to RSDE (1.1) we have the following

2.5 Theorem. Consider (1.1). Assume that condition 1° in Theorem 2.3 holds, and assume that

1° $\sigma(t,0,0) = 0$, $c(t,0,0,z) = 0$,

2° $2x_t\cdot(A(t)x_t + B(t)\beta_t x_t) + \|\sigma(t,x(.),\phi(.))\|^2$
$+\int_Z |c(t,x(.),\phi(.),z)|^2\,\pi(dz) \leq -c_1|x_t|^2$, for all $(x(.),\phi(.)) \in D\times C$;

where $c_1 > 0$ is a constant.

If (x_t,ϕ_t) is a solution of (1.1), then

$$E|x_t|^2 \leq E|x_0|^2 e^{-c_1 t}, \text{for all } t\geq 0;$$

i.e., x_t is exponentially stable in mean square; and there exists a constant $k_1\geq 0$ such that

$$E|\phi_t|^2 \leq k_1 E|x_0|^2(1+e^{-c_1 t}), \text{for all } t\geq 0.$$

Hence, if one denotes that $\phi_\infty = \lim_{n\to\infty}\phi_t$, then

$$E|\phi_\infty|^2 \leq k_1 E|x_0|^2.$$

Furthermore, one has

$$E(|\phi_\infty|^2 - |\phi_t|^2) \leq k''' E|x_0|^2 e^{-k_1 t}.$$

In case condition 2° is substituted by the following condition

2°′ there exists a positive constant $k_5 > 0$ such that

$$2x_t\cdot(A(t)x_t + B(t)\beta_t x_t) + \|\sigma(t,x(.),\phi(.))\|^2$$

$+\int_Z |c(t,x(.),\phi(.),z)|^2 \pi(dz) \geq k_5 |x_t|^2$, for all $(x(.),\phi(.)) \in D \times C$;

and (remember that $\beta \in \widetilde{U}$ implies that β is bounded)

$\|\sigma(t,x(.),\phi(.))\|^2 + \int_Z |c(t,x(.),\phi(.),z)|^2 \pi(dz) \leq k_0(1+\|x\|_t)$,

then

$E|x_t|^2 \geq E|x_0|^2 e^{k_5 t}$, for all $t \geq 0$.

Hence,

$\lim_{t\to\infty} E|x_t|^2 = \infty$.

Theorem 2.5 is a direct consequence of Theorem 2.2 in Chapter 4.

For the more concrete RSDE (1.2) we have the following

2.6 Corollary. Consider RSDE (1.2). If condition 1° in Theorem 2.3 holds, and if $\sigma(t,0,0)=0$, $c(t,0,0,z)=0$, and

$\|\sigma(t,x(.),\phi(.))\|^2 + \int_Z |c(t,x(.),\phi(.),z)|^2 \pi(dz) \leq c_0 |x_t|^2$,

for all $(x(.),\phi(.)) \in D \times C$;

$\eta_0 = \inf_{t\geq 0,\omega}(\eta_1(t,\omega),\cdots,\eta_{r_m}(t,\omega)) > 0$,

$b_M = \sup_{t\geq 0,\omega}(b_{r_1}(t,\omega),\cdots,b_{r_2}(t,\omega)) < \infty$,

where $c_0 \geq 0$ is a constant; and if there exists a constant $\overline{\beta}_0 \geq 0$ such that

(2.8) $\overline{\beta}_0 b_M - 1 + c_0 < 2\eta_0$;

then $\delta_0 = 2\eta_0 + 1 - c_0 - \overline{\beta}_0 b_M > 0$ and for all $\beta \in \widetilde{U}$ satisfying

$0 \leq \beta(t,x(.),\phi(.)) \leq \overline{\beta}_0$

one has

$E\left|x_t^\beta\right|^2 \leq E|x_0|^2 e^{-\delta_0 t}$,for all $t \geq 0$;

and there exists a constant $k_1 \geq 0$ such that

$E|\phi_t|^2 \leq k_1 E|x_0|^2 (1+e^{-\delta_0 t})$,for all $t \geq 0$.

Furthermore,

$E(|\phi_\infty|^2 - |\phi_t|^2) \leq k''' E|x_0|^2 e^{-\delta_0 t}$.

Proof. We only need to verify that the condition 2° in Theorem 2.5 holds. Indeed, now

$2x_t \cdot (A(t)x_t + B(t)\beta_t x_t) + \|\sigma(t,x(.),\phi(.))\|^2$

$+\int_Z |c(t,x(.),\phi(.),z)|^2 \pi(dz) = -2\sum_{i=1}^{r_m}(1+\eta_i(t))x_i^2(t)$

$+2\sum_{i=1}^{r_m-1} x_i(t)x_{i+1}(t) + 2\beta_t x_1(t)\sum_{i=r_1}^{r_2} b_i(t)x_i(t) + \|\sigma(t,x(.),\phi(.))\|^2$

$+\int_Z |c(t,x(.),\phi(.),z)|^2 \pi(dz) \leq [-2(1+\eta_0)+1+\beta_t b_M + c_0]|x_t|^2$.

Hence, Theorem 2.5 applies. ■

Now let us discuss the comparison theorem for solutions to (1.1).

2.7 Theorem. Consider (1.1). Denote $d = r_m$. Assume that

1° $|A(t)| + |B(t)| + |\beta_t| \leq k_0, b_{ij}(t) \geq 0, \forall i,j = 1,\cdots,d$;

where $B(t) = [b_{ij}(t)]_{i,j=1}^d$,

2° $|\sigma_{ik}(t,x,\phi) - \sigma_{ik}(t,y,\psi)|^2 \leq k_N(t)\rho_N(|x_i - y_i|^2)$, as $|x|,|y| \leq N; \forall i,k$;

$\int_Z |c(t,x,\phi,z) - c(t,y,\psi,z)|^2 \pi(dz) \leq k_N(t)\rho_N(|x-y|^2)$, as $|x|,|y| \leq N$;

where $k_N(t) \geq 0, \rho_N(u)$ satisfy the same condition as that in 2° of Theorem 2.1 for each $N = 1,2,...$;

3° $\int_Z |c(t,x,\phi,z) - c(s,y,\psi,z)|^2 \pi(dz) \to 0$,as $|x-y| \to 0, |t-s| \to 0$;

4° $x_i \geq y_i \Longrightarrow x_i + c_i(t,x,\phi,z) \geq y_i + c_i(t,y,\psi,z), 1 \leq i \leq d;$
and

$x + c(t,x,z) \in \overline{R}^d_+.$

Now suppose that $\beta^1(t,x(.),\phi(.)),\ \beta^2(t,x(.),\phi(.)) \in \widetilde{U}$,

$\beta^i(t,x,\phi), i = 1,2$, are both continuous in (x,ϕ) such that

$\beta^1(t,x,\phi) \leq \beta^2(t,x,\phi);$

moreover, one of $\beta^i(t,x,\phi), i = 1,2$, say, $\beta^1(t,x,\phi)$ satisfies the following condition:

5° $\left|\beta^1(t,x,\phi) - \beta^1(t,y,\psi)\right|^2 \leq k_N(t)\rho_N(|x-y|^2 + |\phi-\psi|^2),$

as $|x|, |y| \leq N, |\phi|, |\psi| \leq N$; as $N = 1,2,...,$

where $k_N(t) \geq 0, \rho_N(u)$ satisfy the same condition as that in 2° of Theorem 2.1 for each $N = 1,2,....$

Then

$x^1_i(0) \leq x^2_i(0),\ i = 1,2,...,d,$

implies that

$x^1_i(t) \leq x^2_i(t),\ i = 1,2,...,d,$for all $t \geq 0.$

Furthermore, if

$\sigma(t) = \sigma(t,x,\phi)$ and $c(t,z) = c(t,x,\phi,z)$ do not depend on (x,ϕ),

then for all $t \geq 0$

$d\ \phi^1_i(t) \geq d\phi^2_i(t),\ i = 1,2,...,d.$

Theorem 2.7 can be derived from Theorem 3.4 and 3.5 in Chapter 4 after only small modifications. We omit the details.

Now let us consider a special optimal stochastic population control problem for RSDE (1.1). Denote

$$\widetilde{U}_1 = \left\{ \begin{array}{c} \beta : \beta \in \widetilde{U},\ \overline{\beta}_0 \leq \beta \leq \beta_0, \\ \text{and } \beta \text{ satisfies the condition } 5^\circ \text{ in Theorem 2.7} \end{array} \right\},$$

where $0 < \overline{\beta}_0$ and β_0 are constants. Suppose we want to minimize the following functional

$J(\beta) = E\int_0^T F(x_t)dt,$

among all $\beta \in \widetilde{U}_1$, where $0 \leq T < \infty$ is a arbitrary given constant. Then we have the following

2.8 Theorem. Assume that conditions $1^\circ - 4^\circ$ in Theorem 2.7 holds, and assume that

$\|\sigma(t,x(\cdot),\phi(\cdot)) - \sigma(t,y(\cdot),\psi(\cdot))\|^2$
$+ \int_Z |c(t,x(\cdot),\phi(\cdot),z) - c(t,y(\cdot),\psi(\cdot),z)|^2\, \pi(dz)$
$\leq k_0(|x_t - y_t|^2 + |\phi_t - \psi_t|^2),$
$\|\sigma(t,x(\cdot),\phi(\cdot)) - \sigma(s,x(\cdot),\phi(\cdot))\|^2$
$+ \int_Z |c(t,x(\cdot),\phi(\cdot),z) - c(s,x(\cdot),\phi(\cdot),z)|^2\, \pi(dz) \leq k_0\, |t-s|^2,$

$F(x)$ is a Borel measurable function defined on $\overline{R}^d_+$ such that $F(x)$

$F(x) \leq F(y)$, as $x_i \leq y_i,\ i = 1,2,..,d;\ x,y \in \overline{R}^d_+.$

Let $\beta(t,\omega) = \overline{\beta}_0$. Then

1) $\beta \in \widetilde{U}_1$,
2) $J(\overline{\beta}_0) = \inf_{\beta \in \widetilde{U}_1} J(\beta)$.

Furthermore, suppose now the optimal stochastic population control problem is for RSDE (1.2). If $\overline{\beta}_0$ verifies (2.8), $\eta_i(t), 1 \leq i \leq d$,and $b_i(t), r_1 \leq i \leq r_2$, appeared in coefficients $A(t)$ and $B(t)$ are all bounded, and all conditions in Corollary 2.6 are fulfilled, then the optimal trajectory x_t^0 is also exponentially stable in mean square, i.e. there exists a constant $c_0 > 0$ such that

3) $E\left|x_t^0\right|^2 \leq e^{-c_0 t} E\left|x_0^0\right|^2$,

where $x_t^0 = x_t^{\overline{\beta}_0}$.

The conclusion of Theorem 2.8 can be derived from Theorem 3.20 in Chapter 3 (for 1)), Theorem 2.7 (for 2)) and Corollary 2.6 (for 3)).

Finally, by using the results obtained here, the following conclusions can be drawn: The RSDE (1.1) is a natural and useful mathematical stochastic model for many deterministic systems perturbed by a continuous white noise and a jump random Poisson point noise process. The existence, stability, comparison and convergence of solutions to such systems under mild conditions can be obtained. Furthermore, RSDE (1.2) is also a suitable model for the stochastic population control system. In this model the following conclusions are derived: (i) The size of population depends on the initial size of population, and the fertility rate of females, continuously in some sense (Theorem 2.1 and 2.3). (ii) If one of the last two increases, then so does the size of the population forever as time evolves (Theorem 2.7). (iii) If the stochastic perturbation is not large, and the forward death rate is greater than zero, then it is possible to take a fertility rate of females to make the system exponentially stable in some sense (Corollary 2.6). (iv) If the payoff value functional depends on the size of population monotonically, then the payoff value will take the smallest value, as the fertility rate of the females does (Theorem 2.8).

7.3 Applications to Neurophysiological Control and Others

In this section we will discuss the following RSDE with rectangular boundary, which can be used as a stochastic model for neural network system (Hopfield, 1984, 1985) and also for a population control system (Yu, et al. 1987). (For simplicity we just consider the system without jumps):

$$(3.1)\begin{cases} dx_t = b(t, x_t, \omega)dt + \sigma(t, x_t, \omega)dw_t + d\phi_t, x_0 = x_0 \in \overline{\Theta}, \\ \qquad x_t \in \overline{\Theta}, \text{ for all } t \geq 0, \\ \phi_t \text{ s a } \Im_t - \text{adapte continuous process, finite variational} \\ \qquad \text{on any finite interval with } \phi_0 = 0, \\ \phi_t = \int_0^t n_s d\left|\phi\right|_s, \ \left|\phi\right|_t = \int_0^t I_{\partial\Theta}(x_s) d\left|\phi\right|_s, \\ \qquad n_t \in \aleph_{x_t}, \text{ as } x_t \in \partial\Theta; \end{cases}$$

$$\Theta = \{x = (x_1, ..., x_d) : a_i < x_i < b_i, i = 1, ..., d\},$$

and $-\infty \leq a_i \leq b_i \leq \infty$ are arbitrary given. (3.1) implies many useful cases. For example, if $-a_i = b_i = \infty$, $1 \leq i \leq d$, then $\partial\Theta = \Phi$ (the empty set). Hence, $\phi_t \equiv 0, \forall t \geq 0$. (3.1) becomes the usual SDE (without reflection). If $a_i = 0, \forall 1 \leq i \leq d$; $b_i = \infty, \forall 1 \leq i \leq d$; then (3.1) becomes (1.1), which is the model of stochastic population equation. However, if $-\infty < a_i \leq b_i < \infty, \forall 1 \leq i \leq d$; then (3.1) can be used to study the neurophysiological net problem (Hopfield, 1984, 1985).

In this section we consider the more general case where Θ is any given convex domain in R^d. Thus applying Theorem 2.9 of Chapter 3 one directly gets the following

3.1 Theorem. Assume that
1° $b(t,x,\omega), \sigma(t,x,\omega)$ are $\Im_t^w$-adapted as x is fixed, and there exists a constant $k_0 > 0$ such that

$$|b|^2 + \|\sigma\|^2 \leq k_0(1 + |x|),$$

2° $b(t,x,\omega)$ and $\sigma(t,x,\omega)$ are continuous in x;
3° $2(x-y) \cdot (\, b(t,x,\omega) - b(t,y,\omega)) + \|\sigma(t,x,\omega) - \sigma(t,y,\omega)\|^2$

$$\leq k_N(t)\rho_N(|x-y|^2), \text{ as } |x|, |y| \leq N,$$

where $0 \leq k_N(t)$, it is non-random such that $\int_0^t k_N(s)ds < \infty$, for any $t > 0$, and $\rho_N(u)$ is a concave and strictly increasing function of u in $u \geq 0$ with $\rho_N(0) = 0$ satisfying $\int_{0+} du/\rho_N(u) = +\infty$,for each $N = 1, 2, ...$
Then (3.1) has a pathwise unique strong solution (x_t, ϕ_t).

For the non-random coefficients we can get the existence of a weak solution to (3.1) under assumption that $b(t,x)$ and $\sigma(t,x)$ are bounded, and $\sigma(t,x)$ is uniformly non-degenerate.

3.2 Theorem. Assume that
1° $|b|^2 + \|\sigma\|^2 \leq k_0(1 + |x|^2)$,
2° $\sigma(t,x)$ is uniformly non-degenerate, i.e., there exists a $\delta_0 > 0$ such that for all $\lambda \in R^d$

$$\langle A(t,x)\lambda, \lambda \rangle \geq |\lambda|^2 \delta_0,$$

where $A = \frac{1}{2}\sigma\sigma^*$.
Then (3.1) has a weak solution (x_t, ϕ_t).
Furthermore, if condition 3° in Theorem 3.1 holds, then (3.1) has a pathwise unique strong solution.
Proof. First, we show the result for the case that b and σ are bounded. The proof for this case is similar to that of Theorem 3.17 in Chapter 3. Indeed, one can smooth out b and σ to get b^n and σ^n as in the proof of the above mentioned Theorem and then, apply Theorem 3.1 to get a pathwise unique strong solution (x_t^n, ϕ_t^n) for each RSDE (3.1) with coefficients b^n and $\sigma^n, n = 1, 2, ...$ After a similar discussion, as in the proof of Theorem 3.17 in Chapter 3, a weak solution to (3.1) can be obtained. (Actually, the proof here is even simpler, since there are no jump terms here). Now, applying Theorem 1.3 in Chapter 3, one immediately finds a pathwise unique strong solution of (3.1). For the general case that b and

σ satisfy 1° one can establish the conclusion in completely the same way as in the proof of Theorem 3.8 in Chapter 3. ■

For the stability of solutions to (3.1) we have the following

3.3 Theorem. Assume that conditions 1° - 3° in Theorem 3.1 hold, $0 \in \bar{\Theta}$ and assume that

4° $b(s,0) = 0, \sigma(s,0) = 0, \forall s \geq 0$.

Then 1) $(0,0)$ is a pathwise unique strong solution with $x_0 = 0$;

2) if (x_t, ϕ_t) is the pathwise unique strong solution of (3.1) with $x_0 = x$, then

$\lim_{x \to 0} E \sup_{s \leq t} |x_s|^2 = 0, \ \forall t \geq 0$;

3) furthermore, if

$|b(t,x)|^2 + \|\sigma(t,x)\|^2 \leq k(t) \ |x|^2$,

where $0 \leq k(t)$ is non-random, and $\int_0^T k(t)dt < \infty, \forall 0 \leq T < \infty$, then

$\lim_{x \to 0} E \sup_{s \leq t} |\phi_s|^2 = 0, \ \forall t \geq 0$.

Proof. 1) is true by Theorem 3.1. 2): By Theorem 3.1 again for $x_0 = x$ (3.1) has a pathwise unique strong solution (x_t, ϕ_t). By Ito's formula and 4') of Lemma 2.2 in Chapter 2

$|x_t|^2 \leq |x_0|^2 + \int_0^t (2x_s \cdot b(s,x_s) + \|\sigma(s,x_s)\|^2)ds + 2\int_0^t x_s \cdot \sigma(s,x_s)dw_s$.

Hence by assumption 1° in Theorem 3.1 and by Gronwall's inequality as $0 \leq t \leq T$

$E \sup_{s \leq t} |x_s|^2 \leq k_T E |x_0|^2 e^{k_0' T}$.

2) is proved. 3): By (3.1)

$|\phi_t|^2 \leq 4[|x_t|^2 + |x_0|^2 + \left|\int_0^t b(s,x_s)ds\right|^2 + \left|\int_0^t \sigma(s,x_s)dw_s\right|^2]$.

Applying this inequality and using the additional assumption in 3), one easily derives the conclusion of 3). ■

Moreover, we also have the following

3.4 Theorem. Under conditions 1° – 4° of Theorem 3.3 and $0 \in \bar{\Theta}$, if, in addition,

5° $2x \cdot b(t,x) + \|\sigma(t,x)\|^2 \leq -k_1 |x|^2$,

where $0 < k_1$ is a constant, then

$E|x_t|^2 \leq E|x_0|^2 e^{-k_1 t}$.

Furthermore, if

$|b(t,x)|^2 + \|\sigma(t,x)\|^2 \leq k_2 \ |x|^2$,

where $0 \leq k_2$ is a constant, then

$E|\phi_t|^2 \leq k_3 E|x_0|^2 (1 + e^{-k_3 t})$,

where $k_3 > 0$ is a constant.

Proof. By Ito's formula

(3.2) $E|x_t|^2 = E|x_0|^2 + E\int_0^t (2x_s \cdot b(s,x_s) + \|\sigma(s,x_s)\|^2)ds + 2E\int_0^t x_s \cdot d\phi_s$.

However, by 4') of Lemma 2.2 in Chapter 2 the last term in the right hand side is a decreasing function. Hence by condition 5°

$\frac{d}{dt} E|x_t|^2 \leq -k_1 E|x_t|^2$.

Therefore

$$E|x_t|^2 \leq E|x_0|^2 e^{-k_1 t}.$$

Now by (3.1)

$$|\phi_t|^2 \leq 4[|x_t|^2 + |x_0|^2 + \left|\int_0^t b(s,x_s)ds\right|^2 + \left|\int_0^t \sigma(s,x_s)dw_s\right|^2].$$

Applying this inequality, and using the additional assumption, one easily arrives at the last conclusion. ■

Bibliography

[1] Anulova, S.V.(1978). On process with Levy generating operator in a half-space. *Izv. AH. SSSR ser. Math.* 47:4, 708-750. (In Russian).

[2] Anulova, S.V.(1982). On stochastic differential equations with boundary conditions in a half-space. *Math. USSR-Izv.* 18, 423-437.

[3] Anulova, S.V.(1990). Counter examples: An explosion solution may exist for a stochastic differential equation in a domain with the linear increase of coefficients. *Probab. Th. Appl.* 35:2, 329-331. (In Russian).

[4] Anulova, S.S. and Pragarauskas, H. (1977). On strong Markov weak solutions of stochastic equations. *Lit. Math. Sb.* XVII:2, 5-26. (In Russian).

[5] Barlow, M.T. and Perkins, E. (1984). One-dimensional stochastic differential equations involving a singular increasing process. *Stochastics* 12, 229-249.

[6] Billingsleys, P. (1968). *Convergence of Probability Measures.* John Wiley and Sons, New York.

[7] Bensoussan, A. and Lions, J.L. (1984). *Impulse Control and Quasi-Variational Inequalities.* Gauthier-Villars.

[8] Chaleyat-Maurel, M., El Karoui, N. and Marchal, B. (1980). Reflexion discontinue et systèmes stochastiques. *The Annals of Probab.* 8:6, 1049-1067.

[9] Ekeland, I. (1974). On the variational principle. *J. Math. Anal. Appl.* 47, 324-353.

[10] El Karoui, N. (1975). Processus de reflexion sur R^N. *Seminaire de Probabilité IX, Lect. Notes in Math. 465*, Springer.

[11] El Karoui, N., and Chaleyat-Maurel, M. (1978). Un problem de reflexion et ses applications au temps local et aux equations differentielles stochastiques sur R. Cas continu, Temps Locaux. *Asterisque 52-53*, 117-144.

[12] El Karoui, N., Kapoudjian, C., Pardoux, E., Peng, S. and Quenez, M.C. (1997). Reflected solutions of backward SDEs, and related obstacle problems for PDEs. *The Annals of Probability* 25:2, 702-737.

[13] Elliott, R.J. (1982). *Stochastic Calculus and Application.* Springer-Verlag. N.Y.

[14] Elliott, R.J. and Kohlmann. (1980). The variational principle and stochastic optimal control. *Stochastic* 3, 229-241.

[15] Friedman, A. (1975, 1976). *Stochastic Differential Equations and Applications I, II.* Academdic Press.

[16] Gihman, I.I. and Skorohod, A.V. (1979). *The Theory of Stochastic Processes III*, Springer-Verlag, Berlin.

[17] Gihman, I.I. and Skorohod, A.V. (1982). *Stochastic Differential Equations and their Applications,* Naukova Dumka, Kiev. (In Russian).

[18] Has'minskii, R.Z. (1980). *Stochastic Stability of Differential Equations.* Stjthoff & Noordhoff Interna. Publ. Netherlands.

[19] He, S.W., Wang, J.G. and Yan, J.A. (1992). *Semimartingale Theory and Stochastic Calculus.*

[20] Hida, T. (1980). *Brownian Motion.* Springer.

[21] Hopfield J.J. (1984). Neurons with graded response have collective computational properties like those of two-state neurons. *Proc. Natl. Aca. Sci. USA.* 81, 3088-3092.

[22] Hopfield J.J. and Tank, D.W. (1985). Neural computation of decisions in optimization problems. *Biol. Cybern. 52*, 141-152.

[23] Ikeda, N. and Watanabe, S. (1989). *Stochastic Differential Equations and Diffusion Processes. 2nd edition.* North-Holland.

[24] Ito, K. (1951). *On Stochastic Differential Equations.* Mem. Amer. Math. Soc., 4.

[25] Jacod, J. (1979). Calcul Stochastique et Problemes de Martingales. *L.N.Math. 714.*

[26] Kallianpur, G. (1980). *Stochastic Filtering Theory.* Springer.

[27] Karatzas, I. and Shreve, S.E. (1987). *Brownian Motion and Stochastic Calculus.* Springer-Verlag, Berlin.

[28] Krylov, N.V. (1980). *Controlled Diffusion Processes.* Springer-Verlag. N.Y.

[29] Krylov, N.V. (1995). *Introduction to the Theory of Diffusion Processes.* AMS Providence.

[30] Kunita, H. (1990). *Stochastic flows and Stochastic differential equations.* Cambridge University Press.

[31] Le Gall, J.F. (1983). Applications du temps local aux equations differentielles stochastiques unidimensionnelles. *Lect. Notes Math. 986*, Springer-Verlag, 15-31.

[32] Le Gall, J.F. (1984). One-dimensional stochastic differential equations involving the local time of the unknown processes. *Lect. Notes Math. 1095*, Springer-Verlag.

[33] Lindvall, T. (1973). Weak convergence of probability measures and random functions in the function space $D[0, \infty)$. *Appl.Probab.*, 10, 109-121.

[34] Lions, P.L., Menaldi, J.L. and Sznitman, A.S. (1981). Construction de processus de diffusion reflechis par penalization du domaine. *C.R. Acad. Sci. Paris* I-292, 559-562.

[35] Lions, P.L. and Sznitman, A.S. (1984). Stochastic differential equations with reflecting boundary conditions. *Comm. Pure Appl. Math.* 37, 511-537.

[36] Liptser, R.S. and Shiryaev. (1977, 1978). *Statistics of Random Processes.* I, II. Springer-Verlag. N.Y.

[37] Liptser, R.S. and Shiryaev. (1989). *Theory of Martingales.* Kluwer Academic Publishers. Boston.

[38] Mel'nikov, A.V. (1985). Stochastic equations and Krylov's estimates for semi-martingales. *Stochastics* 10, 81-102.

[39] Menaldi, J.L. (1983). Stochastic variational inequality for reflected diffusion. *Indiana Univ. Math. J.* 32, 733-744.

[40] Menaldi, J.L. and Robins, M. (1983). Processus de diffusion reflechis avec sauts. *C.R. Acad. Sci. Paris* I-297, 533-536.

[41] Menaldi, J.L. and Robins, M. (1985). Reflected diffusion processes with jumps. *The Annals of Probab.* 13:2, 319-342.

[42] Meyer, P.A. (1976). Un cours sur les integrales stochastiques. *Lect. Notes Math. 511*, Springer-Verlag.

[43] Nakao, S. (1972). On the existence of solutions of stochastic differential equations with boundary conditions. *J. Math. Kyoto Univ., 12:1*, 151-178.

[44] Nisio, M. (1981). *Stochastic Control Theory.* ISI LEC. Notes 9, Macmillan India Limited, Delhi.

[45] Novikov, (1973). On moment inequalities and identities for stochastic integrals. Proc. Second Japan-USSR Symp. Prob. Theor., *Lect. Notes in Math. 330*, Springer-Verlag, 333-339.

[46] Øksendal, B. (1992). *Stochastic Differential Equations - An Introduction with Applications.* 3rd Ed.. Springer-Verlag.

[47] Revuz, D. and Yor, M. (1991). *Continuous Martingales and Brownian Motion.* Springer-Verlag.

[48] Rogers, L.C. and Williams, D. (1987). *Diffusions, Markov Processes and Martingales.* V.2: Ito Calculus. John Wiley & Sons, New York.

[49] Saisho, Y.S. (1987). Stochastic differential equation for multi-dimensional domain with reflecting boundary. *Probability Th.& Rel. Fields* 74, 455-477.

[50] Shalaumov, V.A. (1980). On the behavior of a diffusion process with a large drift coefficient in a half-space. *Theor. Probab. Appl. 6, 592-598.*

[51] Situ Rong. (1983). Non-convex stochastic optimal control and maximum principle. *IFAC 3rd Symposium, Control of Distributed Parameter System 1982*, Eds. J.P. Babary and L.L. Letty, Pergamon Press, 401-407.

[52] Situ Rong. (1984a). An application of local time to stochastic differential equations in m-dimensional space. *ACTA Scientiarum Natur. Univ. Sunyatseni*, 3, 1-12.

[53] Situ Rong. (1984b). On strong solutions and stochastic pathwise Bang-Bang control for a non-linear system. *IFAC 9th Triennial World Congress, Hungary, 1984, Proceedings vol. 3, Large-Scale Systems., Decision Making, Mathematics of Control*, Ed. A. Titili, etc.1433-1438.(1985).

[54] Situ Rong. (1985a). On strong solution, uniqueness, stability and comparison theorem for a stochastic system with Poisson jumps. *Lecture Notes in Control and Information Sci.* 75, Distributed Parameter System, Springer-Verlag, 352-381.

[55] Situ Rong. (1985b). Strong solutions and pathwise Bang-Bang control for stochastic system with Poisson jumps in 1-dimensional space. *Proceedings of 5th National Conference on Control Theory and its Applications*, 319-321.

[56] Situ Rong. (1986a). On weak, strong solutions and pathwise Bang-Bang control for non-linear degenerate stochastic system. *IFAC Stochastic Control, USSR, 1986, Proceedings Series, (1987)*, No. 2, Eds. N.K. Sinha and L. Telksnys, Pergamon Press, 145-150.

[57] Situ Rong. (1986b). Strong solutions and pathwise Bang-Bang control for multidimensional non-linear stochastic system with jumps. *4th IFAC Symposium of Control of Distributed Parameter System*, Los Angeles. Calif. USA, June, 1986.

[58] Situ Rong. (1987a). A non-linear filtering problem and its application. *Chin. Ann. Math.* 8B(3), 296-310.

[59] Situ Rong. (1987b). Strong solutions, pathwise control for stochastic nonlinear system with Poisson jumps in n-dimensional space and its application to parameter estimation. *Proc. of Sino-American Statistical Meeting*, Beijing, 389-392.

[60] Situ Rong. (1987c). Local time, Krylov estimation, Tanaka formula, strong comparison theorem and limit theorem for 1-dimensional semimartingales. *Acta Sci. Naturali Univ. Sunyatseni* 10, 14-26.

[61] Situ Rong. (1987d). On non-linear Voterra Ito stochastic system and pathwise control. *Proc. of National Annual Conf. on Control Theory and its Applications*, China, 341-344.

[62] Situ Rong. (1989). On existence theorems of saddle points of stochastic differential games for n-dimensional partial observed stochastic systems. Accepted for *Sino-French Probability and Statistics Conference.*

[63] Situ Rong (1990a). On existence and stability of solutions for reflecting stochastic differential equations with rectangular boundary. *Acta Scientiarum Naturalium Universitatis Sunyatseni* 29:3, 1-11.

[64] Situ Rong (1990b). Stochastic differential equations with jump reflection on half space and pathwise optimal control. *Proc. of National Annual Conference on Control Theory and its Applications*, 679-682.

[65] Situ Rong (1990c). Strong solutions and pathwise control for non-linear stochastic system with Poisson jumps in n-dimensional space. *Chinese Ann. Math.*, Ser. B, 11:4, 513-524.

[66] Situ Rong (1991a). Strong solutions and optimal control for stochastic differential equations in duals of nuclear spaces. *Lect. Notes in Control and Inform. Sci.* 159, Control Theory of Distributed Parameter Systems and Applications, Springer-Verlag, 144-153.

[67] Situ Rong (Rong Situ). (1991b). Theory and application of stochastic differential equations in China. *Contemporary Mathematics*, Vol. 118, AMS, 263-280.

[68] Situ Rong (1991c). Reflecting stochastic differential equations with jumps and stochastic population control. In *Control Theory, Stochastic Analysis and Applications*, Eds. S. Chen and J. Yong, World Scientific, Singapore, 193-202.

[69] Situ Rong (1991d). Stochastic population control problem on white noise measure space. *International Conf. on White Noise Analysis and Applications*, May Nagoya, Japan.

[70] Situ Rong & W.L. Chen. (1992). Existence of solutions and optimal control for reflecting stochastic differential equations with applications to population control theory. *Stochastic Analysis and Applications* 10:1, 45-106.

[71] Situ Rong (1993). Non-linear filtering for reflecting stochastic differential systems. *Proc. of National Annual Conf. on Control Theory and its Appl.*, Ocean Press, 210-213.

[72] Situ Rong (1994). Anticipating stochastic differential equations and optimal control. *Proc. of CSIAM Conf. on Systems and Control*, Shangdong, China, 64-69.

[73] Situ Rong (1995). On backward stochastic differential equations with jumps and with non-Lipschitzian coefficients and applications. *23rd Conf. on Stochastic Processes and their Applications,* June Singapore, Abstracts p.58.

[74] Situ Rong (1995). On existence of strong solutions and optimal controls for anticipating stochastic differential equations with non-Lipschitzian coefficients. *Proc. of 1995 Chinese Control Conference,* Chinese Sci.& Tech. Press, 537-540.

[75] Situ Rong (1996). On comparison theorem of solutions to backward stochastic differential equations with jumps and its applications. *Proc. of the 3rd CSIAM Conf. on Systems and Control*, LiaoNing, China, 46-50.

[76] Situ Rong (1996). On strong solutions and pathwise optimal control for stochastic differential equations in Hilbert space. *Proc. of 1996 Chinese Control Conference*, Shangdong, 396-399.

[77] Situ Rong (1997). One kind of solution for optimal consumption to financial market with jumps by Lagrange multipliers in stochastic control. *Proc. of 1997 Chinese Control Conference,* Wuhan Press, 1205-1209.

[78] Situ Rong (1997). On solutions of backward stochastic differential equations with jumps and applications. *Stochastic Processes and their Applications 66, 209-236.*

[79] Situ Rong (1997). Optimization for financial market with jumps by Lagrange's method. The 2nd International Symposium of Econometrics on Financial Topics, Guangzhou.

[80] Situ Rong (1998). Comparison theorem of solutions to BSDE with jumps and viscosity solution to a generalized Hamilton-Jacobi-Bellman equation. *Conference on Control of Distributed Parameter and Stochastic Systems, June 19-22,* Hangzhou, China.

[81] Situ Rong (1998). Forward and backward stochastic differential systems and financial market with applications to stochastic control. *Proceedings of 1998 Chinese Control Conference,* National Defence University Press, Beijing, 301-305.

[82] Situ Rong (1998). On solutions of reflecting backward stochastic differential equations with jumps. *Conference on Mathematical Finance, Oct.* Shanghai, China.

[83] Situ Rong (1999). Optimization for financial market with jumps by Lagrange's method. *Pacific Economic Review. (*To Appear*).*

[84] Situ Rong and Zeng, A.T. (1999). On solutions of BSDE with jumps and with unbounded stopping terminals I. *Acta Sci. Naturali Univ. Sunyatseni, 38:2,* 1-6.

[85] Situ Rong and Zeng, A.T. (1999). On solutions of BSDE with jumps and with unbounded stopping terminals II. *Acta Sci. Naturali Univ. Sunyatseni.* (To appear).

[86] Situ Rong and Wang Yueping (1999). On solutions of BSDE with jumps, with unbounded stopping times as terminals and with non-Lipschitzian coefficients, and the probabilistic interpretation of solutions to quasi-linear elliptic integro-differential equations. *Applied Maht. & Mechanics.* (To appear).

[87] Situ Rong (1999). Comparison theorem of solutions to BSDE with jumps and viscosity solution to a generalized Hamilton-Jacobi-Bellman equation. In *Control of Distributed Parameter and Stochastic Systems,* Eds.S. Chen, X. Li, J. Yong and X.Y. Zhou, Kluwer, Boston, 275-282.

[88] Situ Rong (1999). On solutions of forward-backward stochastic differential equations with jumps and with non-Lipschitzian coefficients. *Workshop on Infinite Dimensional Analysis and Dirichlet Forms,* March 28 - April 1, WuHan, China.

[89] Situ Rong (1999). On comparison theorems and existence of solutions to backward stochastic differential equations with jumps and with discontinuous coefficients. *The 26th Conference on Stochastic Processes and their Applications,* Beijing.

[90] Skorohod, A.V. (1961, 1962). Stochastic equations for diffusion processes in a bounded region, 1, 2. *Theor. Prob. Appl.*, 6, 264-274; 7, 3-23.

[91] Skorohod, A.V. (1965). *Studies in the Theory of Random Processes.* Addison-Wesley, Reading, Massachusetts.

[92] Skorohod, A.V. (1989). *Asymptotic Methods in the Theory of Stochastic Differential Equations,* AMS Providence.

[93] Stroock, D.W. and Varadhan, S.R.S. (1971). Diffusion processes with boundary conditions. *Comm. Pure Appl. Math. 24,* 147-225.

[94] Stroock, D.W. and Varadhan, S.R.S. (1979). *Multidimensional Diffusion Processes.* Springer-Verlag. N.Y.

[95] Tanaka, H. (1979). Stochastic differential equations with reflecting boundary condition in convex regions. *Hiroshima Math. J. 9,* 163-177.

[96] Tang, S. and Li, S. (1994). Necessary conditions for optimal control of stochastic systems with random jumps. *SIAM J. Control Optim. 32, 1447-1475.*

[97] Veretennikov, A .Ju. (1981). On the strong and weak solutions of one-dimensional stochastic equations with boundary conditions. *Theor. Prob. Appl.,* 26:4, 685-700. (In Russian).

[98] Watanabe, S. (1971). On stochastic differential equations for multi-dimensional diffusion processes with boundary conditions, I, II. *J. Math. Kyoto Univ., 11,* 169-180, 545-551.

[99] Yamada, K. (1994). Reflecting on sticky Markov process with Levy generator as the limit of storage processes. *Stochastic Processes and their Applications 52:1*

[100] Yan, J.A. (1981). *An Introduction to Martingales and Stochastic Integrals.* Shanghai Sci.& Tech. Pub. House. (In Chinese).

[101] Yan, J.A., Peng, S. et al. (1997). *Topics on Stochastic Analysis.* Scien. Press, Beijing. (In Chinese).

[102] Yeh, J. (1973). *Stochastic Processes and Wiener Integral.* Marcel Dekker Inc. N.Y.

[103] Yeh, J. (1995). *Martingales and Stochastic Analysis.* World Scientific.

[104] Yoeurp, Ch. and Yor, M. (1977). Espace orthogonal à une semi-martingale, applications. (A paraître au Z.f. W-theorie).

[105] Yor, M. (1978). Sur la Continuité des temps locaux associés a certaines semi-martingales. *Societé Mathematique de France, Astérisque* 52-53, pp 23-35.

[106] Yu, J.Y., Guo, B.Z. and Zhu, G.T. (1987). The control of the semi-discrete population evolution system. *J. Syst. & Math. Scien.* 7:3, 214-219. (In Chinese).